ANALYTICAL METHODS IN COMBINATORIAL CHEMISTRY

HOW TO ORDER THIS BOOK

BY PHONE: 800-233-9936 or 717-291-5609, 8AM–5PM Eastern Time

BY FAX: 717-295-4538

BY MAIL: Order Department
Technomic Publishing Company, Inc.
851 New Holland Avenue, Box 3535
Lancaster, PA 17604, U.S.A.

BY CREDIT CARD: American Express, VISA, MasterCard

BY WWW SITE: http://www.techpub.com

Cranbury

Analytical Methods in Combinatorial Chemistry

BING YAN

Axys Advanced Technologies, Inc.
South San Francisco, CA

TECHNOMIC PUBLISHING CO., INC.

LANCASTER · BASEL

Analytical Methods in Combinatorial Chemistry
a**TECHNOMIC**® publication

Technomic Publishing Company, Inc.
851 New Holland Avenue, Box 3535
Lancaster, Pennsylvania 17604 U.S.A.

Printed in the United States of America
10 9 8 7 6 5 4 3 2 1

Main entry under title:
Analytical Methods in Combinatorial Chemistry

A Technomic Publishing Company book
Bibliography: p.
Includes index p. 265

Library of Congress Catalog Card No. 99-68040
ISBN No. 1-56676-809-8

Table of Contents

Foreword

"WHY COMBINATORIAL CHEMISTRY, Why Now (and Why You Should Care)" was the title of an article I wrote near the beginning of the combinatorial chemistry explosion phase in the mid-1990s. While it's easy now to understand why the time was ripe for the use of combinatorial chemistry, this is an idea that initially took root slowly. For many chemists the goal of synthesizing large numbers of compounds smacked of production, not what a medicinal chemist did. It's taken some time to accept that the goal of efficient, homogeneous library synthesis carries all the challenge of natural product synthesis and more, because making a library based on any molecule is more difficult than is making the compound itself. While combinatorial builds on the armamentarium of synthetic methods, the invention of tools is necessary because we have the goal of making library synthesis and screening as easy as possible. With the new millennium comes recognition that combinatorial chemistry is, most importantly, a new way for synthetic chemists to be valuable.

Combinatorial chemistry is likewise a new way for analytical chemists to be valuable. While most of us can take one NMR spectrum, none of us can take tens of thousands of NMR spectra on a human timescale. Similarly challenging, given a microtiter plate with 96 samples and the structure expected for each sample, how would one determine the absolute concentration of compound in each sample? Multiply this by thousands, and the magnitude of the analytical challenges quickly comes into focus.

This book by Dr. Yan, the first of its kind, fills an important need in the field of combinatorial chemistry. The chemical analysis devices of the past are insufficient to meet the needs of today, and there will be a lot of creativity required to fill those needs. Understanding both the work and the philosophies that underlie analytical combinatorial chemistry will help to stimulate that creativity and to engage it productively.

ANTHONY W. CZARNIK
Illumina
San Diego, California, USA

Preface

A TREMENDOUS NEED for analytical methodologies in various phases of combinatorial chemistry is the main driving force for me to write this book. Although several books on combinatorial chemistry and solid-phase organic synthesis are available, no book specifically dealing with analytical issues in combinatorial chemistry has been published up to now. This book intends to show that the lack of analytical methods in combinatorial chemistry is no longer true. This book also wishes to bring everyone's attention to the fact that major breakthroughs are badly needed to resolve many analytical issues in combinatorial chemistry, particularly the incompatibility between parallel synthesis and the serial analysis and purification.

This book contains most analytical methods that have been applied to solving various problems in broad areas of combinatorial chemistry, that is, chemistry optimization, library synthesis, lead selection, and lead optimization. The focus of combinatorial synthesis technology in this book is primarily solid-phase organic synthesis (SPOS). Only limited examples related to solid-phase peptide synthesis (SPPS) are included with an emphasis on their implication for the small molecule synthesis. Methods for analysis of the cleaved libraries are clearly applicable for libraries made by other methods, such as solution-phase and liquid-phase synthesis.

This book is organized according to the operation process of combinatorial chemistry starting with the description of the basic properties of the unique analytical sample solid supports and their effects on analysis and solid-phase organic reactions in Chapter 2. Chapters 3 through 5 describes analytical methods routinely applied to the tedious process of reaction optimization. These techniques are various FTIR and Raman methods, MS, NMR, UV-Vis, fluorescence, and other analytical methods. Methods for the reaction optimization stage emphasize on-support analysis, because chemical changes and reaction yield on resin are the most relevant information for multi-step synthesis. Then, the discussions move to the analytical control of the library quality in Chapter 6. In contrast to the reaction optimization stage, the high-throughput in-solution analysis is more relevant at this stage because biological *in vitro* assays are mostly performed in solution. The various methods for selecting active lead compounds from combinatorial libraries and the ways to reveal their structure or identity are discussed next in Chapter 7. Standard high-throughput in-solution assays are the main methods for selecting active compounds from libraries containing discrete compounds. The selection of lead compounds from pooled combinatorial libraries relies on on-resin assays and various affinity selection methods. These methods are still in the development stage and may make impacts for the future development of combinatorial chemistry. Structure elucidation methods on a single bead have great potential, as the future highly miniaturized assays will allow activity evaluation from fmol to pmol amount of compounds, which is much below a single bead's loading. The applications of analytical methods in the optimization of the *in vitro* activity and physicochemical properties of lead compounds are then illustrated in Chapter 9, and the concept of "lead compound" is also discussed. In every chapter, the methods are summarized by providing the main advantages and weaknesses of the current technologies and future needs. Some promising new technologies and future perspective are presented in chapter 9.

Although I kept adding references until the submission of this book, it is still impossible to include all the publications that deserve to be included. I apologize to authors whose works are not included here. I foresee the second edition will be sooner than normal due to the rapid development in this field. Any comments, criticisms, and suggestions are welcome and appreciated.

Because analytical challenges accompany a combinatorial chemist at every step of his or her endeavor, this book serves as a valuable companion and a comprehensive reference book for chemists involved in combinatorial chemistry. This book also serves as an introduction to lead beginners to enter the field of combinatorial chemistry knowing every analytical issue and possible solution. For those who are interested in analytical chemistry, this book tries to include as many research possibilities and directions as possible in order to stimulate vigorous research in this field. This book is also a timely textbook for university and graduate school students after their acquaintance with basic analytical techniques. It will introduce graduate students and college seniors to the updated applications of modern analytical technologies and to the rapidly growing field of combinatorial chemistry.

Bing Yan
October, 1999
South San Francisco

Acknowledgments

IT HAS BEEN a great experience to write this book using all of my weekends and evenings in a year, because the literature source is huge. Therefore, I am very grateful to all the people who gave me encouragement and support during the past year. I am particularly grateful to Lina Liu for sacrificing her spare time to help me prepare the manuscript. I also want to thank Drs. Mark Bradley (University of Southampton), Qun Sun (Purdue Pharma, L.P.), William Fitch (Affymax), and Yen-Ho Chu (Ohio State University), and Qing Tang (Novartis) for their critical reading of the manuscript and their expert opinions and suggestions. I especially thank Dr. Anthony Czarnik for the inspiration and encouragement to make me take over the lifetime challenge in combinatorial chemistry. I am grateful for the support from many chemists around me in Novartis as well as from many other companies and universities. I thank all the authors and publishers, especially the American Chemical Society, for allowing me to use figures from their publications in this book. I am grateful to Dr. Steven Strauss from Technomic for recognizing the importance of this topic and to Susan Farmer, Lori Eby, and other members of Technomic's staff for editorial assistance and the production of the book. This book is dedicated to the eighth birthday of my son Patrick.

1

Analytical Issues in Combinatorial Chemistry

1.1 COMBINATORIAL CHEMISTRY

Combinatorial chemistry research has undergone enormous development recently (Figure 1.1). These developments have been the subject of many reviews (*Reviews:* Jung & Beck-Sickinger, 1992; Pavia et al., 1993; Felder, 1994; Janda, 1994; Gordon et al., 1994; Czarnik, 1995; Lyttle, 1995; Thompson & Ellman, 1996; DeWitt & Czarnik, 1996; Still, 1996; Balkenhohl et al., 1996; Lam et al., 1997; Fenniri, 1996; Trautwein et al., 1997; Wentworth Jr. & Janda, 1998. *Books:* Bunin, 1998; Gordon & Kerwin, 1998; Obrecht & Villalgordo, 1998; Terrett, 1998; Czarnik & DeWitt, 1997; Devlin, 1997; Moos et al., 1997; Wilson & Czarnik, 1997; Abelson, 1996; Chaiken & Janda, 1996; Cortese, 1996; Epton, 1996; Jung, 1996; Geysen et al., 1995; Houghten, 1995; Merrifield, 1993).

Combinatorial chemistry can be defined (Czarnik, 1998) as a subfield of chemistry that aims to combine a small number of chemical reagents, in all combinations defined by a given reaction scheme, to yield a large number of well-defined products in a form that is easy to screen for properties of interest. It can be considered as an umbrella term that covers a wide array of topics, including the combinatorial and high-throughput parallel synthesis and all the associated technologies for handling, analyzing, and screening these libraries and obtaining useful molecules from them. Combinatorial chemistry has come to be regarded as a tool rather than an end in itself. Its success is measured in terms of lead compounds and new chemical entities discovered in various areas, such as molecular recognition, catalysis, drug development, material sciences, agrochemical, and other applications.

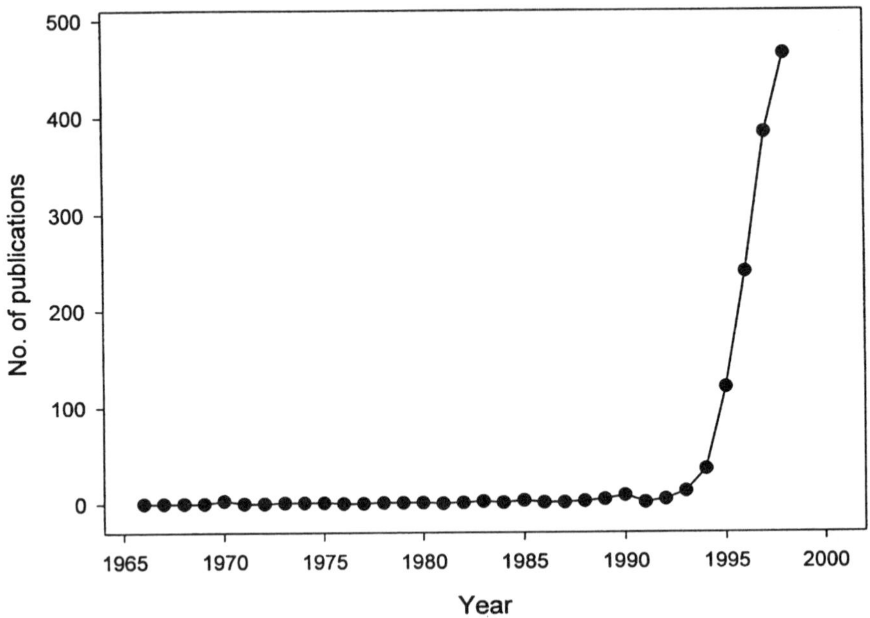

Figure 1.1 *Number of publications in combinatorial chemistry based on a search result in SciFinder for the term "combinatorial chemistry."*

In drug discovery, several factors have catalyzed the dramatic growth of combinatorial chemistry. First, the rapid development in genetics and molecular biology can now provide ever-increasing numbers of therapeutic target proteins for testing. Second, to meet the need for screening biological targets, significant progress has been made in high-throughput screening technologies since the beginning of the 1990s. It is possible to test more than 100,000 compounds in a period of months. This has made clear that the synthesis of molecules has become the rate-limiting step in the drug discovery process. Third, the huge cost and the lengthy process to develop a marketable drug have become formidable so that all pharmaceutical companies are eagerly searching for more efficient and cost-effective ways of drug discovery.

Combinatorial synthesis is a technology that allows the rapid synthesis of large numbers of compounds with diverse features (structural and physicochemical properties). The term "combinatorial chemistry" used to be more accurately reserved for the synthesis of compound libraries containing an extremely large number of compounds with all possible combination of substituents to distinguish them from "parallel synthesis," where the compound libraries consist of a smaller number of individual or discrete compounds, but it is now used indiscriminately to embrace both methodologies. Combinatorial libraries can be made on solid supports, soluble polymer supports, or in solution. As seen from Figure 1.1, the widespread interest in combinatorial chemistry is a quite recent phenomenon.

By creating vast new sources of organic compounds, the efficient screening machines will identify active compounds for defined targets—a process known as lead generation. Once active compounds are found, more focused libraries around those structures provide valuable structure-activity relationship data via what is known as lead optimization. Note that current lead generation and lead optimization are only based on *in vitro* activity. Physicochemical and other properties are optimized in later developmental stages. Ideally, compounds with good physicochemical properties should also be selected as lead compounds, and their *in vitro* activity can be optimized in lead optimization.

In the lead identification phase, the goal is to make the most diverse organic compound libraries in the shortest time. The molecular diversity of a compound library means the greatest variation in molecular structure, size, shape, polarity, and polarizability—therefore, a wide variation in *in vitro* activity and physicochemical properties such as LogP, solubility, permeability, stability, and bioavailability. The lead selection should be based on all of these criteria. Unfortunately, due to technical and philosophical reasons, lead compounds are currently selected only on the basis of *in vitro* activity alone. There is no reason to discard a compound that has a low *in vitro* activity, but has good physicochemical properties in the lead discovery stage.

1.2 SYNTHESIS METHODS

Few advances in chemistry have had as profound an effect on biotechnology as solid-phase supported synthesis. First developed by Merrifield (1963) for peptide synthesis, the method has already been adapted to small organic molecules (Fridkin et al., 1966; Leznoff, 1974; Crowley & Rapoport, 1976; Leznoff, 1978; Frechet, 1981; Hodge, 1988; Brown, 1997) and is now routinely used in laboratories worldwide. Although SPOS has been continuously studied by several academicians during the 1970s and 1980s, the technology flourished only after the seminal publications of Bunin and Ellman (1992) and of De Witt et al. (1993).

Solid-phase chemistry has advantages that make it an extremely important combinatorial methodology. Because the synthesis is conducted on a solid support, the removal of excess reagent and byproducts is accomplished easily by copious washing of the polymer. A large excess of reagents can be added and then washed away after the reaction terminates. This allows each reaction step to be driven very close to completion. If, on the other hand, the reaction efficiency were to be much less than 100%, then a fraction of unreacted chains would remain incomplete at the end of each reaction step. Multiply this by any reasonable number of steps, and the resulting combinatorial library will contain a large portion of undesired products.

Two formats for library synthesis are high-throughput parallel synthesis on polymer supports or in solution and combinatorial synthesis on polymer support using methods such as "split-and-pool" synthesis (also known as split synthesis). Parallel synthesis makes discrete compounds and, if pure, provides substance for unambiguous testing. The number of compounds and diversity made through this method are, however, limited. The split synthesis is a promising technique to cover much more chemical diversity. It is able to make a library with millions of compounds while each bead contains only one compound. In a sense, split synthesis also makes discrete compounds at a single bead level. Because of the huge number of compounds, it is not practical at present to assay discrete compounds released from every bead. Future development in miniaturization of HT screening will enforce this format. At present, compounds from split libraries are usually assayed in the form of mixtures. The screening of pooled compounds in solution, however, may give ambiguous results. On-bead screening has also been applied when combined with tagging schemes to identify lead compounds.

After years of widespread practice of SPOS, difficulties associated with this technology became more evident. First, the synthetic methods required to make organic molecules on a polymer support have not been optimized. The shortage of SPOS literature, the insufficient knowledge of solid support and its effect on SPOS, and the lack of effective ways of monitoring many diverse organic reactions on support have significantly prolonged the optimization of combinatorial reactions. Further problems include the following:

- Reactions are frequently thought to be retarded on solid phase owing to the lack of understanding of solid-phase reaction kinetics and fundamental effects of support on organic reactions.
- Heterogeneous reagents are typically less effective.

- Resin supports display significant solvent-dependent swelling and shrinking properties, which can severely affect reaction rates and site accessibility.
- The optimal solvent for resin swelling may not be the optimal solvent for the desired reaction.
- There is an attachment step and a cleavage step.

Because of these reasons, the size of the typical library has been decreasing for the past several years, and there has been an upswing of solution-phase synthesis in 1997–1999.

Even though a lot of work is needed to perfect the SPOS methodology, SPOS is still the foundation of the now most widely practiced form of combinatorial chemistry and remains the method of choice for most combinatorial library synthesis. In addition to solid-phase synthesis, the synthesis of small molecule combinatorial libraries has also been pursued using both liquid-phase (on soluble polymers see Bayer & Mutter, 1972; Bonora et al., 1990; Han et al., 1995) and solution-phase chemistry (Storer, 1996; Garr et al., 1996; special issue of Tetrahedron, 1998).

In liquid-phase combinatorial synthesis (LPCS), the key is a soluble, linear homopolymer [polyethylene glycol monomethyl ether (MeO-PEG)], which also serves as a terminal protecting group for the library of compounds synthesized. Two properties that are inherent in this homopolymer's structure provide the necessary elements for it to be attractive in a combinatorial format. First, due to its helical structure, MeO-PEG has a strong propensity to crystallize; thus, as long as the polymer remains unaltered during the construction of the library, purification by crystallization can be accomplished at each stage of the synthesis. Second, MeO-PEG has remarkable solubilizing effects in a variety of aqueous and organic solvents. This property can be used to advantage if the homopolymer is treated as a reagent and is used in large excess. Another virtue of the polymer's good solubility is that all manipulations in the LPCS, including split synthesis, can be carried out under homogeneous conditions. Furthermore, reaction progress can be monitored by NMR. A drawback is that intermediate separation and purification are not as simple as solid-phase synthesis.

Solution-phase combinatorial synthesis has undergone some recent progress. This field is basically an amalgam of synthetic/medicinal chemistry, parallel processing, and laboratory automation. A wide range of reactions is potentially available for solution-phase library generation. In contrast to solid-phase synthesis, no additional steps for attachment to or detachment from supports are required, and, thus, only the particular reaction of interest needs to be developed. The reactions and techniques to be employed in solution-phase library generation are already familiar to all organic chemists. Because the chromatographic purification is impossible in high-throughput synthesis, intermediate purification and removal of excess reagent and solvent are the major hurdles. These make the multistep synthesis nearly impossible. In "split-and-pool" libraries, each bead contains a compound that is physically separated, and tagging is possible to code for the active bead, while this is impossible for solution-pooled compounds. This book focuses on SPOS. Analytical methods for cleaved compounds are clearly also applicable to libraries from solution-phase and liquid-phase synthesis.

1.3 ANALYTICAL CHALLENGES

The rapid development of combinatorial chemistry poses challenges to many disciplines. Challenges facing analytical chemists are in at least four major areas:

1.3.1 Properties of Solid Supports

Although there are many advantages to carry out organic synthesis on solid support, the effects of these supports on reactions' outcomes cannot be ignored. The effects range from basic swelling and solvent compatibility to site-site interaction in the matrix and alteration of reaction kinetics and the entrapped impurities.

Many of these effects have been inadequately studied. The prolonged reaction optimization time and the recent trends to the reduced library size and moving away from SPOS can partially be attributed to the lack of knowledge on the effects of solid supports.

1.3.2 Reaction Optimization Stage

Synthetic route development in combinatorial synthesis remains a great hurdle and time-limiting factor in most library synthesis projects. It is estimated that reaction development generally can take months, and the library synthesis likely takes only weeks. It is not feasible to purify intermediates in multistep synthesis and library products if the libraries are big, so investing work at the route development stage becomes essential. Although it is impractical to characterize all of the products formed during a combinatorial synthesis, significant credibility could be gained by using the same chemical manipulations to synthesize a single representative compound and small trial libraries and to measure the success rate. A good quality control (QC) of building blocks and reagents and analysis of yield, byproducts, and minimization of side reactions are also crucial.

On-support analytical methods are ideal for monitoring reactions at this stage because chemical changes and the reaction yield on solid support are the most relevant information. To "cleave and analyze" is time-consuming, laborious, and destructive. Some synthetic intermediates are not even stable enough to cleavage conditions.

1.3.3 Library Synthesis Stage

After reactions are optimized and trial libraries are rehearsed, a final large library will be generated on polymer supports or in solution. The quality control of this final collection of compounds is a major challenge due to the large number of compounds that are each present in very small amounts. Should QC be done on support or after cleavage? Because most *in vitro* assays are performed in solution, it is advantageous and more relevant to characterize the library in solution. The high-throughput analytical methods are needed for such characterization. Furthermore, the high-throughput methods unavoidably require automatic data interpretation capability, especially for high-throughput NMR methods. It is highly desirable, at least for discrete libraries, to obtain the concentration and purity as the structure is confirmed. Frankly, the serial format of analysis and purification protocols will never meet the needs of parallel synthesis. A major breakthrough is urgently needed.

1.3.4 Lead Selection and Optimization Stage

Should the selection assay be done on support or in solution? Routine HT screening methodology is still the main method for screening discrete compound libraries in solution. If the target biomolecule is soluble, then on-bead screening can be more efficient and has been used. Unfortunately, the utility of on-bead screening is limited by the fact that not all targets are soluble. Additionally, attachment to a solid support can distort the structure of the protein. In-solution screening is clearly more favorable and unambiguous. Another issue is how to balance the unambiguity in single-compound evaluation with the time-efficient, but more complicated mixture evaluation. The split synthesis actually makes discrete compounds at the level of single bead. Fewer than one-third of compounds made on a single low loading TentaGel bead are enough for a current screening assay (96-well format). As the screening methods become more miniaturized, significantly less materials will be needed for evaluation.

An important issue is that the "lead" selection criteria need to be redefined. A new chemical entity must possess a combination of several optimal properties. Taking a drug as an example, a combination of optimal potency, bioavailability, permeability, and physicochemical properties (not to mention stability and feasibility for synthesis and formulation) is required. It is improper to select lead only based on *in vitro* activity and to disregard the compounds that have good physicochemical characteristics and bioavailability, but that scored low in *in vitro* activity assay.

It has been almost six years since SPOS was reintroduced into combinatorial chemistry. Many progresses have been made in analytical methods development at various stages of combinatorial chemistry. Several reviews on various aspects of this topic have appeared recently (Egner & Bradley, 1997; Gallop & Fitch, 1997; Balkenhohl et al., 1997; Shapiro et al., 1997; Fitch, 1998; Yan, 1998; Yan & Gremlich, 1999; Swali et al., 1999; Fitch, 1999). A relative comprehensive compilation of developments in analytical methodologies is presented in the following chapters.

1.4 REFERENCES

Abelson, J. N. *Combinatorial Chemistry.* 1996, Academic Press, San Diego. ISBN 0121821684.

Balkenhohl, F., Bussche-Hunnefeld, C., Lansky, A., Zechel, C. Combinatorial synthesis of small organic molecules. *Angew. Chem. Int. Ed. Engl.* 1996, 35, 2288.

Balkenhohl, C., von dem Bussche-Hunnefeld, F., Lansky, A., Zechel, C. Combinatorial chemistry. *Fresenius J. Anal. Chem.* 1997, 359, 3.

Bayer, E., Mutter, M. Liquid phase synthesis of peptides. *Nature.* 1972, 237, 512.

Bonora, G. M., Scremin, C. L., Colonna, F. P., Garbesi, A. HELP (high efficiency liquid phase) new oligonucleotide synthesis on soluble polymeric support. *Nucleic Acids Res.* 1990, 18, 3155.

Brown, R. Recent developments in solid-phase organic synthsis. *Contemp. Org. Syn.* 1997, 4, 216.

Bunin, B. A. *The Combinatorial Index.* 1998, Academic Press, San Diego. ISBN 0121413403.

Bunin, B. A., Ellman, J. A. A general and expedient method for the solid-phase synthesis of 1,4-benzodiazepine derivatives. *J. Am. Chem. Soc.* 1992, 114(27), 10997–8.

Chaiken, I. M., Janda, K. D. *Molecular Diversity and Combinatorial Chemistry. Libraries and Drug Discovery.* 1996, Americal Chemical Society, Washington, DC. ISBN 0841234507.

Cortese, R. *Combinatorial Libraries: Synthesis, Screening and Application Potential.* 1996, Walter de Gruyter, New York. ISBN 311014395X.

Crowley, J. I., Rapoport, H. Solid-phase organic synthesis novelty or fundamental concept? *Acc. Chem. Res.* 1976, 9, 135.

Czarnik, A. W. Combinatorial organic synthesis using Parke-Davis's DIVERSOMER method. *Chemtracts—Org. Chem.* 1995, 8, 13.

Czarnik, A. W. Combinatorial chemistry—What's in it for analytical chemists? *Anal. Chem.* 1998, 378A.

Czarnik, A. W., DeWitt, S. H. *A Practical Guide to Combinatorial Chemistry.* 1997, American Chemical Society, Washington, DC. ISBN 084123485X.

Devlin, J. P. *High Throughput Screening: The Discovery of Bioactive Substances.* 1997, Marcel Dekker, Inc., New York. ISBN 0824700678.

DeWitt, S. H., Czarnik, A. W. Combinatorial Organic Synthesis Using Parke-Davis's Diversomer Method. *Acc. Chem. Res.* 1996, 29, 114.

DeWitt, S. H., Kiely, J. S., Stankovic, C. J., Schroeder, M. C., Reynolds, D. M., Ravia, M. R. "Diversomers": An approach to nonpeptide, nonoligomeric chemical diversity. *Proc. Natl. Acad. Sci. U. S. A.* 1993, 90, 6909.

Egner, B. J., Bradley, M. Analytical techniques for solid-phase organic and combinatorial synthesis. *Drug Discovery Today.* 1997, 2, 102.

Epton, R. *Innovation and Perspectives in Solid Phase Synthesis & Combinatorial Libraries.* 1996, Mayflower Scientific Limited, Birmingham.

Felder, E. R. The challenge of preparing and testing combinatorial compound libraries in the fast lane, at the front end of drug development. *Chimia.* 1994, 48, 531.

Fenniri, H. Recent advances at the interface of medicinal and combinatorial chemistry. Views on methodologies for the generation and evaluation of diversity and application to molecular recognition and catalysis. *Current Med. Chem.* 1996, 3, 343.

Fitch, W. L. Analytical methods for quality control of combinatorial libraries. *Annual Reports in Combinatorial Chemistry and Molecular Diversity.* 1998, 1, 59.

Fitch, W. L. Analytical methods for quality control of combinatorial libraries. *Molecular Diversity.* 1999, 4, 39.

Frechet, J. M. J. Synthesis and applications of organic polymers as supports and protecting groups. *Tetrahedron.* 1981, 37, 663.

Fridkin, M., Patchornik, A., Katchalski, E. Use of polymers as chemical reagents. *J. Am. Chem. Soc.* 1966, 88, 3164.

Gallop, M. A., Fitch, W. L. New methods for analyzing compounds on polymeric supports. *Curr. Opin. Chem. Biol.* 1997, 1, 94.

Garr, C. D., Peterson, J. R., Schultz, L., Oliver, A. R., Underiner, T. L., Cramer, R. D., Ferguson, A. M., Lawless, M. S., Patterson, D. E. Solution phase synthesis of chemical libraries for lead discovery. *J. Biomol. Screening.* 1996, 1, 179.

Geysen, H. M., Armstrong, R. W., Kay, B. K. *Chemical Diversity: The Synthetic Techniques of Combinatorial Chemistry.* 1995, American Chemical Society, Washington, DC.

Gordon, E. M., Barrett, R. W., Dower, W. J., Fodor, S. P. A., Gallop, M. A. Applications of combinatorial technologies to drug discovery. 2. Combinatorial organic synthesis, library screening strategies, and future directions. *J. Med. Chem.* 1994, 37, 1385.

Gordon, E. M., Kerwin, J. F. J. *Combinatorial Chemistry and Molecular Diversity in Drug Discovery.* 1998, John Wiley & Sons, New York. ISBN 0471155187.

Han, H., Wolfe, M. M., Brenner, S., Janda, K. D. Liquid-phase combinatorial synthesis. *Proc. Natl. Acad. Sci. U. S. A.* 1995, 92, 6419.

Hodge, P. in *Synthesis and Separations Using Functional Polymers.* 1988, Sherrington, D. D., Hodge, P. Eds. Wiley, Chichester, Chapter 2.

Houghten, R. A. *Combinatorial Libraries: The Theory and Practice of Combinatorial Chemistry.* 1995, ESCOM, Leiden.

Janda, K. D. Tagged versus untagged libraries: methods for the generation and screening of combinatorial chemical libraries. *Proc. Natl. Acad. Sci. U. S. A.* 1994, 91, 10779.

Jung, G. *Combinatorial Peptide and Nonpeptide Libraries: A Handbook.* 1996, VCH, Weinheim, New York. ISBN 3527293809.

Jung, G., Beck-Sickinger, A. G. Multiple peptide synthesis methods and their applications. *Angew. Chem. Int. Ed. Engl.* 1992, 31, 367.

Lam, K. S., Lebl, M., and Krchnak, V. The “One-Bead-One-Compound” combinatorial library method. *Chem. Rev.* 1997, 97, 411.

Leznoff, C. C. The use of insoluble polymer supports in organic chemical synthesis. *Chem. Soc. Rev.* 1974, 3, 65.

Leznoff, C. C. The use of insoluble polymer supports in general organic synthesis. *Acc. Chem. Res.* 1978, 11, 327.

Lyttle, M. H. Combinatorial chemistry: A conservative perspective. *Drug Dev. Res.* 1995, 35, 230.

Merrifield, R. B. Solid phase peptide synthesis. I. The synthesis of a tetrapeptide. *J. Am. Chem. Soc.* 1963, 85, 2149–2154.

Merrifield, R. B. *Life During a Golden Age of Peptide Chemistry: The Concept and Development of Solid-Phase Peptide Synthesis.* 1993, American Chemical Society, Washington, DC. ISBN 0841218420.

Moos, W. H., Pavia, M. R., Kay, B. K., Ellington, A. D. *Annual Reports in Combinatorial Chemistry and Molecular Diversity.* 1997, ESCOM, Leiden. ISBN 9072199235.

Obrecht, D., Villalgordo, J. M. *Solid-Supported Combinatorial and Parallel Synthesis of Small-Molecular-Weight Compound Libraries.* 1998, Tetrahedron Organic Chemistry Series, V. 17. Pergamon Press. ISBN 0080432573.

Pavia, M. R., Sawyer, T. K., Moos, W. H. The generation of molecular diversity. *Bioorg. Med. Chem. Lett.* 1993, 3, 387.

Shapiro, M., Lin, M., Yan, B. On-resin analysis in combinatorial chemistry, in ACS Book: *A Practical Guide to Combinatorial Chemistry,* Czarnik, A. W., DeWitt, S. H. Eds., 1997, 123.

Still, W. C. Discovery of sequence-selective peptide binding by synthetic receptors using encoded combinatorial libraries. *Acc. Chem. Res.* 1996, 29, 155.

Storer, R. Solution-phase synthesis in combinatorial chemistry: applications in drug discovery. *Drug Discovery Today.* 1996, 1, 248.

Swali, V., Langley, G. J., Bradley, M. Mass Spectrometric analysis in combinatorial chemistry. *Curr. Opinion in Chem. Biol.* 1999, 3, 337.

Terrett, N. K. *Combinatorial Chemistry.* 1998, Oxford University Press, New York. ISBN 0198502206.

Tetrahedron, 1998. 54. No. 16. Special issue.

Thompson, L. A., Ellman, J. A. Synthesis and applications of small molecule libraries. *Chem. Rev.* 1996, 96, 555.

Trautwein, A., Bauser, M., Fruchtel, J., Richter, H., Jung, G. *Nachr. Chem. Tech. Lab.* 1997, 45, 158.

Wentworth Jr., P., Janda, K. D. Generating and analyzing combinatorial chemistry libraries. *Curr. Opinion in Biotech.* 1998, 9, 109.

Wilson, S. R., Czarnik, A. W. *Combinatorial Chemistry. Synthesis and Applications.* 1997, John Wiley & Sons, Inc., New York. ISBN 047112687X.

Yan, B. Monitoring the progress and the yield of solid phase organic reactions directly on resin supports. *Acc. Chem. Res.* 1998, 31, 621.

Yan, B., Gremlich, H. U. Role of Fourier transform infrared spectroscopy in the rehearsal phase of combinatorial chemistry: a thin layer chromatography equivalent for on-support monitoring of solid-phase organic synthesis. *J. Chromatogr. B.* 1999, 725, 91.

2

An Examination of the Analytical Sample: Resin Support

BEFORE THE SYNTHESIS of a combinatorial library, the sequence of reactions to be used needs to be optimized to give the highest yield and fewest side products. Unlike the synthesis in solution-phase or on soluble polymers, solid-phase synthesis requires a major effort for reaction optimization or reaction validation. One of the major obstacles for adapting reactions to solid support is the lack of knowledge about the solid support and the effects of the support on organic reactions. The consequence is the prolonged reaction optimization process, poor reaction yields, the formation of side products, and even failed syntheses. This chapter examines properties of solid-phase samples and analytical studies targeted to understand these properties as well as the effects of swollen resin on solid-phase organic reactions. These solid supports are an important platform on which to build combinatorial libraries and also analytical samples to be analyzed during the reaction optimization process. The success of many analytical methods relies on the physicochemical and swelling properties of solid supports.

In the past three decades, resin properties have sometimes been topics of investigations in the context of solid-phase peptide synthesis (SPPS) (Merrifield, 1984; Meldal, 1997) and solid-phase organic synthesis (SPOS) (Crowley et al., 1973; Overberger & Sannes, 1974; Pittman & Evans, 1973; Kraus & Patchornik, 1978; Guyot, 1988; Hodge, 1997). Although the effects of solid supports on SPOS have not been fully evaluated, the effects on SPPS have been studied more thoroughly. For combinatorial synthesis, the practice of SPOS has undergone rapid expansion both in scale and in depth. However, so far, insufficient attention has been paid to the effects of resin properties in combinatorial synthesis. As a result, unwanted products and impurities are produced unnecessarily along with target products.

SPOS is carried out predominantly on beaded polymer resins or on grafted polyethylene surfaces. The former has been the major form of support used in solid-phase combinatorial and parallel synthesis and, therefore, is the focus of this chapter. The reaction conditions and the outcome of the solid-phase reaction are usually different from those of the solution-phase counterpart due to the entirely different microenvironment of the solid supports. The physical and chemical characteristics and chemical compatibility of the solid support often determine the success of SPOS. Although the effects of resin solvation on SPPS reactions may be useful for the optimization of SPOS reactions, differences between SPPS and SPOS must be noted. In this chapter, examples in both categories will be given in order to clarify properties of the support. Instead of considering all SPOS reactions, this chapter will cite selected examples to illustrate some key properties and effects posed by solid supports. Interested readers are referred to several other reviews on related topics (Merrifield, 1984; Meldal, 1997; Crowley et al., 1973; Overberger & Sannes, 1974; Pittman & Evans, 1973; Kraus & Patchornik, 1978; Guyot, 1988; Hodge, 1997).

Resin-based solid-phase synthesis is actually the gel-phase synthesis, because it is the highly solvated resin that provides a reaction matrix. Therefore, the following sequence of topics will be considered:

- physical properties of resins
- resin swelling and solvation
- effects of the solvated resin on solid-phase reactions

2.1 PHYSICAL PROPERTIES OF RESINS

Polystyrene-based resin is the most popular solid support for solid-phase combinatorial and parallel synthesis. Its physical properties have a direct impact on the design of the library, the optimization of reaction conditions, and the synthetic scheme. These properties include the following:

- resin type
- size distribution
- loading and the reactive site distribution
- thermal stability
- chemical stability

2.1.1 Resin Type

2.1.1.1 POLYSTYRENE RESINS

Two main classes of polystyrene (PS) supports, lightly and highly crosslinked PS resins, have been used in SPOS. The most popular PS support in both SPPS and SPOS is the 1% DVB crosslinked gel-type PS resin prepared by suspension polymerization. Other lightly crosslinked (2–5% DVB) gel-type resins and the highly crosslinked (5–60% DVB) macroporous resins have also been used. Gel-type resins have no permanent porosity, but swell to varying degrees in many organic solvents to form micropores—the space between crosslinks occupied by the swelling solvent. These spaces contain solutions of polymer segments, not just solvent. Therefore, they are referred to as microporous. After removing the solvent, the solvent-induced porous structure and sometimes also the chemically fixed porosity remain preserved. In macroporous resins, permanent pores or holes of various sizes are formed. Beads consist of conglomerates of microspheres (100–200 nm) that look like cauliflower florets, and each microsphere shows smaller nuclei (10–30 nm) fused together. Small pores are distributed between the nuclei.

PS resin is highly hydrophobic. Due to the absent spacer, reactions on PS resin tend to be significantly affected by the hydrophobic PS matrix, which may disfavor reactions involving polar and charged species. For example, reactions involving charged, highly polar reagents or aqueous solution may be retarded. Such unfavorable situations can be corrected with phase-transfer catalyst (see section 2.3.2).

2.1.1.2 PS-PEG RESINS

PS-PEG (polyethylene glycol) resins differ from PS resins in that they contain dominantly (up to 70%) PEG spacers (50–60 ethylene oxide units) in their composition (Rapp, 1996; Zalipsky et al., 1994). The chemical structures of PS and PS-PEG resins are shown in Figure 2.1. PEG units are immobilized on functionalized PS resin by means of, for example, anionic graft copolymerization. A PS resin bead with a diameter of 44 μm shows an increase in its diameter to 80 μm after being grafted with PEG to the extent that the PEG/PS ratio is 3/1 by weight. PS-PEG shows almost no solvent preference for resin swelling (Figure 2.2). This is due to the good miscibility of PEG chains with a majority of solvents. A structural account of the good solvent compatibility of PS-PEG resins was given by Meldal (1997). Because of the vicinal arrangements of carbon-oxygen bonds throughout the chain, PEG assumes helical structures with gauche interaction between the polarized bonds. There are

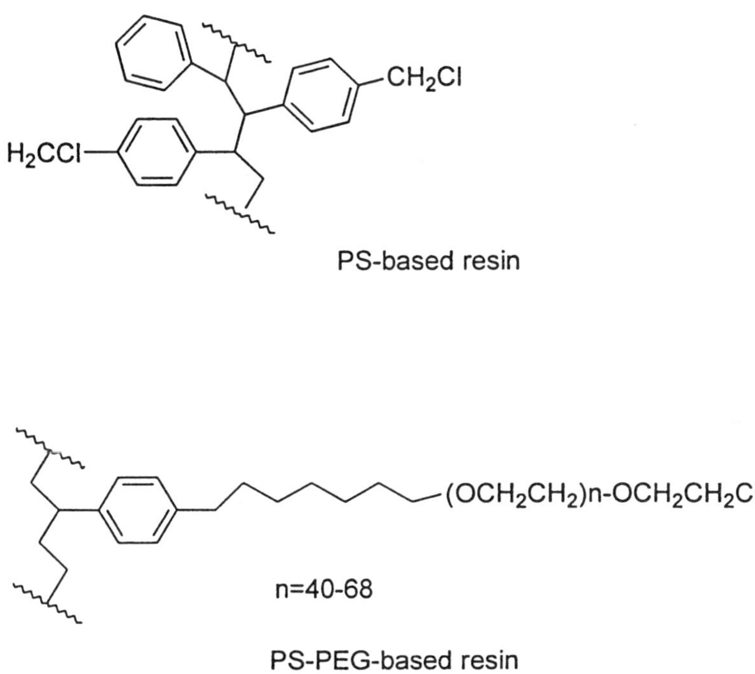

Figure 2.1 *Comparison of the chemical structures of PS- and PEG-PS-based resins.*

three possible helical arrangements of PEG that allow the low-energy gauche conformations of vicinal oxygen-carbon bonds. One is largely hydrophobic with the oxygen atoms in the interior of the helix; the second is of intermediate polarity; and the third is quite hydrophilic, exposing the oxygen atoms to the environment. The PEG molecule, therefore, has an amphipathic nature and is solvated well both by polar and nonpolar solvents.

Therefore, the properties of the PEG chains determine the mechanical, physicochemical behavior of the resin. Because the reactive sites in TG resin, one of PEG-PS resins, are located at the end of long, flexible spacers and are, therefore, well separated from the PS backbone, they are less affected by the hydrophobic PS matrix.

2.1.1.3 OTHER RESIN AND NON-RESIN SUPPORTS

Kieselguhr (or diatomaceous earth), a porous, inert, inorganic, and highly heterogeneous material prepared as granules from natural

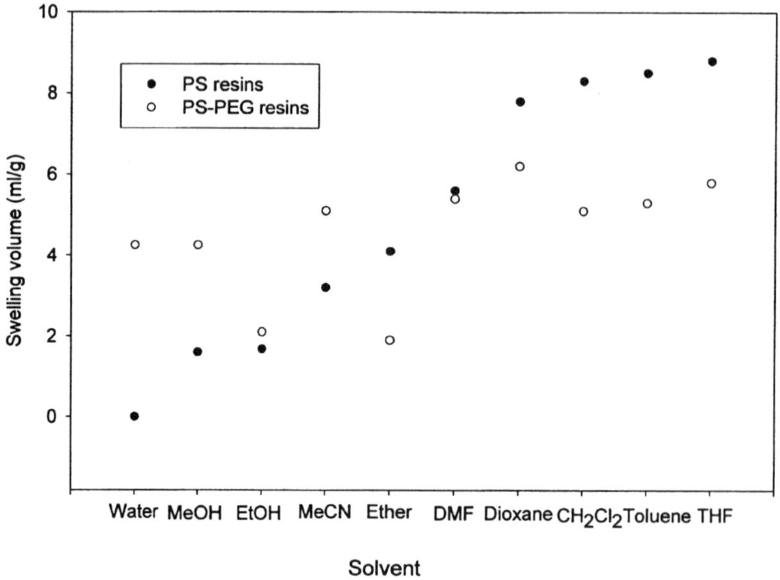

Figure 2.2 *Swelling property of PS and PS-PEG resins in different solvents. Data were from Rapp (1996).*

sediments, was used as a matrix for dimethylacrylamide polymerization to afford a resin that was stable to flow (Atherton et al., 1981).

Polypropylene shaped into arrays of pins or small caps attachable to pins (Bray et al., 1990) has been modified by grafting. Next, the surface was polymerized to form a polydimethylacrylamide coating. These rigid pins have been used for the simultaneous synthesis of up to 96 peptides. Microtube™ from IRORI (San Diego, CA) is a similar surface-modified polymer support used for SPOS (Li et al., 1998). Several other materials have been derivatized and used for solid-phase synthesis including beads made from sintered polyethylene, cellulose fibers (cotton, paper, Sepharose, and LH-20), porous highly crosslinked polystyrene or polymethacrylate, controlled pore glass, and silica. For more complex amounts of SPOS supports, see two perspectives in J.O.C. (1999).

2.1.2 Size Distribution

Resin beads vary in diameter, and the most commonly used sizes are 100–200 mesh (150–75 μm) and 200–400 mesh (75–37 μmm). In general, current PS-PEG resins have a more uniform size distribution than PS resins.

Macro PS-PEG resin (400–800 μm) is now commercially available and increases the capacity/bead to 100 or 1000 fold (nmol/bead), whereas the regular PS-PEG resin beads have diameters of 90–130 μm and 50–200 pmol/bead capacity. The compounds synthesized in the macro beads are enough for biological assays and for analytical studies such as HPLC and NMR on the materials released from a single macro bead.

2.1.3 Loading and the Site Distribution

Measured as molar equivalents of reactive groups per gram of resin, the loading capacities are typically 0.5–1.2 mmol/g and 0.2–0.4 mmol/g for PS and PS-PEG resins, respectively. Higher loading resins have been made in both categories. Grafted crowns or Microtubes have a low loading largely depending on the surface area. The most popular method for determining the loading capacity is to convert a functional group into a Fmoc derivative and then cleave the Fmoc groups for quantitation. Sometimes, elemental analysis is used if the resin contains elements such as N, Cl, Br, S, and I (Frechet & Schuerch, 1971; N'Guyen & Boileau, 1979; Frechet et al., 1979a & b; Mitchell et al., 1978; Forman et al., 1995; Yan et al., 1998). Grafted polymers such as crowns and Microtubes cannot be quantitated by the elemental analysis method, but their loading can be determined using the Fmoc-derivatization or electrochemical methods (Li et al., 1998).

Although manufacturers report the loading capacities of commercial resins and information on the method used for loading determination, the loading value of a resin can change over time because the reactive groups are not inert to storage and handling. Therefore, it is critical to determine the resin loading before the synthesis and to use the same batch of resin throughout the entire process of chemistry development and library synthesis.

Because resins are uniformly functionalized, the relative amounts of reactive groups on the bead surface and its interior can be estimated by volume calculations. If the "depth" of the surface is defined as 0.1 μm, the total amount of reactive sites on the surface is about 0.6% of the total reactive sites for a bead with a diameter of 100 μm. Therefore, more than 99% of these reactive sites are in the interior of the resin bead.

Autoradiography of a microtomed section of resin containing radiolabeled peptide showed random site location (Merrifield, 1969). Furthermore, the uniform distribution of reactive sites extends to beads of various sizes. For example, it has been found that resin beads labeled with radioactive peptide showed constant specific activity even though differing in diameter by a factor of five (Crowley et al., 1973).

2.1.4 Thermal Stability

Because many organic reactions require high temperatures, the thermal stability of resin beads is an important issue for SPOS. Dry resin exists in a glassy state, and an increase in temperature will transform the resin into a rubbery state. The temperature characterizing this transition is called the glassy temperature (T_g), and T_g values for polystyrene resins copolymerized with 5, 10, and 20% DVB are 102, 116, and 133° C, respectively (Sanetra et al., 1985). Polystyrene-based resins used for SPOS are commonly 1% DVB crosslinked PS, which should have a T_g of < 100° C. In highly crosslinked resins, the glassy transition virtually disappears, and the specific heat continuously increases with increasing temperature up to the threshold of thermal decomposition.

It is unknown if there is any effect of the glassy/rubbery state transition on the solvation status of resin at high temperatures and on the kinetics and efficiency of the resin-supported organic reactions. Some high temperature reactions have been carried out well above the T_g, for example, at 130° C (Sun & Yan, 1998), suggesting that this transition does not affect polymer-supported organic reactions. However, quantitative studies of the temperature effects are needed to ensure that other high-temperature organic reactions can be transferred to solid supports. On the other hand, organic reactions sometimes need to be carried out at low temperatures, and the effect of resin structure and solvation at very low temperature needs further study. Finally, it has been noted that another type of rubbery state can be obtained by swelling the resin with solvent and then drying it so that only a small amount of solvent remains (Wieczorek et al., 1982).

2.1.5 Chemical Stability

The DVB crosslinked polystyrene matrix is very stable. However, the functional groups on the resin are not. More studies are needed to completely understand their stability. For such studies, more quantitative analytical methods need to be developed. From single-bead FTIR studies, it has been found that trityl chloride resin is not stable and there always exists a low level of hydrolyzed product. By aldehyde quantitation studies, it has been found that the loading of aldehyde groups declines slowly with time. Other reactive functional groups may behave similarly during storage.

2.2 RESIN SWELLING AND SOLVATION

Gel-type (low crosslinked PS) resins swell to two to six times their original volume depending on the solvent used. In this type of resin, solvation is crucial to form a gel-phase for synthesis. The swelling of macroporous resins is more complicated, because they contain both pores and a polymer phase. The pores fill with solvent, and the polymer phase may swell as in gel-phase resins, except that the crosslink density is higher.

The mutual interactions between the solvent and resin bead result in significant effects on both. On one hand, as beads swell, their reactive chains become more mobile so that their reactivity and their influence on the accessibility of incoming reagent change. On the other hand, the mobility of solvent or reagent molecules is greatly reduced when they enter the bead microenvironment. These two aspects are further explained below.

2.2.1 Effects of Swelling on Resin

Regen (1975) studied the influence of solvent on the motion of molecules "immobilized" on PS matrices. The influence of solvent on the rotational motion of 2,2,6,6-tetramethyl-4-piperidinol-1-oxyl covalently bound to crosslinked PS, silica, and a PS ion-exchange resin has been examined by electron paramagnetic resonance spectroscopy. The degree of swelling of the polymer supports, as defined by the solvent employed, is an important factor in determining the mobility of the attached spin label. In the investigation, it was found that PS (Diagram 2.1) underwent extensive swelling in carbon tetrachloride with an accompanying increase in motion of the nitroxide moiety. When it was swollen in ethanol, slow nitroxide tumbling and little solvent uptake were observed (Figure 2.3). Examination of the influence of solvent on the ion-exchange resin (Diagram 2.1) revealed exactly the reverse effects. Only ethanol and 2-propanol swollen the ion-exchange resin significantly, and only with these solvents, a large increase in the motion of the spin label was observed (Figure 2.3). Evidently, the introduction of charged sites along the polymer backbone makes the lattice more compatible with ethanol and less compatible with carbon tetrachloride.

A.

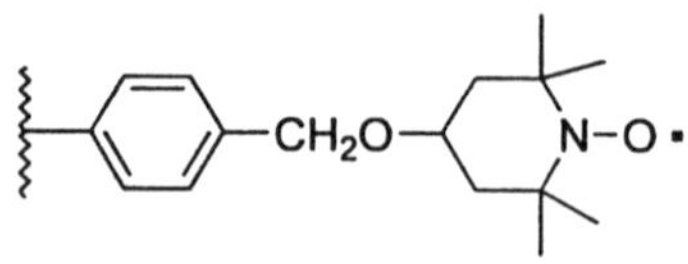

1

nitroxide-PS-2% DVB

B.

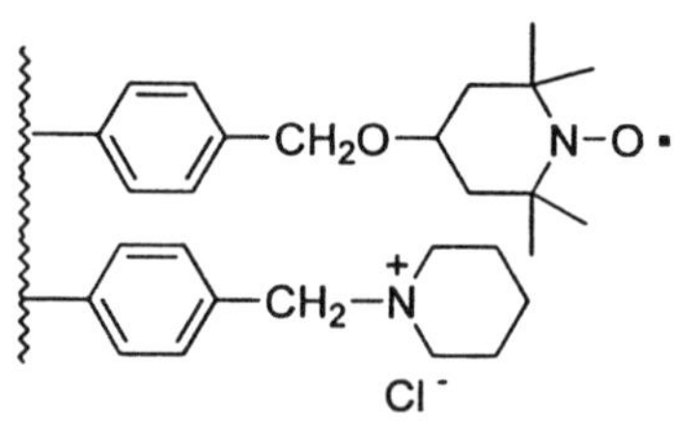

2

nitroxide-PS-2% DVB ion-exchange resin

Diagram 2.1

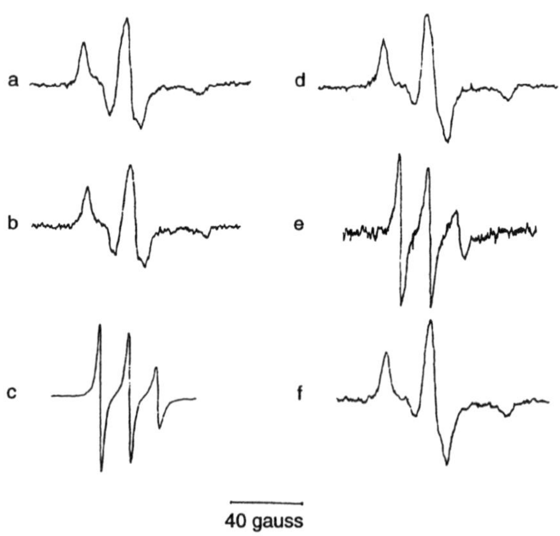

Figure 2.3 *EPR spectra of: (a) 1 in the dry state, (b) 1 swollen in ethanol, (c) 1 swollen in carbon tetrachloride, (d) 2 in the dry state, (e) 2 swollen in ethanol, (f) 2 swollen in carbon tetrachloride. (Reproduced with permission from* J. Am. Chem. Soc., *1975, 97, 3108–3112. Copyright© 1975 American Chemical Society.)*

Movement restrictions and the chain conformation in various swollen crosslinked PS gels were studied by ^1H NMR line-shape analysis (Schneider et al., 1985). The MAS ^1H NMR spectra of PS with 2–10% DVB or ethylene dimethylacrylate, swollen to equilibrium in CCl_4, $CDCl_3$ or benzene, were measured. The spectra were analyzed as a folding of a narrow line-shaped function approximated by the MAS NMR spectrum and an orientation-dependent dipolar broadening function. As indicated by the NMR spectra, the rate of conformational transition depends on solvent, temperature, and degree of crosslinking, but remained to be very rapid even at 10% crosslinking. It has been reported that resin structure, especially the tether length, is the dominant factor influencing ^1H NMR line widths, with the solvent playing a secondary role (Keifer, 1996).

Solvation, crucial for SPOS, is particularly important for PS resins. Although ^1H NMR line width of PS-PEG resins varies little with solvent selection (Figure 2.2), the reactivity of PS-PEG resins may be also influenced by the solvent, because the compatibility of the resin and the solvent with reagent and the transition state are also important. For example, the reaction rate of PS-PEG resins is not always higher than PS resins under identical conditions (Li & Yan, 1998).

2.2.2 Effects of Swelling on Reagent and Solvent Molecules

Regarding the effect of support on small molecules, the limited diffusion of reactant or reagent molecules into the bead interior and the reduced mobility of small molecules within the bead are particularly important.

If diffusion of soluble molecules into the active sites of the beads is rate limiting, this can result in the supported reactant displaying significant size selectivity. This is illustrated by the work of Grubb and Kroll (1971) on the hydrogenation of olefins. A catalyst was prepared by equilibrating 2% crosslinked polystyrene beads in which 8% of the repeat units were benzyldiphenylphosphine residues with a twofold excess of tris-(triphenylphosphine)-chlororhodium(I). When used as a catalyst with benzene as the solvent at 25° C under 1 atm of hydrogen, the relative rates of hydrogen uptake, with hexene arbitrarily set to be 100, were cyclohexene 39; cyclooctene 15; cyclododecene (cis-and trans-mixture) 8.8; and Δ^2-cholestene 1.2. For a similar soluble catalyst, the relative rates of hydrogen uptake, again relative to hexene arbitrarily set to be 100, were all in the range of 71–100. Therefore, when using a supported catalyst, the rate of reduction is inversely correlated to the molecular size of the olefin.

In the example above, the effect of the solid support on reaction rate is particularly pronounced because most of the catalytic sites likely involve two or more polymer-supported phosphine ligands, which will serve as extra crosslinks and restrict diffusion through the matrix. Furthermore, many of the active sites will be on these crosslinks. In most other reaction systems, access to the active sites is less of a problem. An example was shown in a single-bead FTIR study of the surface reaction rate vs. the rate in the bead interior for a series of esterification reactions (Yan et al., 1996a). No difference in rates on the bead surface and in its interior was found for those reactions. However, when the reaction is slowed down due to incompatibility or other problems such as an aqueous reaction on a polystyrene resin, a different reaction rate on the bead surface and in the bead interior may occur.

Solvent and reagent molecules, when entering the resin bead, undergo changes in their mobility and, perhaps, reactivity. Ford and coworkers (1984) studied the ^{13}C spin-lattice relaxation times (T_1) and NOEs of toluene in crosslinked polystyrene gel beads 16–18% ring substituted with benzyltri-n-butylphosphonium chloride groups. The ^{13}C spin-lattice relaxation time (T_1) is particularly useful because its measurement requires no isotope labeling and relaxation by the direct dipolar mechanism is intramolecular, allowing interpretation in terms of rotational correlation times. Toluene inside and outside the bead gave separate ^{13}C peaks for each carbon atom. The T_1 values and rate constants for exchange of toluene between the sites were measured by selective T_1 experiments with concurrent relaxation and exchange. The T_1 values of all protonated carbon atoms of toluene inside the bead were reduced to 0.05 to 0.33 times their values in low-viscosity toluene solution. The T_1 results are explained by slower rotational diffusion of toluene in the crosslinked polystyrene gels. The self-diffusion coefficient of toluene in the gel is estimated from the exchange rate constants to be 0.02 times that in toluene solution. Self-diffusion coefficients of toluene in swollen crosslinked PS beads with 5.7–40% crosslinking and in linear PS solution were determined by pulsed-field-gradient spin-echo (PGSE) NMR experiments (Pickup et al., 1986). A comparison of solvent diffusion in the crosslinked and the non-crosslinked systems at similar concentrations shows remarkably close agreement. These results suggest that the diffusion rate in the crosslinked polymer beads is not significantly affected by the presence of crosslinks and is mainly determined by the concentration of solvent in the bead. Self-diffusion coefficients of toluene, acetonitrile, chloroform, and dichloromethane in swollen 1–20% crosslinked PS beads were determined by application of a new model to magnetization-transfer ^{13}C NMR data for exchange of solvent in and out of the beads (Ford et al., 1987). In 1% crosslinked beads, the self-diffusion coefficient of dichloromethane is reduced only by a factor of two compared with bulk liquid. The poorest swelling solvent acetonitrile shows a reduction in its self-diffusion coefficient by a factor of 15. By comparing interactions of cyclohexane, tetrahydrofuran, and acetonitrile with swollen PS beads using ^{13}C NMR spectroscopy, it was found that the mobility of solvent molecules was affected by the crosslinking density in the case of good swelling solvents, whereas little dependence of the crosslinking density was seen for poor solvents (Ogino & Sato, 1995). These NMR studies show that different solvents have different self-diffusion coefficients in swollen beads depending on their swelling ability, beads crosslinking density, and specific interactions of each solvent with resin. Kurth and coworkers (Wilson et al., 1998) studied solvent and reagent accessibility of PS resin crosslinked with α, ω-bis-4(vinylbenzyl) ethers of mono-, di-, tetra-, and hexa-9-ethylene glyco. They found that the swelling of 2, 10, and 20% crosslinked PS-PEG beads was higher or comparable to 2% PS beads. Fluorescence quenching of polymer-bound dansyl groups by $Et_3O{\cdot}BF_4$ indicated that the most rapid intra-resin diffusion of this electrophilic reagent occurred in toluene. PS-PEG beads exhibited faster quenching than 2% PS beads in poor swelling solvents but were slower to quench than 2% PS beads in toluene. It is expected that the diffusion of reagent molecules into resin will depend on the molecular properties of the reagent and the properties of the resin.

2.2.3 Swelling Ability of Solvents

Regen (1974) used the spin-labeled technique in examining the mobility of a nitroxide bound to crosslinked PS resin (A in Diagram 2.1) in the solvent swollen state. His results established that the choice of swelling solvent has a substantial influence on the physical nature of the resin-bound nitroxide and the dominant role of solvent is to define the degree of swelling of the PS lattice (Figure 2.4). A distinct trend is that the rotational correlation time of the spin label decreases as the degree of swelling of the polymer increases. Those solvents that swell PS matrices the most will allow for the greatest mobility of the substrates bound to them. The solvation of resin by single-solvent and mixed-solvent systems can be correlated to the Hildebrand solubility (δ) and hydrogen-bonding solubility (δ_h) parameters (Fields & Fields, 1991). A systematic study of the correlation between resin solvation and solvent properties revealed the importance of the sum of solvent

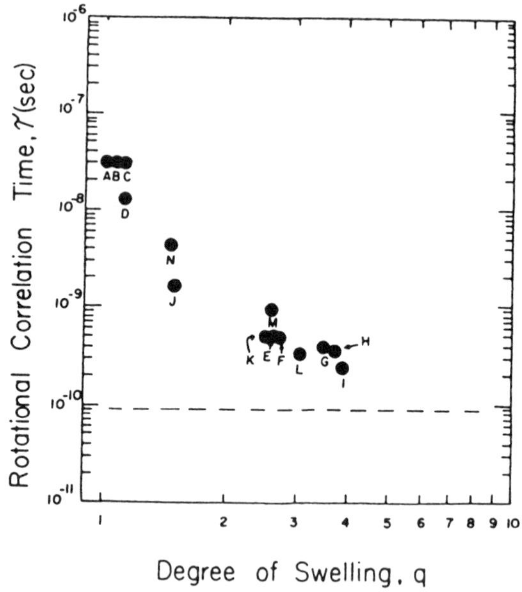

Figure 2.4 *Plot of τ as a function of q for 1 swollen in (A) no solvent, (B) ethanol, (C) 2-propanol, (D) acetonitrile, (E) ethyl acetate, (F) N, N-dimethylformamide, (G) carbon tetrachloride, (H) dichloromethane, (I) benzene, (J) ethanol in benzene (67% v/v), (K) ethanol in benzene (33% v/v), (L) ethanol in benzene (17% v/v), (M) 4% cross-linked swollen in benzene, and (N) 12% crosslinked resin swollen in benzene; the dashed line (------) represents the value of τ measured for a benzene solution of nitroxide on a non-crosslinked polymer. (Reproduced with permission from* J. Am. Chem. Soc., *1974, 96, 5275–5276. Copyright© 1974 American Chemical Society.)*

electron acceptor (AN) and electron donor (DN) numbers (Cilli et al., 1996). This method measures the capacity of the solvent to act as a Lewis acid or Lewis base, a donor or acceptor of electron pairs, such as AN and DN numbers, respectively. According to this theory, the solvent effect upon a solute molecule is generally considered as an acid-base type interaction. AN represents the electrophilic property of the solvent and is related to the relative chemical shift of ^{31}P in triethylphosphine in the particular solvent, with hexane as a reference solvent on the one hand and triethylphosphine oxide-$SbCl_5$ in 1,2-dichloroethane on the other, to which the acceptor of 0–100 has been assigned. In contrast, the DN number is representative of the basic (nucleophilic) property of the solvent molecule and is defined as the molar enthalpy for the reaction of the donor with $SbCl_5$ as reference acceptor in 1,2-dichloroethane. A

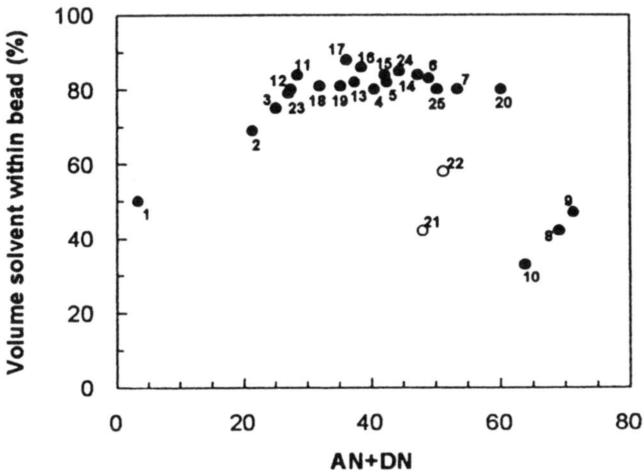

Figure 2.5 *Swelling of resin as a function of solvent (AN+DN) value. Symbols l-25 represent different solvent systems (see Cilli et al., 1996 for details). (Reproduced with permission from* J. Org. Chem., *1996, 61, 8995. Copyright© 1996 American Chemical Society.)*

plot of swelling versus (AN+DN) displays bell-shaped curves with maximum solvation region characteristic of each class of resin, which, in turn, depends on the nature and amount of resin-bound peptide sequence (Figure 2.5). The occurrence of a maximum solvation region in this plot is important for both SPPS and SPOS methodologies, because it facilitates the selection of the most appropriate solvent system for each step of the synthesis cycle. However, experimental select of solvent is now used instead of theoretical prediction.

Beads must remain swollen throughout the reaction. When the loading of reactive groups is high, the swelling properties can change considerably as the original functional groups are transformed into others, especially when these transformations involve a change in polarity. An extreme case is the reaction of Merrifield resin with tertiary amines to give quaternary ammonium salts, where the polymer changes from having relatively non-polar groups to having ionic groups. Such dynamic changes in resin-swelling properties are discussed next.

2.2.4 Dynamic Resin Solvation

Swelling is a macroscopic term to describe resin solvation, which is the association of solvent molecules with the resin spacer, linker, and the bound-substrate to create an enlarged bead. The proper solvation of the bound substrate and external reagent is crucial for the reaction to occur, and the success of SPPS and SPOS depends on the accessibility of the growing resin-bound substrate to reagent. Resin-bound reaction products are expected to have physicochemical properties that differ from the initial polystyrene resin, especially if the attached compound introduces a component of different polarity to the resin. This additional component might require a solvent of different polarity to ensure optimal solvation and, hence, accessibility.

Although studies of the properties of most resin-bound organic substances are still lacking, resin-bound peptides have been well-characterized. During the course of a multi-step synthesis, both the size of bead and its swelling characteristics will change. Merrifield (1969) showed that the diameter of a dry peptide resin bead on which a peptide with a MW of 5957 was assembled increased from 40 to 94 μm. Thus, the relative volume increase was five-fold. There was no indication that the resin matrix became filled with peptide, even at a peptide/resin weight ratio of 4:1, or that the efficiency of the synthesis became limited by increased steric hindrance. In CH_2Cl_2, the swelling ratio was 1.84 for unloaded resin and gradually decreased as the peptide content increased. It became constant at 1.33 after the peptide content reached 2.5 g per gram of resin. In DMF, the ratio was 1.49 for unloaded resin and increased to 1.84 for the same peptide-loaded resin.

In SPPS, many peptides have not been prepared in adequate yields due to unknown causes. Some of the problems include incomplete coupling and deprotection steps, the formation of truncated peptides, and the deletion sequences. To gain further understanding of these synthetic problems, Marshall and coworkers (Hancock et al., 1973) investigated the source of truncated and deletion sequences in SPPS and found that dynamic solvation was a major factor. In another example, MAS NMR characterization of the SPPS product of the difficult sequence formed by the 10-residue peptide NH_2-Thr-Glu-Gly-Glu-Ala-Arg-Gly-Ser-Val-Ile-OH (Dhalluin et al., 1997) showed that resin swelling influenced the outcome of the reaction at critical synthetic steps. A detailed secondary structure determination of the growing peptide on the resin beads, based on NOE analysis and the ^1H and ^{13}C chemical shift deviations, indicated an extended structure on the whole length of the sequence for peptides with less than eight residues (Figure 2.6). For 9- and 10-residue peptide resin, spectral broadening indicated aggregation of peptide. The addition of H-bond accepting DMSO disrupted peptide aggregation caused by inter-peptide chain H-bonding, improved the ^1H NMR spectral linewidth, and increased reaction yield (36% to 61% for 10-residue peptide and 58% to 67% for

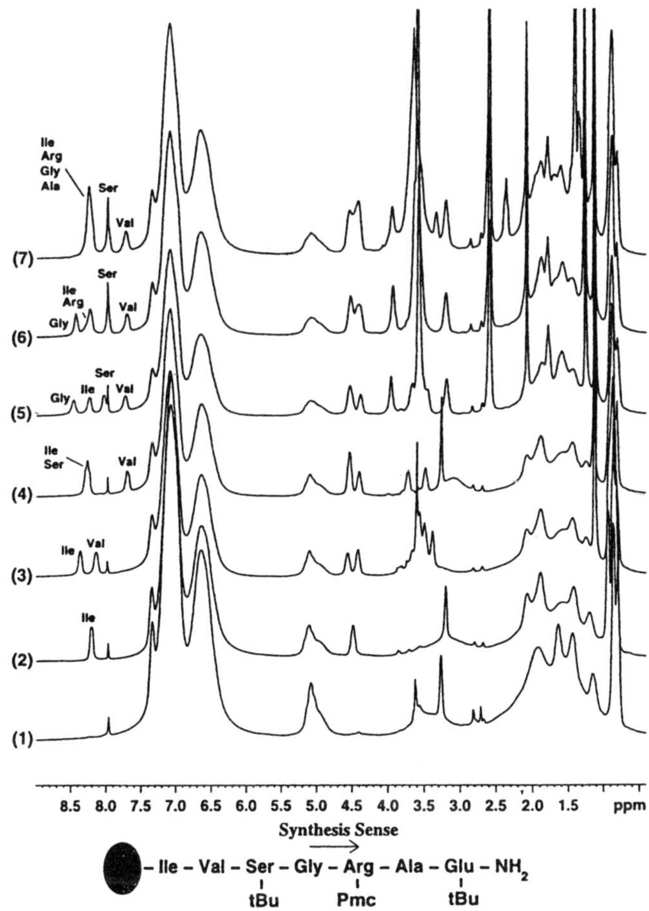

Figure 2.6 *[1]H 1D MAS spectra for the first seven peptidyl-resins (resin-Ile-Val-Ser(Ot-Bu)-Gly-Arg(Pmc)-Ala-Glu(Ot-Bu)-NH$_2$) swollen in DMF-d$_7$ at room temperature and 4 kHz spinning rate. The assignment of the NH resonances at each synthesis step is shown. (Reproduced with permission from* J. Am. Chem. Soc., *1997, 119, 10494–10500. Copyright© 1997 American Chemical Society.)*

9-residue peptide) at the same time (Figure 2.7). This is because the disruption of peptide aggregates allowed the remaining residual free amino functions to be more accessible to the deprotecting and acylating reagents, resulting in better coupling efficiency at critical synthetic steps.

According to the Merrifield strategy, SPPS begins in an apolar environment due to the PS matrix, and the environment evolves gradually toward a more polar nature when the peptide chain grows. If the

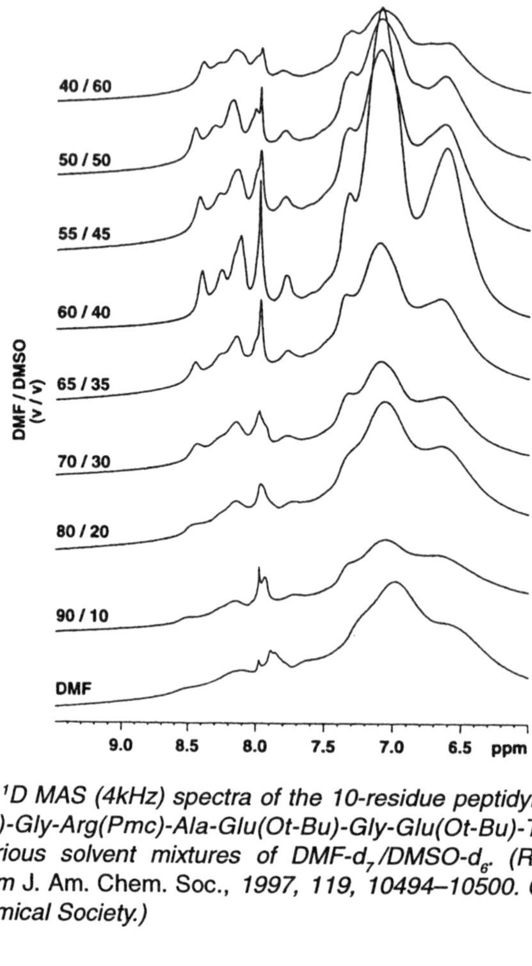

Figure 2.7 *^1H ^1D MAS (4kHz) spectra of the 10-residue peptidyl-resin (resin-Ile-Val-Ser(Ot-Bu)-Gly-Arg(Pmc)-Ala-Glu(Ot-Bu)-Gly-Glu(Ot-Bu)-Thr(Ot-Bu)-NH_2) swollen in various solvent mixtures of DMF-d_7/DMSO-d_6. (Reproduced with permission from* J. Am. Chem. Soc., *1997, 119, 10494–10500. Copyright© 1997 American Chemical Society.)*

nature of the solvent does not evolve in a parallel way, aggregation and crosslinking of the peptide-complex (depending on the sequence, the side chain protection groups, and the resin loading) may result in and lead to an incomplete solvation of the peptide resin, shrinkage of the resin, reduced reagent penetration, and finally a decrease in coupling yield. In summary, chain mobility is essential for optimal yields and may be enhanced by the use of polar solvent to swell the peptide-resin; and DMSO, when added to DMF, may be used to disrupt peptide aggregation. These studies of SPPS are relevant to SPOS where dynamic solvation problems need to be considered in combinatorial or parallel synthesis.

2.3 EFFECTS OF THE SOLVATED RESIN ON SOLID-PHASE REACTIONS

The resin support has significant effects on the physical and chemical nature of the attached reactive groups and entrapped reagent molecules. These effects are incompletely understood so far, but some aspects are emerging as follows:

- the effects of solvated resin on SPPS
- resin support as another "solvent phase"
- support effects on SPOS reaction kinetics
- site separation and site interaction
- support effects on product purity

2.3.1 The Effects of Support Solvation on SPPS

The influence of the chain length on the coupling reaction in SPPS was investigated by Hagenmaier (1970) for the synthesis of the dodecapeptide $(Leu\text{-}Ala)_6$ on Merrifield resin (2% DBV, 200–400 mesh). The method of Dorman (1969) was used to determine the efficiency of the coupling reaction. The first six steps of synthesis gave almost quantitative yields after repetition of the coupling reaction using fresh BOC-amino acid plus dicyclohexylcarbodiimide. In later coupling steps, yield decreased steadily with increasing chain length and was more pronounced for leucine than for alanine due to greater steric hindrance in the case of leucine. Carrying out the coupling reaction for a third time did not increase the yield, which suggests that aggregation of the elongated peptide chains prevented reagent access as discussed in the previous section (Dhalluin et al., 1997).

In a study comparing various solid supports for SPPS, five different polymers [polystyrene-1% DVB (I), polystyrene macroporous (II), Kel-F-g-styrene (III), polyacrylamide (IV), and controlled pore glass (V)] were tested in order to evaluate the influence of the chemical and physical conditions of the support on the reaction (Albericio et al., 1989). The Fmoc-(60–67)-uteroglobin-OH-protected segment was coupled onto H-(68–70)-uteroglobin previously assembled on each of the solid supports (Diagram 2.2). Microporous polystyrene (I) and polyacrylamide (IV) appear to be the most convenient supports for segment coupling. This correlates very well with the smaller line widths at half-height observed for supports I and IV in gel-phase ^{19}F NMR spectra in $CDCl_3$ (Figure 2.8) or DMF and in ^{13}C NMR spectra of the same materials observed in $CDCl_3$. The faster coupling rates observed using (I) (2 h,

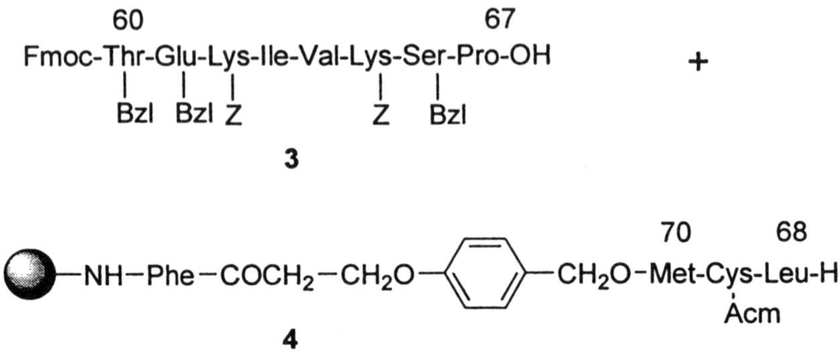

Diagram 2.2

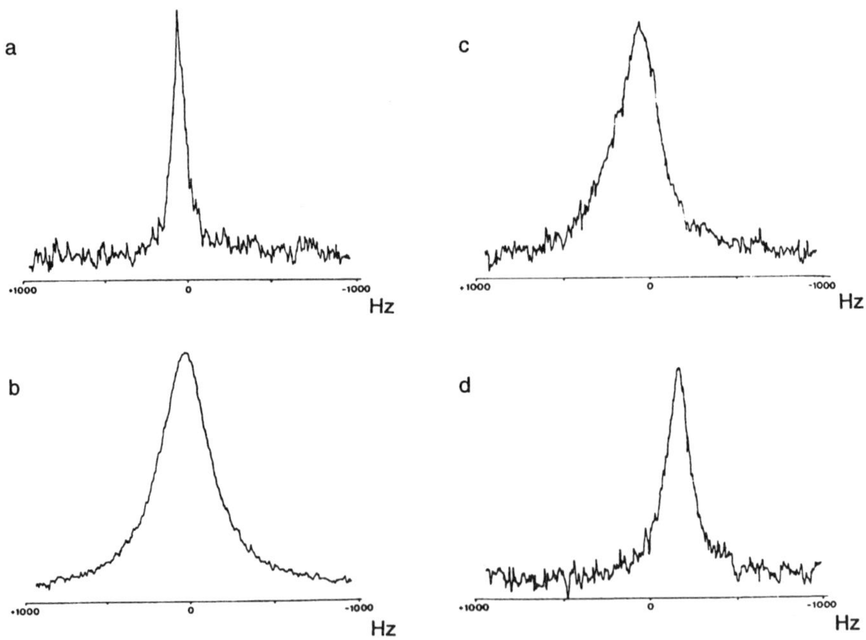

Figure 2.8 *^{19}F NMR signal of Boc-5-flurotryptophan (BFT) bound to support I (a), II (b), III (c), and IV (d) swollen in $CDCl_3$. Spectra were run at 1888 MHz, with a 30-kHz spectral width using 90° pulses and an 8-s recycling time. (Reproduced with permission from* J. Org. Chem., *1989, 54, 360–366. Copyright© 1989 American Chemical Society.)*

31% yield and 7 h, 100%) as compared to IV (2 h, 24% and 7 h, 87%) are in agreement with the T_1 measurements (I: 0.95 $CDCl_3$, 0.71 DMF; IV: 0.61 $CDCl_3$, 0.50 DMF) that show a clear difference between resin I and the remaining resins.

2.3.2 Resin Support as Another "Solvent Phase"

Evidence has accumulated to start revealing the nature of the resin support as a solvent phase.

2.3.2.1 SELECTIVE ABSORPTION OF REACTANT AND REAGENT MOLECULES

Polymer-supported reaction occurs when the substrate enters the resin. One factor determining reactivity is the polarity of the resin. On an anion-substituted macroreticular resin, the selective absorption of the polar substrate into the resin increased the rate of reactions involving polar reactants (Cainelli et al., 1982).

Polymer supports can selectively absorb reactants and create a more concentrated reactant pool inside the resin bead. For the reduction reaction shown in Scheme 2.1, both the starting material and the reagent were absorbed into a polymer support, so that the reaction yield and the stereochemical selectivity were greatly improved compared to control experiments with no polymer present. The effect was

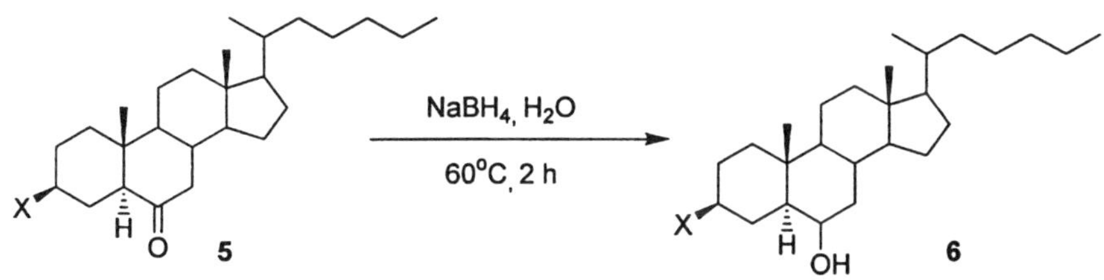

1. X = OH yield: 0% (no polymer), 65%(2% DVB anionic polymer)

2. X = H 63% (no polymer), 82% (2% DVB anionic polymer)

Scheme 2.1

further enhanced when a catalytic amount of tetra-n-butylammonium bromide (TBAB) was added. Due to the structural differences between the gel-type PS resins and the macroporous PS resins, only the latter improved the stereochemistry (Briggs & Hodge, 1988).

Regen and Dulak (1977) investigated the cosolvent-like property of PS-PEG resins. Illustrated in Scheme 2.2, the polar PS-PEG resin significantly enhanced the rates and yields of these aqueous reactions, and two possible mechanisms for this resin effect might be (1) the polymer may provide a common phase where a high effective concentration of the reactants is possible; (2) the microenvironment within the resin phase may influence starting material and transition state stability.

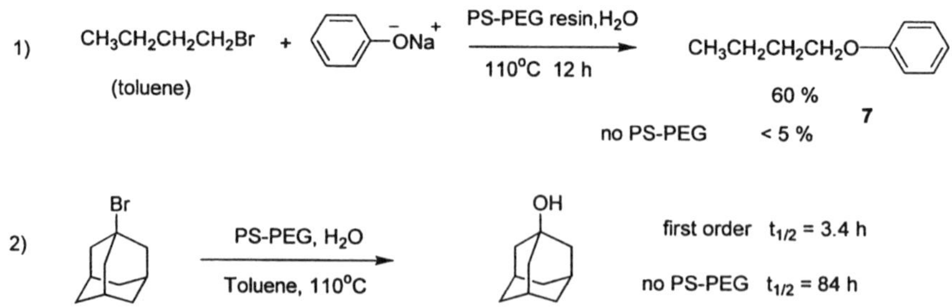

8

Scheme 2.2

2.3.2.2 THE EFFECT OF PHASE-TRANSFER CATALYST (PTC) ON SPOS REACTIONS

Without using PTC, the reactions shown in Scheme 2.3 gave 30% yield; with PTC, yields of 88–98% were achieved (Frechet et al., 1979a). PTCs have also facilitated a range of S_N2 reactions on Merrifield resin using amines, acids, salt, alcohol, and phenols (N'Guyen & Boileau, 1979). Three series of aqueous SPOS reactions (nucleophilic substitutions on Merrifield resin; nucleophilic addition of CN^- to formylpolystyrene resin; and reactions involving polymeric nucleophiles) were studied by Frechet and coworkers (Frechet et al., 1979b). These results demonstrated remarkable success for phase-transfer catalysis. These results indicated that the swollen resin can be viewed as another "solvent phase" in terms of their effects on organic substrates and reactions.

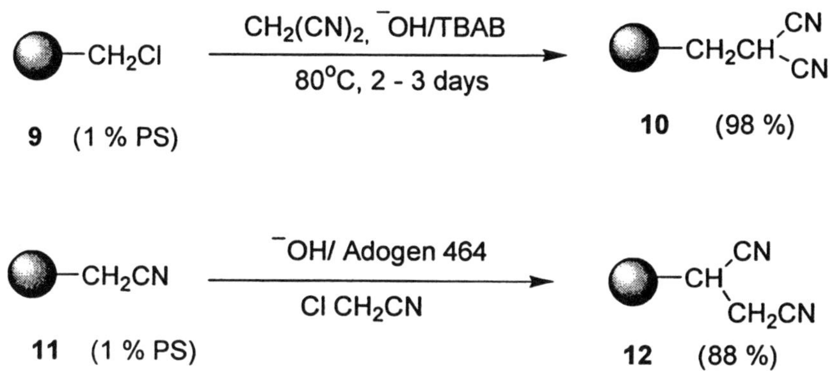

Scheme 2.3

2.3.3 Support Effects on SPOS Reaction Kinetics

The stereotypical view of a resin bead is a rigid, non-solution-like matrix in which organic reagents have to overcome a steric barrier in order to diffuse into the resin bead. Although the resins are usually heavily solvated (swollen resin beads contain 80% solvent), reactions are undoubtedly affected by the proximity of the attached reactant to the spacer and the polymer matrix. Synthesis using soluble polyethyleneglycol (PEG) polymers was sought in order to take advantages of certain benefits of both solution and solid-phase synthesis. An alternative approach has been the development of PS-PEG resins designed to combine useful features from both insoluble PS support and the soluble PEG support.

Kinetics of a series of organic reactions (Scheme 2.4) on both PS and PS-PEG resins have been determined using single-bead FTIR microspectroscopy (Li & Yan, 1998). It is generally assumed that reactions proceed faster on PS-PEG resin because it provides a more "solution-like" matrix compared to PS resins. The so-called "solution-like" matrix means that the attached reactant is floating in a mobile PEG phase within the PS-PEG resin instead of being attached to an "immobile" PS phase. This hypothesis may explain some results of solid-phase peptide synthesis and some organic reactions involving polar reagents. For example, oxidation of an alcohol (reaction 1 in Scheme 2.4) is more rapid on PS-PEG than on PS resin. However, of the four reaction types studied in this investigation, two reactions exhibited higher reaction rate when carried out on PS resins than that on PS-PEG resins.

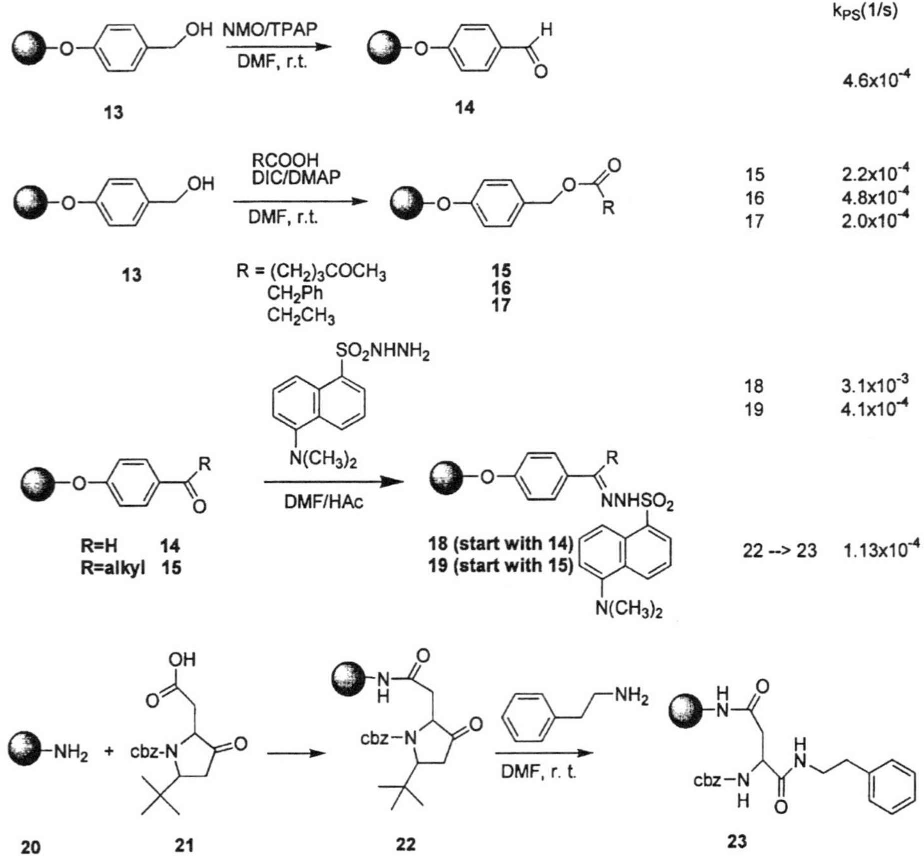

	k_{PS}(1/s)	k_{TG}(1/s)	k_{TG}/k_{PS}
	4.6×10^{-4}	1.8×10^{-3}	3.9
15	2.2×10^{-4}	2.3×10^{-4}	1.0
16	4.8×10^{-4}	4.2×10^{-4}	0.9
17	2.0×10^{-4}	2.2×10^{-4}	1.1
18	3.1×10^{-3}	1.8×10^{-3}	0.6
19	4.1×10^{-4}	1.9×10^{-4}	0.5
22 --> 23	1.13×10^{-4}	6.26×10^{-6}	0.055

Scheme 2.4

Table 2.1 Comparison of reaction rates on PS- and TG-based resins.

Reaction Scheme	k_{PS} (1/s)	k_{TG} (1/s)	k_{TG}/k_{PS}
1	4.6×10^{-4}	1.8×10^{-3}	3.9
2 (R = 3)	2.2×10^{-4}	2.3×10^{-4}	1.0
2 (R = 4)	4.8×10^{-4}	4.2×10^{-4}	0.9
2 (R = 5)	2.0×10^{-4}	2.2×10^{-4}	1.1
3 (R = H)	3.1×10^{-3}	1.8×10^{-3}	0.6
3 (R = 3)	4.1×10^{-4}	1.9×10^{-4}	0.5
4 (step 2)	1.13×10^{-4}	6.26×10^{-6}	0.055

ESR studies have shown that the rate of movement of the pendant group is high in swollen PS resin. Merrifield (1984) pointed out that this extensive chain flexibility plays an important role in the rate and extent of the reactions in solid-phase peptide synthesis. In our previous kinctics studies (Yan et al., 1996a; Yan et al., 1996b; Sun and Yan, 1998; Li and Yan, 1998), we found that, except for trityl-like linkers, the proximity of the reactive group to the spacer and the PS backbones in general poses little steric hindrance. This may be due to the fact that full solvation and the dynamic fluctuation of the PS matrix allow free access of reagent into the resin bead. The steric hindrance assumption cannot explain the results in Table 2.1. What then determines the relative reaction rate on different resin supports?

Czarnik (1998) summarized some key observations, including ours, on this subject and pointed out that solid-phase synthesis supports are like solvents. The effect of polymer matrix on the reaction rate may be very similar to the effect of solvent on the rate of a solution reaction. A swollen bead is the ultimate microreactor for SPOS. Although 80% of the bead is solvent, spacers and matrix exert significant effects on incoming reagents and reactants depending on their molecular properties, such as polarity. The solvent-like support effect and the effect of phase-transfer catalysts on the solid-phase reaction discussed in the previous section help substantiate this hypothesis. A complete understanding of how the structure of a support affects the reactivity of an attached reactant awaits further investigations.

2.3.4 Site Separation and Site Interaction

Although solvent swelling can improve resin accessibility to reagents, it also creates some unwanted effects, such as site-site interactions. Intuitively, the site-site distance should increase with swelling of the resin, but, actually, the dynamic distance between reactive sites becomes smaller so that site-site interactions occur. Pendant moieties on PS resins do not maintain the separation implied by percent substitution and the well-defined bond lengths in polystyrene. The expected site-separation and "pseudo dilution" do not hold after resin swelling. Because of the long PEG spacers, site-site interaction is much more extensive in PS-PEG resins. Because the interactions between reactive groups inside the resin can alter the SPOS reactions, it is critical to fully understand the site-separation and site-interaction effects in PS resins. In this section, the sequence of discussion will be as follows:

- site separation effect
- site interaction effect
- factors affecting site separation/interaction

2.3.4.1 SITE SEPARATION EFFECT

Cyclic peptides are usually prepared from linear peptides by intramolecular cyclization reactions. The carboxylic end of the peptide is usually activated by the formation of an active ester, anhydrides, azides, or acid chlorides, and the free terminal amino end allowed to react with the active terminal carbonyl group at high dilution. Because of intermolecular condensation occurring under these conditions, linear oligopeptides are formed in addition to the desired cyclic peptide, so that solution-phase synthesis results in reaction mixtures from which cyclic peptides are usually isolated in relatively low yields.

Fridkin et al. (1965) used the following two polymer reagents in the preparation of the peptide esters: crosslinked poly-4-hydroxy-3-nitrostyrene and a branched copolymer of DL-lysine and 3-nitro-L-tyrosine. Cyclo(Gly-Gly) was synthesized on these two polymers in 75% yield (DMF, r.t. 12 h). Cyclo (tetra-L-Ala) was also synthesized on the polymer (r.t. 12 h) with yields of chromatographically pure peptide of 50–60% compared to 40% obtained from solution-phase synthesis from tetra-L-Ala ester (pyridine, 100° C, 12 h).

Flanigan and Marshall (1970) also reported solid-phase synthesis of cyclic peptides at higher overall yields with a minimization of racemization during activation and subsequent cyclization. Five cyclic peptides were synthesized on polymer supports with yields of 40–60%.

Camps et al. (1971) observed hydroxyl band splitting (3580 and 3460 cm^{-1}) and pointed out that in 2% DVB PS, the hydroxyl groups might be divided into free form at 3580 cm^{-1} and the associated (hydrogen-bonded) hydroxyl groups at 3460 cm^{-1}. The experimental proof was obtained later by demonstrating that the addition of the hydrogen bond-accepting solvent DMSO converted the free hydroxyl band at

3580 cm^{-1} to the hydrogen-bonded band at 3460 cm^{-1}. Alternatively, increased steric hindrance around the hydroxyl group converted the band at 3460 cm^{-1} to the free hydroxyl band at 3580 cm^{-1} (Yan and Sun, 1998).

Carboxyl resin exhibits two carbonyl bands in the IR spectrum, a strong band at 1694 cm^{-1} and a weak band at 1727 cm^{-1}. Letsinger and coworkers (1964) pointed out that the former may be attributable to the hydrogen-bonded carboxyl groups and the latter to the free carboxyl groups.

The most important factor governing intrapolymeric reaction is the frequency of encounters between reactive substituents. Unlike the diffusion-limited rate of collisions between monomeric reagents in solution, the encounter rate of reactive groups in a polymer is limited by the potentially more restricted conformational motions of the polymer backbone. Mazur and Jayalekshmy (1976, 1979) carried out a crude measurement of the encounter frequency of substituents focusing on the chemistry of *o*-benzyne (1,2-dehydrobenzene) covalently bound to a PS support (Scheme 2.5). While monomeric benzyne undergoes very rapid dimerization in solution, this pathway was found to be greatly retarded for the polymer-bound analog. *o*-Benzyne was first

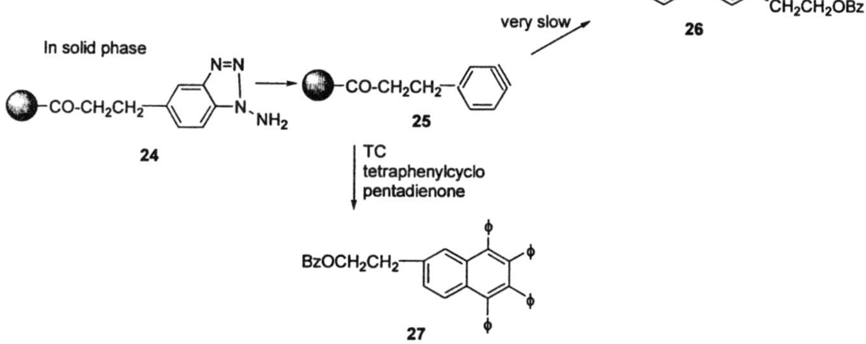

Scheme 2.5

generated by oxidation of 1-aminobenzotriazole derivatives bound to various PS supports. Upon oxidization, the benzyne intermediate was converted to a mixture of isomeric aryl acetates. This reaction was not observed for monomeric 1-aminobenzotriazole, where dimerization was the predominant pathway. The frequency of encounters between the polymer-bound benzyne intermediates allows the relatively slow acetate formation to compete with the encounter-controlled dimerization process. Delayed trapping of the benzyne intermediate by a diene through Diels-Alder reaction demonstrated that the frequency of encounter between the intermediates bound to a 2% crosslinked PS resin must lie between 2.6×10^{-2} and 1.1×10^{-3}. This prolonged lifetime for the resin-bound benzyne molecules indicated considerable site separation within the bead.

Scheme 2.6 shows an intrapolymeric glycyl transfer or intraresin nucleophilic reaction (Rebek Jr. & Trend, 1979). The empirical value of one to two minutes for the half-life of this compound, assuming a second-order process for its disappearance, was derived from an acid anhydride quenching study. If acyl transfer between sites of the compound do not occur at every encounter, this half-life corresponds to a lower limit for the encounter frequencies of reagents attached to such polymers. This figure agrees well with that derived by Mazur and Jayalekshmy (1976, 1979) for the dimerization rate of polymer-bound benzyne.

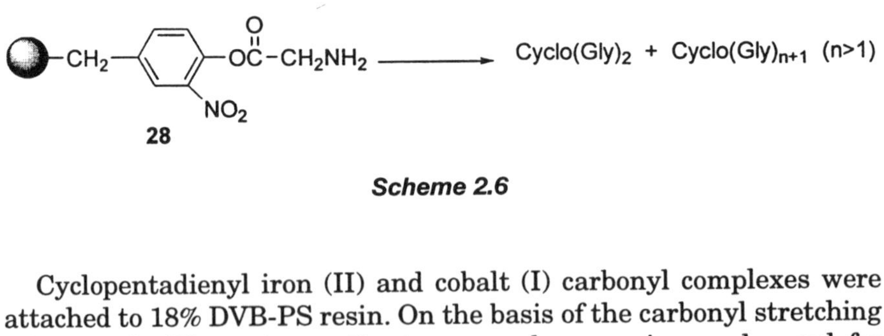

28

Scheme 2.6

Cyclopentadienyl iron (II) and cobalt (I) carbonyl complexes were attached to 18% DVB-PS resin. On the basis of the carbonyl stretching vibration in IR spectrum, the iron complex remains unchanged for months at room temperature under exclusion of light and air. In contrast, solution phase C_5H_5-$Fe(CO)_2H$ has been observed to be a highly unstable species, which rapidly loses H_2 to form binuclear $(C_5H_5)_2Fe_2(CO)_4$. The absence of any μ-CO absorption, typical of such a binuclear particle, from the polymer-supported iron compound substantiates the view that attachment to a crosslinked polymer can suppress any metal-metal interactions and that the unusual stability of the polymer-supported iron hydride species is a consequence of their mutual isolation by the polymer network (Gubitosa et al., 1977).

Similarly, photolysis of $C_5H_5Co(CO)_2$ in solution leads to a succession of di- and tri-nuclear clusters such as $(C_5H_5)_2Co_2(CO)_2$, $(C_5H_5)_2Co_2(CO)_3$ and $(C_5H_5)_3Co_3(CO)_3$ characterized by IR absorption around 1970, 1800, and 1670 cm^{-1}. However, when polymer-bound $C_5H_5Co(CO)_2$ is irradiated in petroleum ether (a poor swelling solvent) suspension at ambient temperature with an unfiltered medium-pressure Hg or Xe lamp, a disappearance of the absorption at 1960 and 2020 cm^{-1} was observed, indicating the photodissociation of the CO ligands. No trace of any μ-CO band is observed at wavelength below 1900 cm^{-1}. These data clearly indicate the absence of any CO-bridged, di- or polynuclear species (Gubitosa et al., 1977).

Taking advantage of the site-site separation effect, Leznoff has performed a series of successful monofunctionalization reactions of diols and dialdehyde (Leznoff & Wong, 1972; Wong & Leznoff, 1973a; Wong & Leznoff, 1973b).

2.3.4.2 SITE INTERACTION EFFECT

Titanocene with various loading was attached to 20% DVB PS resin, and the hydrogenation of the 1-hexene was studied by polymer-supported Cp_2TiCl_2 at room temperature. The rate of hydrogenation in $mLmin^{-1}$ $mmol^{-1}$ of Ti indicated the percentage of the titanium centers that were active. As the loading increased, the percentage of metal centers close enough to interact increased, and, consequently, the number of potential active sites eventually decreased due to site-site interaction and destruction. The rate of hydrogenation of 1-hexene as a function of the loading of Ti on the polymer beads was determined, and the result was a bell-shaped curve with a maximum (Figure 2.9) (Grubb et al., 1977). This indicated that the rate of catalysis increased as the Ti loading increased until site-site interaction caused a loss of active sites.

A relationship between resin-bound mercapto groups was studied (Wulff and Schulze, 1978). Because mercapto groups can be selectively re-oxidized to disulfides (see Scheme 2.11 later), their spatial relationship can be quantitatively determined on the basis of the re-oxidized portion. The reoxidation of the SH-groups was carried out with iodine in methanol. Polymers with less crosslinking exhibited higher degrees of reoxidation (69%). Remarkably, a 33% reoxidation has been observed even in highly-crosslinked polymers, suggesting that the polymer chains retain considerable flexibility, making reactions possible between widely distributed SH-groups. A higher degree of oxidation of mercapto groups in resin swollen in benzene (a good swelling

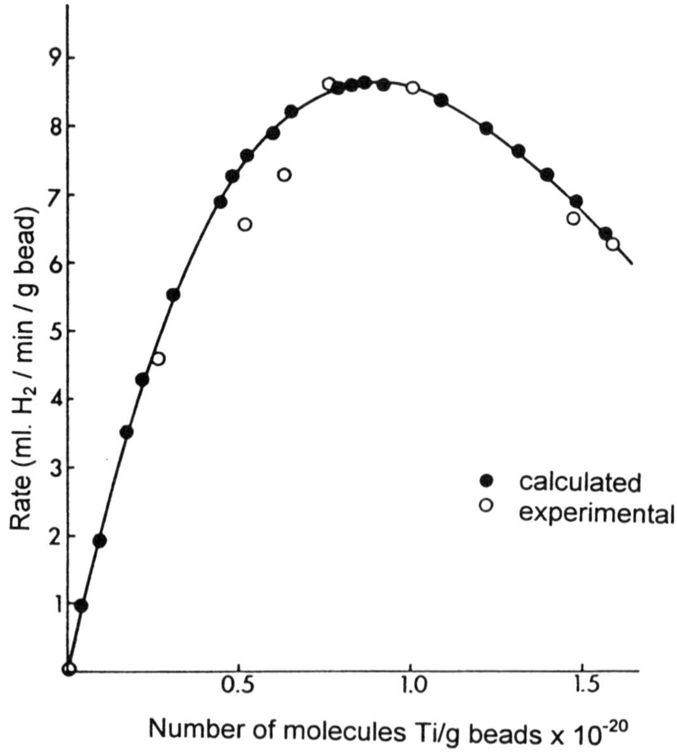

Figure 2.9 *Rate of hydrogenation of 10-hexene as a function of the loading of Ti on the polymer beads. (Reproduced with permission from* J. Am. Chem. Soc., *1977, 119, 4517–4518. Copyright© 1977 American Chemical Society.)*

solvent for PS resins) indicated that the increased mobility of polymer chains in better swollen resin promotes site interaction (Wulff & Schulze, 1978).

Crowley and Rapoport (1970) carried out a series of studies of Dieckmann cyclization reactions. In the Dieckmann cyclization (Crowley et al., 1973) shown in Scheme 2.7, the cyclized product was formed in 19% yield. This dimerization undoubtedly occurred because the individual moieties were not effectively separated from each other during the reaction. The best interpretation for this site-site interaction is the resin flexibility during the reaction.

The utilization of a polymeric phenylthiomethyllithium reagent for the homologation of alkyl iodides was reported (Crosby & Kato, 1977). The use of 1,4-diiodobutane provided an appropriate system for the

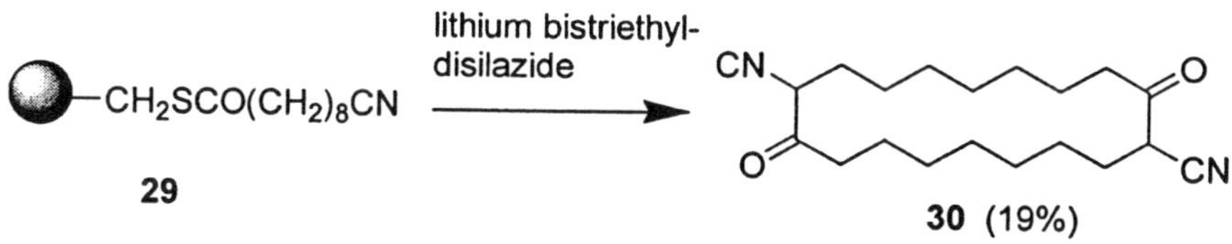

Scheme 2.7

study of intraresin reactions of polymer-bound ionic functional groups. The ratio of 1,5- to 1,6-diiodoalkanes obtained by the two-step (Scheme 2.8) homologation sequence was a direct indication of the interaction of reactive sites.

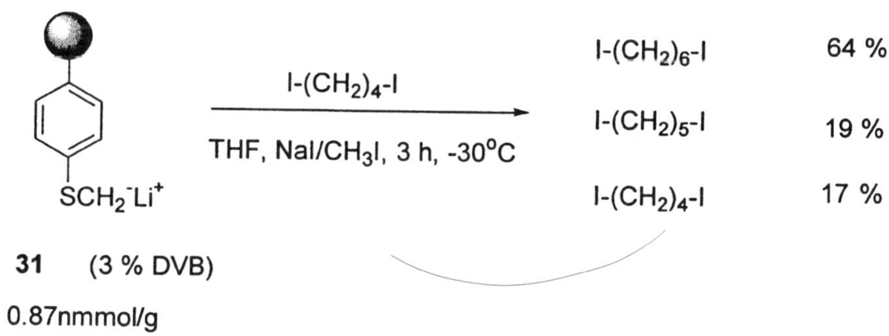

Scheme 2.8

Taking advantage of the site-site interaction effect, a mixed condensation product was synthesized by purposefully attaching two different kinds of reactants onto resin (Scheme 2.9) (Kraus & Patchornik, 1971).

2.3.4.3 FACTORS AFFECTING SITE SEPARATION

The equilibrium between site-site separation and site-site interaction can be explained on the basis of favorable kinetics relationships between the desired reactions and competitive reactions. When the less kinetically favorable reactions are attempted with polymer-bound reagent, the extent of site-site interactions will increase.

Resin swelling has greatly increased site-site interactions. Two decomposition pathways of the resin-bound carbonic benzoic anhydride were investigated under various conditions. Intrastrand reaction

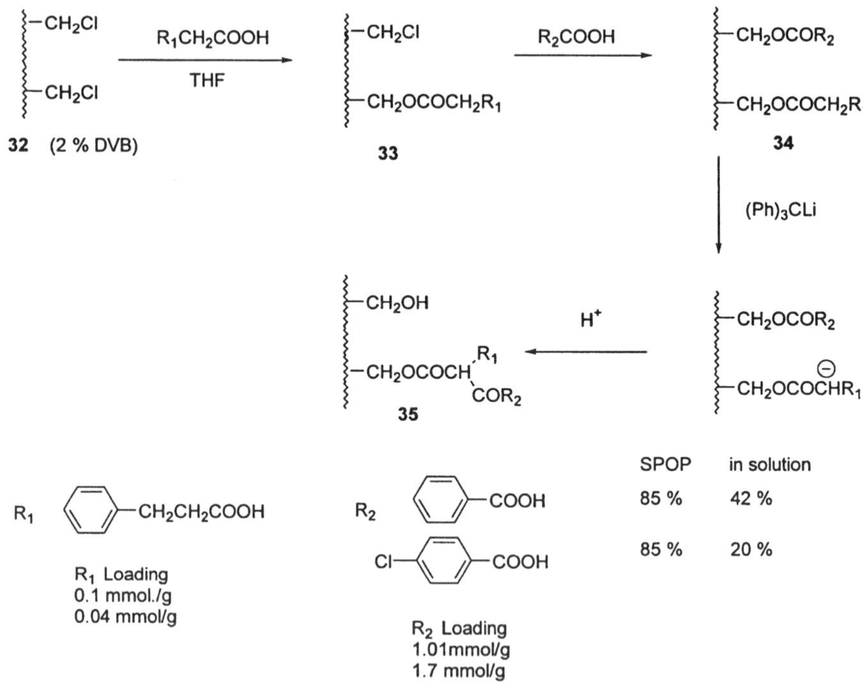

Scheme 2.9

resulted in the formation of the benzoate ester on the resin (A), while interstrand gave the benzoic anhydride and polymeric carbonate (B) (Scheme 2.10). The latter pathway was favored in solvent-swollen resin and the former in dry resin (Martin et al., 1978). A higher degree of oxidation of mercapto groups in resin swollen in benzene indicated the better mobility of polymer chains in better-swollen resin promotes site interactions as shown above (Wulff & Schulze, 1978).

Regen and Lee (1977) used a method introduced by Collman and colleagues (1972), which consists of IR measurements of the 1:1 and 2:1 complex formed between 2% DVB crosslinked styryl diphenylphosphine and $Co(NO)(CO)_3$. A high degree of site isolation promotes the formation of the 1:1 complex (IR band of NO at 1755 cm^{-1}), and a high degree of site interaction promotes the formation of the 2:1 complex (IR band at 1710 cm^{-1}) between the resin-bound phosphine and $Co(NO)(CO)_3$. Starting with the 1:1 complex, IR indicated that addition of very small quantities of the good swelling solvent m-xylene to the resin resulted in a rapid increase in site-site interaction and formation

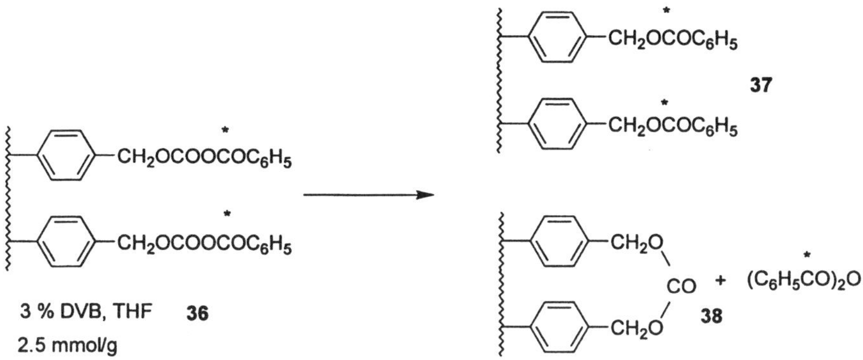

3 % DVB, THF **36**
2.5 mmol/g

	37	38
100°C, 10 h	59%	25%
25°C, 1 year	68%	20%
90 °C in dioxane 10 h	15 %	84 %
37 °C in dioxane, 15 h, TEA	17 %	82 %

Scheme 2.10

of the 2:1 complex at the expense of the 1:1 complex (Figure 2.10). In the absence of solvent or in the presence of an excess of a poor swelling solvent such as n-hexadecane, no change was observed in the IR spectrum from the starting 1:1 complex. These results clearly establish that bead swelling is an important factor in site isolation in low-crosslinked PS resins.

It is interesting to compare results from different laboratories on this topic. For example, Grubb and Kroll (1971) reported observation of site isolation on a macroreticular 20% crosslinked PS resin using a poor swelling solvent, cyclohexane. In contrast, Scott et al. (1977) found significant site-site interaction in the same resin when a good swelling solvent, CH_2Cl_2, was used. These results indicated the influence of solvent in determining site isolation on highly crosslinked resins. The intensity of the free hydroxyl band at 3580 cm^{-1} increases as the microenvironment of the hydroxyl group becomes bulkier (Figure 2.11 a & b). By controlling both the steric hindrance and the local concentration, total site isolation can be achieved as shown in Figure 2.11.

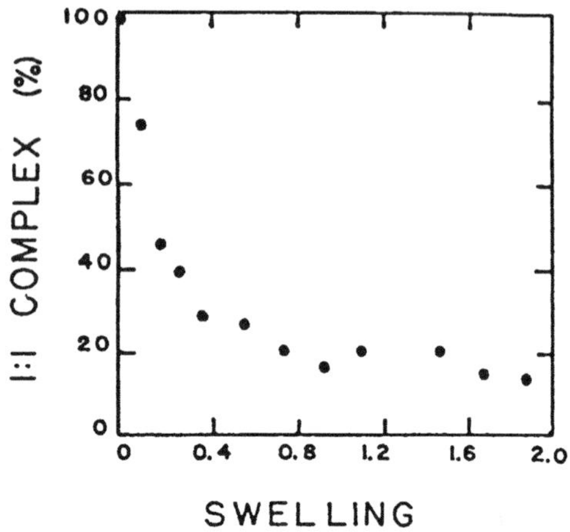

Figure 2.10 *Plot of percent of 1:1 complex remaining as a function of degree of swelling, g of* m*-xylene/g of dry resin, of the resin after a reaction time of 240 h at 70° C as estimated through relative IR band intensities. (Reproduced with permission from Macromolecules, 1977, 10, 1418–1419. Copyright© 1977 American Chemical Society.)*

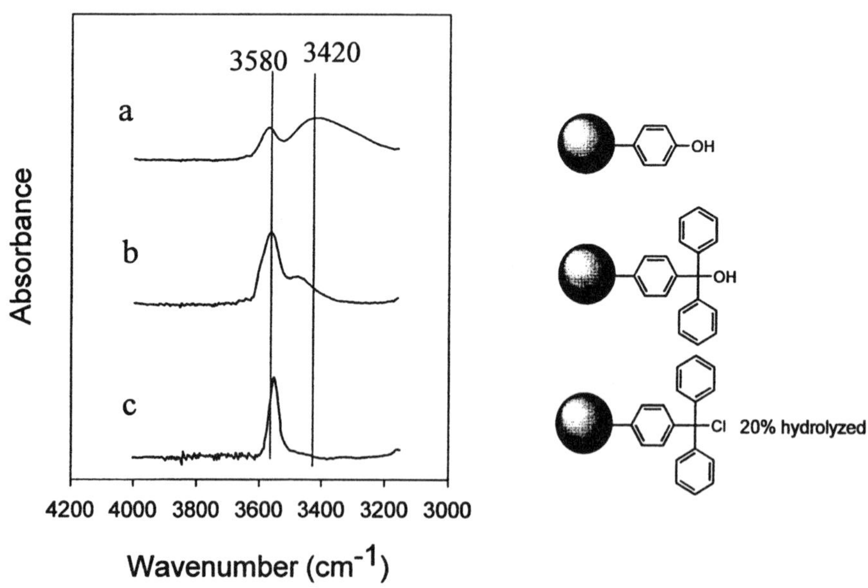

Figure 2.11 *IR spectra taken from a single flattened bead for resins shown on the right. (a) Wang resin; (b) Tritylhydroxyl resin; (c) Tritylchloride resin with ~20% hydrolysis (to tritylhydroxyl resin).*

Site-site interaction can also be controlled by altering the reacting group/reactive site ratio. In the reaction of crosslinked chloromethyl PS with 1,4-butanedithiol, the site interaction was controlled by using a different ratio of starting materials (Scheme 2.11) (Wulff and Schulze, 1978). The interstrand reacting product was reduced from 97% to 26% when the excess of dithiol was raised from threefold to 11-fold. The reaction of Wang resin with a series of diacid chloride (Scheme 2.12) showed both the dynamic nature of the resin (lack of diacid chloride length dependence) and the effect of the molecular ratio. The interstrand product was reduced from 72% to 54% when the excess of diacid chloride was raised from twofold to 10-fold. (Yan & Sun, 1998).

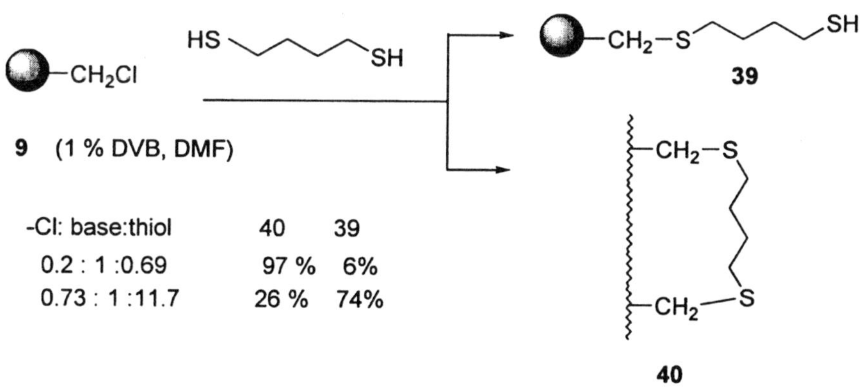

Scheme 2.11

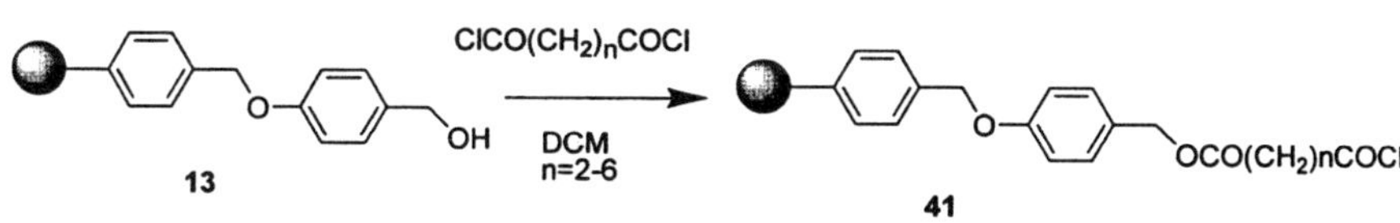

Scheme 2.12

2.3.5 Support Effects on Product Purity

Because SPOS resin itself is a synthetic product, the quality of resins from different commercial sources or different batches from the same source may vary. Some resins may contain residues of covalently or non-covalently bound synthetic intermediates (such as carboxylpolystyrene resin); others may have undergone partial hydrolysis during handling (such as tritylchloride resin) or may have changed their loading value during storage (such as formylpolystyrene). The result of any of these problems could be impurities in the final cleaved synthetic product. Because it is often impractical to purify combinatorial libraries or products from combinatorial synthesis so far, the impact of the resin on the quality of synthetic product is immense. In order to minimize some of these problems, MacDonald et al. (1995) used prewashing to eliminate entrapped resin impurities, and this approach is recommended for the routine removal of non-covalently bound impurities. They also suggested the use of mild cleavage conditions to liberate the desired compounds without causing resin breakdown or impurity cleavage.

2.4 SUMMARY

SPOS is highly sensitive to the microenvironment of the resin and other reaction conditions. When designing the synthesis of a combinatorial library, the influence of the resin after solvation and swelling must be carefully considered. This ensures the quality of the final library that is often too large to be completely characterized. The physicochemical and swelling properties of resin supports have direct impact on many spectroscopic analysis methods such as NMR.

Selective solvation of resin alters the local reactivity and accessibility of the bound substrate as well as mobility of the entrapped reagent. Resin solvation changes during the course of the reaction if the attached substrate changes polarity or other physical and chemical properties. In addition, swollen beads can selectively absorb compounds and thereby influence the reaction kinetics. The use of a phase-transfer catalyst can enhance the rate of reaction on resin and illustrates the solvent-like property of resin supports. Swollen resin can allow sites to interact with each other in 1% DVB resins in varying degrees depending on solvent, resin, and reactivity of the pendant groups so that total site isolation seems only achievable by controlling several factors simultaneously, such as decreasing loading, increasing steric hindrance, and increasing the degree of crosslinking.

The control of these properties during SPOS for combinatorial chemistry involves the proper selection of resin and solvent, alternating solvents to accommodate dynamic solvation of the resin, optimizing kinetics when changing solid supports, and carefully controlling resin impurities. This will lead to a synthesis of a high-quality combinatorial library. The study of solid support effects on SPOS is an area that needs more attention. Also needed is clarification of the influence of high and low temperatures, resin compatibility with various organic molecules, and dynamic solvation in SPOS.

2.5 REFERENCES

Albericio, F., Pons, M., Pedroso, E., Giralt, E. Comparative study of supports for solid-phase coupling of protected-peptide segments. *J. Org. Chem.* 1989, 54, 360.

Atherton, A., Brown, E., Sheppard, R. C. A physically supported gel polymer for low pressure, continuous flow solid phase reactions. Application to solid phase peptide synthesis. *J. Chem. Soc. Chem. Commun.* 1981, 1151.

Bray, A. M., Maeji, N. J., Geysen, H. M. The simultaneous multiple production of solution phase peptides, assessment of the Geysen method of simultaneous peptide synthesis. *Tetrahedron Lett.* 1990, 31, 5811.

Briggs, J. C., Hodge, P. Potassium borohydride reductions of ketones absorbed into polymer supports: Stereochemical effects. *J. Chem. Soc. Chem. Commun.* 1988, 310.

Cainelli, G., Contento, M., Manoscalchi, F., Plessi, L. Polymer-supported reagents: Polar effects in substrate selectivity. *J. Chem. Soc. Chem. Commun.* 1982, 725.

Camps, F., Castells, J., Ferrando, M. J., Font, J. Organic syntheses with functionalized polymers: I. Preparation of polymeric substrates and alkylation of esters. *Tetrahedron Lett.* 1971, 20, 1713.

Cilli, E. M., Oliveira, E., Marchetto, R., Nakaie, C. R. Correlation between solvation of peptide-resins and solvent properties. *J. Org. Chem.* 1996, 61, 8992.

Collman, J. P., Hegedus, L. S., Cooke, M. P., Norton, J. R., Dolcetti, G., Marquardt, D. N. Resin-bound transition metal complexes. *J. Am. Chem. Soc.* 1972, 94, 1789.

Crosby, G. A., Kato, M. J. The utilization of a polymeric phenylthiomethyllithium reagent for the homologation of alkyl iodides and its application for the study of intraresin reactions of polymer-bound functional groups. *J. Am. Chem. Soc.* 1977, 99, 278.

Crowley, J. I., Harvey III., T. B., Papoport, H. Solid phase synthesis. Evidence for and quantification of intra-resin reactions. *J. Macromol. Sci.-Chem. A.* 1973, 1117.

Crowley, J. I., Rapoport, H. Cyclization via solid phase synthesis. Unidirectional Dieckmann products from solid phase and benzyl triethylcarbinyl pimelates. *J. Am. Chem. Soc.* 1970, 92, 6363.

Czarnik, A. W. Solid-phase synthesis supports are like solvents. *Biotech. Bioeng. (Combin. Chem.)* 1998, 61, 77.

Dhalluin, C., Boutillon, C., Tartar, A., Lippens, G. Magic angle spinning nuclear magnetic resonance in solid-phase peptide synthesis. *J. Am. Chem. Soc.* 1997, 119, 10494.

Dorman, L. C. A non-destructive method for the determination of completeness of coupling reactions in solid phase peptide synthesis. *Tetrahedron Lett.* 1969, 2319.

Fields, G. B., Fields, C. G. Solvation effects in solid-phase peptide synthesis. *J. Am. Chem. Soc.* 1991, 113, 4202.

Flanigan, E., Marshall, G. R. Synthesis of cyclic peptide on dual function supports. *Tetrahedron Lett.* 27, 1970, 2403.

Ford, W. T., Ackerson, B. J., Blum, F. D., Periyasamy, M., Pickup, S. Self-diffusion coefficients of solvents in polystyrene gels. *J. Am. Chem. Soc.* 1987, 109, 7276.

Ford, W. T., Periyasamy, M., Spivey, H. O. Mechanisms of polymer-supported catalysis. 7. Carbon-13 NMR relaxation of toluene in cross-linked polystyrene phase-transfer catalyst gels. *Macromolecules.* 1984, 17, 2881.

Forman, F. W., Sucholeiki, I. Solid-phase synthesis of biaryls via the stille reaction. *J. Org. Chem.* 1995, 60, 523.

Frechet, J. M. J., de Smet, M. D., Farrall, M. J. Application of phase-transfer catalysis to the chemical modification of cross-linked polystyrene resins. *J. Org. Chem.* 1979a, 44, 1774.

Frechet, J. M. J., de Smet, M. D., Farrall, M. J. Chemical modification of crosslinked resins by phase transfer catalysis: Preparation of polymer-bound dinitriles and diamines. *Tetrahedron Lett.* 1979b, 137.

Frechet, J. M. J., Schuerch, C. Solid-phase synthesis of oligosaccharides. I. Preparation of the solid support. Poly[p-(1-propen-3-ol-1-yl)styrene]. *J. Am. Chem. Soc.* 1971, 93, 492.

Fridkin, M., Patchornik, A., Katchalski, E. J. A synthesis of cyclic peptides utilizing high molecular weight carriers. *J. Am. Chem. Soc.* 1965, 87, 4646.

Grubb, R., Lau, C. P., Cukier, R., Brubaker, Jr., C. Polymer attached metallocenes evidence for site isolation. *J. Am. Chem. Soc.* 1977, 99, 4517.

Grubb, R. H., Kroll, L. C. Catalytic reduction of olefins with a polymer-supported Rhodium(I) catalyst. *J. Am. Chem. Soc.* 1971, 93, 3062.

Gubitosa, G., Boldt, M., Brintzinger, H. H. Supported cyclopentadienyl metal cabonyl complexes. 1. Mononuclear iron(II) and cobalt(I) to a polystyrene support. *J. Am. Chem. Soc.* 1977, 99, 5174.

Guyot, A. Synthesis and structure of polymer supports. In *Syntheses and Separations Using Functional Polymers.* Sherrington, D. C., Hodge, P. Eds. 1988, 1–42.

Hagenmaier, H. The influence of the chain length on the coupling reaction in solid phase peptide synthesis. *Tetrahedron Lett.* 1970, 283.

Hancock, W. S., Prescott, D. J., Vagelos, P. R., Marshall, G. R. Solvation of the polymer matrix. Source of truncated and deletion sequences in solid phase synthesis. *J. Org. Chem.* 1973, 38, 774.

Hodge, P. Polymer-supported organic reactions: what takes places in the beads? *Chem. Soc. Rev.* 1997, 26, 417.

Keifer, P. A. Influence of resin structure, tether length, and solvent upon the high-resolution ^{1}H NMR spectra of solid-phase-synthesis resins. *J. Org. Chem.* 1996, 61, 1558.

Kraus, M. A., Patchornik, A. The directed mixed ester condensation of two acids bound to a common polymer backbone. *J. Am. Chem. Soc.* 1971, 93, 7325.

Kraus, M. A., Patchornik, A. Polymeric carriers as immobilizing media—Fact and fiction *Israel J. Chem.* 1978, 17, 298.

Letsinger, R. L., Kornet, M. J., Mahadevan, V., Jerina, D. M. Organic and biological chemistry II (Reaction on polymer supports). *J. Am. Chem. Soc.* 1964, 86, 5163.

Leznoff, C. C., Wong, J. Y. The use of polymer supports in organic synthesis. The synthesis of monotrityl ethers of symmetrical diols. *Can. J. Chem.* 1972, 50, 2892.

Li, R., Xiao, X., Czarnik, A. W. Merrifield MicroTube™ reactors for solid phase synthesis. *Tetrahedron Lett.* 1998, 39, 8581.

Li, W., Yan, B. Effects of polymer supports on the kinetics of solid-phase organic reactions: A comparison of polystyrene- and TentaGel-based resins. *J. Org. Chem.* 1998, 63, 4092.

MacDonald, A. A., DeWitt, S. H., Ghosh, S., Hogan, E. M., Kieras, L., Czarnik, A. W., Ramage, R. The impact of polystyrene resins in solid-phase organic synthesis. *Mol. Diversity.* 1995, 1, 183.

Martin, G. E., Shambhu, M. B., Shakhshir, S. R., Digenis, G. A., Polymer-bound carbonic-carboxylic anhydride functions. Preparation, site-site interactions, and synthetic applications. *J. Org. Chem.* 1978, 43, 4571.

Mazur, S., Jayalekshmy, P. Pseudodilution, the solid-phase immobilization of benzyne. *J. Am. Chem. Soc.* 1976, 98, 6710.

Mazur, S., Jayalekshmy, P. Chemistry of polymer-bound *o*-bnzyne. Frequency of encounter between substituents on cross-linked polystyrenes. *J. Am. Chem. Soc.* 1979, 101, 677.

Meldal, M. Properties of solid supports. *Methods in Enzymol.* 1997, 289, 83.

Merrifield, R. B. The role of the support in solid phase peptide synthesis. *Brit. Polym. J.* 1984, 16, 173.

Merrifield, R. B. Solid-phase peptide synthesis. *Adv. Enzymol.* 1969, 32, 221.

Mitchell, A. R., Kent, S. B. H., Engelhard, M., Merrifield, R. B. A new synthetic route to *tert*-butyloxycarbonylaminoacyl-4-(oxymethyl)phenylacetamidomethyl-resin, an improved support for solid-phase peptide synthesis. *J. Org. Chem.* 1978, 43, 2845.

N'Guyen, T. D., Boileau, S. Phase transfer catalysis in the chemical modification of polymers. Part II. *Tetrahedron Lett.* 1979, 28, 2651.

Ogino, K., Sato, H. NMR analysis of interaction between styrene-divinylbenzene gel beads and small molecules. *J. Polymer Sci. B.* 1995, 33, 189.

Overberger, C. G., Sannes, K. N. Polymer as reagents in organic synthesis. *Angew. Chem. Int. Ed. Engl.* 1974, 13, 99.

Pickup, S., Blum, F. D., Ford, W. T., Periyasamy, M. Transport of small molecules in swollen polymer beads. *J. Am. Chem. Soc.* 1986, 108, 3987.

Pittman Jr., C. U., Evans G. O. Polymer-bound catalysts and reagents. *Chem. Tech.* 1973, 560.

Rapp, W. PEG grafted polystyrene tentacle polymers: Physico-chemical properties and application in chemical synthesis. In G. Jung (Ed.), *Combinatorial Peptide and Nonpeptide Libraries: A Handbook.* VCH, Weinheim, 1996, 425–464.

Rebek Jr., J., Trend, J. E. On the rate of site-site interactions in functionalized polystyrenes. *J. Am. Chem. Soc.* 1979, 101, 737.

Regen, S. L. Influence of solvent on the motion of molecules "immobilized" on polystyrene matrices and on glass surfaces. *J. Am. Chem. Soc.* 1975, 97, 3108.

Regen, S. L. Influence of solvent on the mobility of molecules covalently bound to polystyrene matrices. *J. Am. Chem. Soc.* 1974, 96, 5275.

Regen, S. L., Dulak, L. Solid phase cosolvents. *J. Am. Chem. Soc.* 1977, 99, 623.

Regen, S. L., Lee, D. P. Influence of solvent on site isolation in polystyrene resins. *Macromolecules.* 1977, 10, 1418.

Sanetra, R., Kolarz, B. N., Wlochowicz, A. Determination of the glass transition temperature of poly(styrene-co-divinylbenzene) by inverse gas chromatography. *Polymer.* 1985, 26, 1181.

Schneider, B., Doskocilova, D., Dybal, J. Motional restrictions and chain conformation in various swollen crosslinked polystyrene gels from ^{1}H NMR line-shape analysis. *Polymer.* 1985, 26, 253.

Scott, L. T., Rebek, J., Ovsyanko, L., Sims, C. L. Organic chemistry on the solid phase. Site-site interactions on functionalized polystyrene. *J. Am. Chem. Soc.* 1977, 99, 625.

Sun, Q., Yan, B. Single bead IR monitoring of a novel benzimidazoles synthesis. *Bioorg. Med. Chem. Lett.* 1998, 8, 361.

Wieczorek, P. P., Ilavsky, M., Kolarz, B. N., Dusek, K. J. Mechanical behavior and structure of single beads of homogeneous and macroporous styrene-divinylbenzene copolymers. *Appl. Polymer Sci.* 1982, 27, 277.

Wilson, M. E., Paech, K., Zhou, W.-J., Kurth, M. J. Solvent and reagent accessibility within oligo(ethylene glycol) ether (PEG) cross-linked polystyrene beads. *J. Org. Chem.* 1998, 63, 5094.

Wong, J. Y., Leznoff, C. C. The use of polymer supports in organic synthesis. II. The syntheses of monoethers of symmertrical diols. *Can. J. Chem.* 1973a, 51, 2452.

Wong, J. Y., Leznoff, C. C. The use of polymer supports in organic synthesis. II. Selective chemical reactions on one aldehyde group of symmetrical dialdehydes. *Can. J. Chem.* 1973b, 51, 3756.

Wulff, G., Schulze, I. Directed cooperativity and site separation of Mercapto groups in synthetic polymers. *Angew. Chem. Int. Ed. Engl.* 1978, 17, 537.

Yan, B., Fell, J. B., Kumaravel, G. Progression of organic reactions on resin supports monitored by single bead FTIR microspectroscopy. *J. Org. Chem.* 1996a, 61, 7467.

Yan, B., Jewell Jr., C. F., Myers, S. W. Quantitatively monitoring of solid-phase organic synthesis by combustion elemental analysis. *Tetrahedron.* 1998, 54, 11755.

Yan, B., Sun, Q. Crucial factors regulating site interactions in resin supports determined by single bead IR. *J. Org. Chem.* 1998, 63, 55.

Yan, B., Sun, Q., Wareing, J. R., Jewell, C. F. Real-time monitoring of the catalytic oxidation of alcohols to aldehydes and ketones on resin support by single-bead Fourier transform infrared microspectroscopy. *J. Org. Chem.* 1996b, 61, 8765.

Zalipsky, S., Chang, J. L., Albericio, F., Barany, G. Preparation and applications of polyethylene glycol-polystyrene graft resin supports for solid-phase peptide synthesis. *Reactive Polymers.* 1994, 22, 243.

3

Solid-Phase Reaction Optimization Using FTIR Methods

IN THE REHEARSAL phase, optimal reaction conditions need to be selected and implemented rapidly. Currently, the products and intermediates are mostly chemically (e.g., using TFA) or photochemically cleaved from the resin and then purified for analysis with conventional spectroscopic methods. The indirect information obtained from the compound derived from cleavage is used to judge the intended chemistry on the solid support. However, the requirement for cleavage of the compound from the resin for analysis renders the method inherently destructive. The "cleave and analyze" method is a time-consuming, expensive, and laborious process. Some synthetic intermediates are not stable enough for the "cleave and analyze" protocol. It is especially a waste of time and sample for characterizing intermediates in multistep synthesis. Although photochemical linkers simplify this procedure, the cleavage yield is usually low, and the types of reactions that can be carried out on such labile linkers may be limited. The cleavage of diverse compounds from resin supports relies on the chemical linker used during the synthesis. Furthermore, the isolated yield is usually different from the on-resin yield that is more important to know for multistep synthesis. It would be much more advantageous to be able to analyze the intermediates or products while still bound to the resin. These demands have consequently stimulated the development of on-support analytical methods. FTIR is a method that analyzes synthetic product exclusively on support.

Recent progresses in analytical methods associated with combinatorial chemistry have been summarized in several reviews (Egner & Bradley, 1997; Gallop & Fitch, 1998; Balkenhohl et al. 1997; Deben et al., 1998; Shapiro et al., 1997; Yan, 1998; Yan & Gremlich, 1999). This chapter is specifically limited to reviewing the application of FTIR techniques with an emphasis on single-bead FTIR technique in the monitoring of solid-phase organic synthesis (SPOS) and the impact of these methods on the rehearsal phase of combinatorial chemistry. Applications are illustrated with examples in this overview. Experimental details can be found in original papers listed in the references section.

3.1 COMPARISON OF FTIR AND RAMAN TECHNIQUES

3.1.1 Techniques

The unavailability of thin-layer chromatography (TLC) in SPOS has caused severe analytical problems in the routine quality control of the solid-phase reactions. Consequently, various FTIR methods have risen in importance to meet the challenge. Peptide resins have been studied using the macro FTIR method by holding milligram amounts of swollen beads between salt windows (Narita et al., 1984, 1988, 1989) or by the FT-Raman method (Larsen et al., 1993). Analysis of solid-phase organic reactions has also been performed on 5–10 mg of resin beads using various techniques such as the KBr pellet method (Frechet & Schuerch, 1971; Crowley & Rapoport, 1980; Blanton & Salley, 1991; Chen et al., 1994; Moon et al., 1994; Reggelin & Brenig, 1996; Deben et al., 1997), diffuse reflectance infrared Fourier transform spectroscopy (DRIFTS) (Chan et al., 1997), photoacoustic FTIR spectroscopy (PAS) (Gosselin et al., 1996), and, alternatively, by FTIR microspectroscopy on a single resin bead (Yan et al., 1995; Yan & Kumaravel, 1996; Yan et al., 1996a, b; Pivonka et al., 1996; Pivonka & Simpson, 1997).

Organic reactions selected to assemble a combinatorial library should preferably be robust and well behaved. Therefore, the resin-bound structures to be monitored are rarely unknowns. Consequently, the major analytical task is to confirm the intended chemistry rather than a full structural elucidation. Because FTIR is a technique sensitive to organic functional group changes, it is well suited for this purpose. The principle of monitoring reactions by IR is based on the functional group inter-conversions via chemical reaction or by the appearance or disappearance of functional groups carried by building blocks or protecting groups introduced or removed during the reaction. The functional group to be monitored need not be directly involved in the reaction that is monitored. For the rehearsal of a library synthesis on solid supports, building blocks to be used in solid-phase reaction can even be selected to contain an IR detectable group at a remote site. Many FTIR and Raman methods have been used to analyze SPOS. These techniques are described briefly in the following:

1. FTIR microspectroscopy. This method records spectrum from a small sample (such as a single bead) in the transmission or reflectance mode. A microscope accessory is required for the measurement. A liquid nitrogen-cooled MCT detector is used to enhance the sensitivity. In the transmission mode, several beads are spread on a diamond window and flattened by hand-press or with a compression cell (SpectraTech, Shelton, CT). The IR beam is focused on a flattened single bead using the view mode of the microscope. The blank area surrounding the bead is cut off using an adjustable aperture. A spectrum is recorded with 32 scans (< 1 min). Then, the focus is moved to the nearby blank area of the same size on the diamond window to record the background.
2. Beam condenser. This simple optical accessory is used to reduce the IR radiation from a typical 8-mm beam to 2 mm at the sample plane. This allows the analysis of smallest quantity of pressed resin beads (50–100 beads) without using the KBr dilution. A diamond compression cell is used to flatten beads and for the measurement. The same diamond cell without beads is then used to record a background spectrum.
3. Attenuated total reflection (macro- and micro-ATR). Under certain conditions, infrared radiation passing through a prism made of a high refractive index infrared transmitting material (ATR crystal) will be totally reflected internally. If a sample is brought in contact with the totally reflecting surface of the ATR crystal, an evanescent wave in the less dense medium extends beyond the reflecting interface, and the evanescent wave will be attenuated in regions of the infrared spectrum where the sample absorbs energy. The intensity of the wave decays exponentially with the distance from the surface of the ATR crystal. The distance, which is on the order of micrometers, makes ATR generally insensitive to sample thickness, allowing for the analysis of thick or strong absorbing samples. Micro-ATR measurement, which is based on the use of a micro ATR crystal in an ATR objective, on the partial surface from a single bead has been reported (Yan et al., 1996a).
4. Macro and single-bead FT Raman spectroscopy. FT Raman spectroscopy is based on inelastic light scattering, in which scattered photons exchange energy with the sample. Most commonly, the

scattered photon loses energy to a vibration mode of the sample molecule, leading to a downward frequency shift. This Raman shift is equal in energy to the light absorbed by the same molecule in an IR absorption experiment if the mode is active toward both techniques. Because of the differences in the selection rules, both FTIR and FT Raman spectra must be measured to obtain the complete vibration characteristics of totally symmetric molecules or functional groups. For an asymmetric molecule, an FT Raman spectrum provides information similar to that in an FTIR spectrum except the intensity difference.

The factor contributing to the intensity of Raman band is the change of polarizability during the vibration. Therefore, strong FT Raman signals can be obtained for functional groups with low polarity and high polarizability, such as C≡C, C=C, N=N, S-H, C≡N, C=N, C=S, C-S, and symmetric vibrations of groups with high degree of symmetry, such as NO_2, SO_2, CO_2.

The excitation light used in FT Raman is in the near-infrared region (1064 nm). The spectra produced are free of the fluorescence interference that is usually typical for spectra obtained by visible laser excitation. FT Raman spectra can be acquired in macro-mode on 0.2-mg beads or in micro-mode, such as single-bead FT Raman microspectroscopy.

5. KBr pellet method. This is a widely used sampling technique for solid samples. The method consists of mixing finely ground (0.5 μm average particle size) sample with a pure, dry spectroscopic grade of KBr powder, usually at a concentration of about 1% sample in KBr, and transferring the mixture to a die where it is pressed until the KBr particles coalesce into a clear disk. The disk is measured using the regular FTIR instrument in which the diameter of the IR beam is 8 mm. For resin samples, although polystyrene-PEG resins can be ground into small particles, polystyrene resin beads cannot. The sample particle size in the resin KBr pellet is 50–150 μm rather than the required 0.5 μm. This causes problems such as light scattering in resin spectrum.
6. Diffuse reflectance infrared Fourier transform spectroscopy (DRIFT). When infrared radiation is directed onto the surface of a solid sample, two types of energy reflectance can occur. One is specular reflectance, and the other is diffuse reflectance. The specular component is the radiation that reflects directly off the sample surface (it is the energy that is not absorbed by the sample). Diffuse reflectance is the radiation that penetrates into the sample and then emerges. A diffuse reflectance accessory is designed so that the

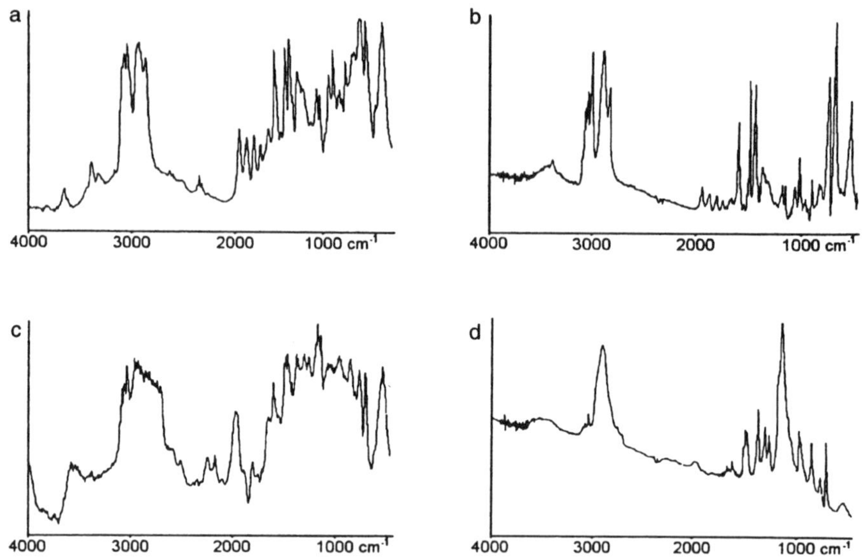

Figure 3.1 *Comparison of DRIFTS and KBr spectra of NovaBiochem HL resin (a,b) and NovaBiochem TG resin (c,d) (Reproduced with permission from* Tetrahedron Lett. *1997, 38, 2822–2823. Copyright© 1997, Elsevier Science Ltd).*

diffuse reflected energy is optimized and the specular component is minimized. The optics collect the scattered radiation and direct it to the infrared detector. Typical spectra of PS and PS-PEG resins recorded with DRIFT method as compared with KBr method are shown in Figure 3.1.

7. Photoacoustic spectroscopy. When modulated infrared radiation is absorbed by a sample, the substance heats and cools in response to modulated infrared energy impinging on it. The heating and cooling is converted into a pressure wave that can be communicated to a surrounding gas and can be detected by an acoustic detector (essentially a sensitive microphone in the enclosed sample chamber). In such measurements, the acoustic detector replaces the infrared detector of the spectrometer. A typical spectrum recorded with this method is compared with the KBr method in Figure 3.2.

Resin beads contain 99% of functional groups in the bead interior. A transmission measurement is a method predominantly measuring the bead interior. Among the methods mentioned above, 1, 2, and 5 are transmission analysis methods. These methods obtain chemical information from overall substance attached to the whole bead, while

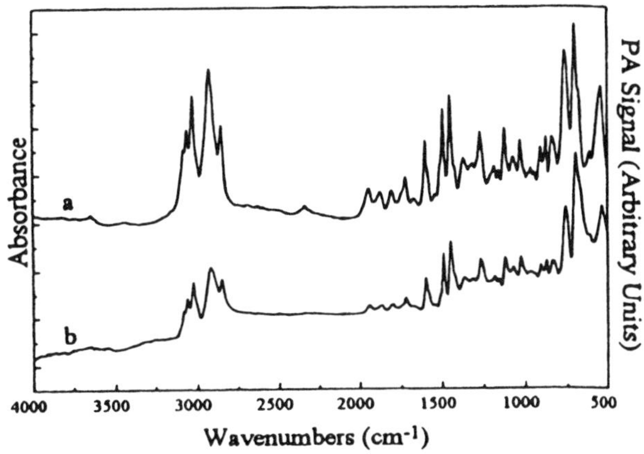

Figure 3.2 *FTIR spectra of Merrifield's resin (a) PA-FTIR; (b) FTIR of KBr pellet (Reproduced with permission from* J. Org. Chem. *1996, 61, 7980. Copyright© 1996, American Chemical Society).*

methods 3, 6, and 7 are only sampling molecules attached on the bead surface. Method 4 is based on the analysis of light scattering mostly from the outer layer of the bead.

3.1.2 Analysis of Resin-Bound Compounds

In order to evaluate these methods in the analysis of resin-based organic molecules, resin-bound molecules in Diagram 3.1 were analyzed by various methods (Yan et al., 1999).

1. Oxime resin (**1** in Diagram 3.1). The spectrum of oxime resin (Figure 3.3a) contains several interesting features: a sharp hydroxyl band at 3503 cm^{-1}, a weak carbon-nitrogen double bond stretch at 1660 cm^{-1}, and asymmetric and symmetric stretching vibrations of the nitro group at 1523 and 1345 cm^{-1}. All of these signals have been identified by FTIR methods except the water interference about 3400 cm^{-1} and the more overlapping fingerprint region bands in the ATR method. KBr method is subject to interference and baseline tilting. FT Raman spectroscopy failed to unambiguously detect bands at 3503 and 1523 cm^{-1}. However, the N=O symmetric stretching of the nitro group at 1345 cm^{-1} is stronger in the FT Raman spectrum than in the IR spectrum.
2. Fmoc-cys(t-Buthio)-Wang resin (**2** in Diagram 3.1). Figure 3.3b shows the spectra of Fmoc-Cys(t-Buthio)-Wang resin obtained by

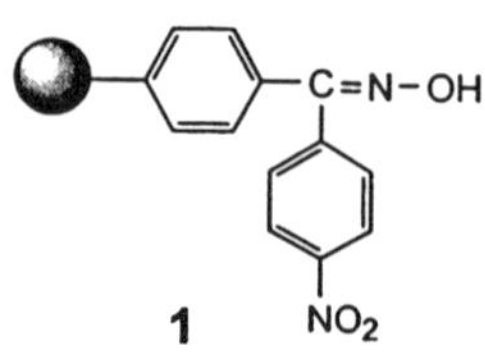

1

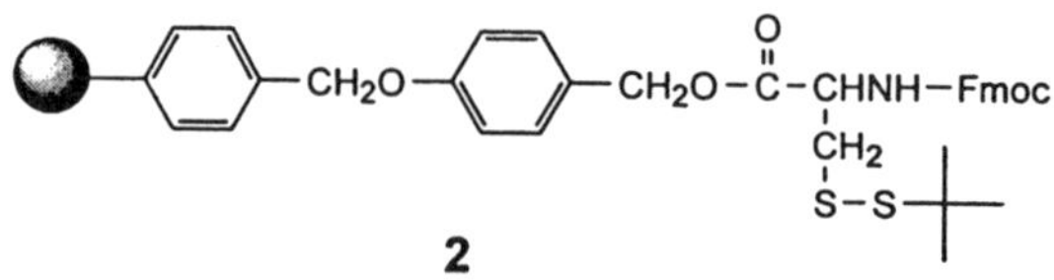

2

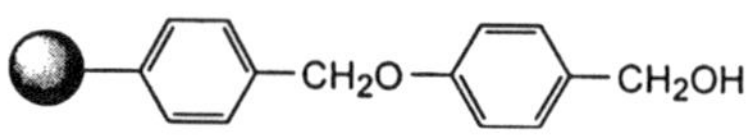

3

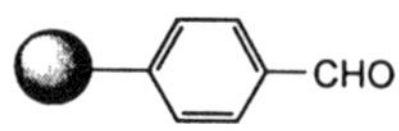

4

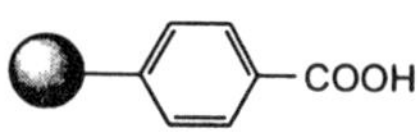

5

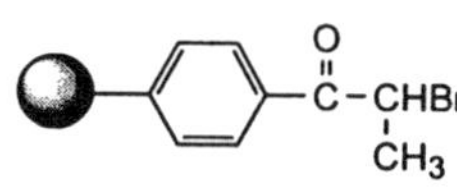

6

Diagram 3.1

various methods. Characteristic features in this compound are an ester carbonyl, a Fmoc carbonyl, an amide NH, and a disulfide bond. The two carbonyl bands overlap at 1727 cm^{-1} and NH stretch is at 3420 and 3350 as a doublet. These features are easily observed by FTIR methods, but not by FT Raman. The vibration of S-S is between 450 and 550 cm^{-1}, which is not detected by FTIR methods. The vibration of S-S bond was detected by the FT Raman method (Figure 3.3b inset).

3. Wang resin (**3** in Diagram 3.1). Wang resin (Figure 3.3c) contains a benzylic hydroxyl group. A sharp band at 3573 cm^{-1} and a broad band at 3449 cm^{-1} observed by all methods with the exception of the FT Raman method are attributed to the free and intraresin hydrogen-bonded hydroxyl groups (Yan & Sun, 1998). A strong CH_2-OH bending vibration at 1232 cm^{-1} is also observed by all methods except FT Raman. This is due to the different selection rule in Raman scattering. Vibrations of polar groups are not easily observable.
4. Formylpolystyrene resin (**4** in Diagram 3.1). The spectrum of formylpolystyrene resin is characterized by a band at 2713 cm^{-1} attributed to the aldehyde (O=)C-H stretch and a band at 1702 cm^{-1} attributed to the carbonyl group (Figure 3.3d). All FTIR methods

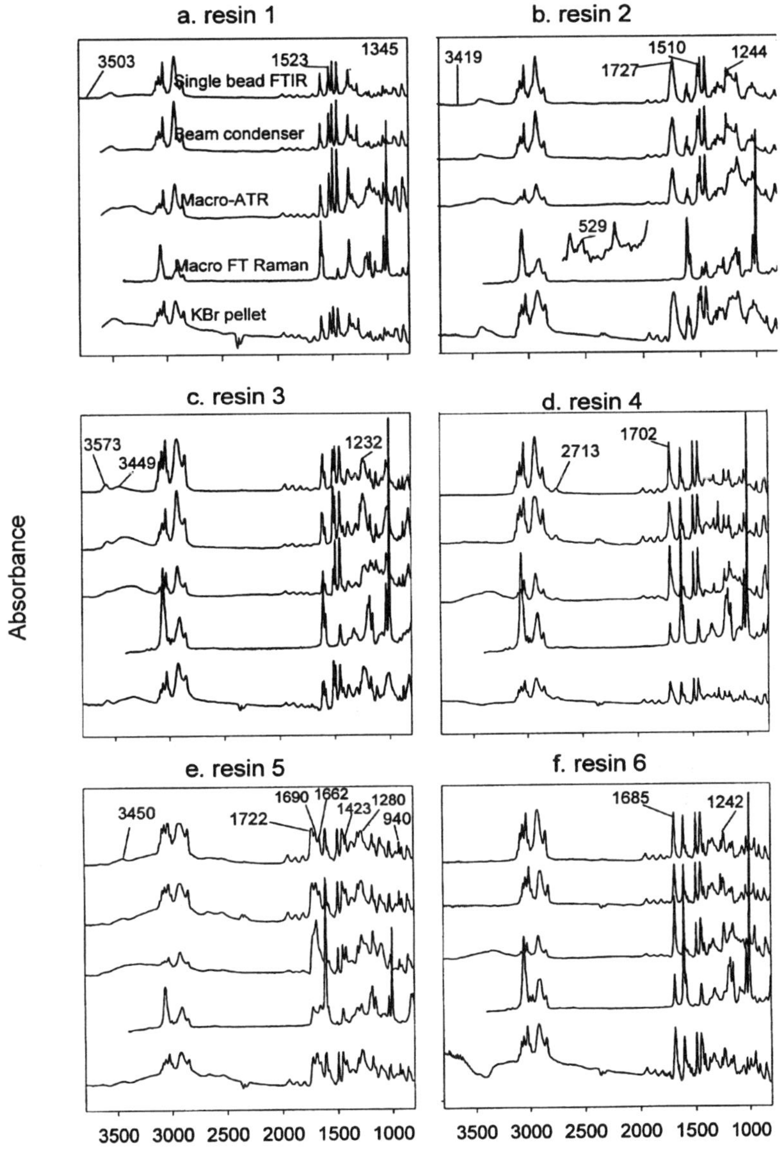

Figure 3.3 *Spectra of resin-bound organic compounds (Diagram 3.1) obtained by various FTIR and FT Raman methods (see panel a). Resins are (a) oxime resin; (b) Fmoc-cys(t-buthio)-Wang resin; (c) Wang resin; (d) formylpolystyrene resin; (e) carboxylpolystyrene resin; (f) brominated Wang resin.*

can detect these signals except for interference and S/N ratio problems in cases of ATR and KBr pellet. In FT Raman spectrum, these signals are relatively weaker.

5. Carboxylpolystyrene resin (**5** in Diagram 3.1). Spectra of carboxylpolystyrene resin are shown in Figure 3.3e. In the liquid or in solution at concentrations over 0.01 M, carboxylic acids exist as dimers due to strong hydrogen bonding. Because of the strong bonding, a free hydroxyl stretching vibration near 3500 cm^{-1} is observed only in very dilute solution in nonpolar solvents in which a mixture of monomer and dimer is observed. In the spectrum of resin-bound carboxylic acid, the O-H shows a broad band from 3300 to 2500 cm^{-1}, a feature of hydrogen-bonded carboxyl groups. A small sharp band at 3450 cm^{-1} is also observed in spectra recorded by single-bead, beam-condenser, and KBr pellet methods. This band is an indication of the presence of a portion of free carboxyl groups. This "site isolation" effect is further supported by the frequencies of the carbonyl group. The band at 1722 cm^{-1} was attributed to the free carboxyl groups, and the band at 1690 cm^{-1} was attributed to the associated carboxyl groups (Letsinger et al., 1964). The origin of the band at 1662 cm^{-1} is not so clear. Because this band remains after esterification of the resin, it may come from the intermediate in the resin preparation, which is a ketoester, or an amide formed with an amine impurity during the resin preparation process. In addition to O-H and the carbonyl groups, the -OH in-plane bending at 1423 cm^{-1}, the C-O stretching at 1280 cm^{-1}, and the O-H out-of-plane bending at 940 cm^{-1} are all observed. The carbonyl vibrations were weakly observed by the FT Raman method. In addition, the above-mentioned vibrations are not observed, except the band at 1280 cm^{-1}.
6. Brominated Wang resin (**6** in Diagram 3.1). Brominated Wang resin (Figure 3.3f) showed a characteristic peak at 1685 cm^{-1} for the carbonyl group and a strong band at 1242 cm^{-1} for the Φ-C(=O) stretching and bending vibrations (Figure 3.3f) besides other features. Spectra from the single-bead and beam-condenser methods clearly display these signals. ATR and KBr methods also show these features, except the interference to which these methods are usually subjected. The macro FT Raman method showed a weak band for the benzene-conjugated carbonyl at 1685 cm^{-1} and a C-Br vibration below 600 cm^{-1} (not shown), but failed to clearly detect the Φ-C(=O) stretching and bending vibration at 1242 cm^{-1}.

Table 3.1 Comparison of various FTIR and raman techniques.

	Rapid Analysis	Less Sample	High S/N & Resolution	Less Artifact & Interference	Low Cost	Total
Single bead	5	5	5	5	2	22
beam condenser	5	3	5	5	4	22
ATR	5	3	4	3	4	19
Raman	4	2	3	5	1	15
KBr	2	1	4	1	5	13
DRIFT	5	1	2	4	4	16
Photoacoustic	5	1	3	4	3	16

Note: Various methods were compared for features listed in these five entries. Their relative rankings in these aspects were scored. The highest score is 5 and the lowest is 1.

3.1.3 Comparison of Techniques

Spectra obtained with FTIR and FT Raman techniques presented in Figure 3.3 are compared along with the published data on DRIFT (Chan et al., 1997) and photoacoustic technique (Gosselin et al., 1996) in terms of information content and quality, resolution and artifact, sample requirement, speed, ease of operation, and the instrument cost (Table 3.1).

1. Information content and quality. In general, all FTIR methods provide useful information on the solid-phase samples, but the quality and the extent of the revealed information vary. From Figure 3.3, it is clear that single-bead FTIR and the beam condenser methods are subject to less artifact interference and, therefore, offer more information on the resin-bound compounds. This demonstrated that transmission mode always gives highest quality. Spectra obtained by the ATR method exhibit lower resolution in the fingerprint region compared with those obtained by single-bead and beam-condenser methods. Moisture interference at 3100–3500 cm^{-1} is significant in ATR spectra. The KBr method is the only destructive method among all methods in this investigation. It is not possible to grind the beads. The whole beads in the pellet cause light scattering and, thus, baseline tilting. Moisture interference is also prominent. Because of the different selection rules for Raman spectroscopy, many important vibrations are not detectable by FT Raman method, such as OH, NH, aliphatic C=O, C=N and C-O vibrations, only considering the limited examples studied in this

work. However, FT Raman can detect some IR-inactive or weak vibration modes, such as the stretching of S-S (Figure 3.3b). Thus, FT Raman can be considered as a complementary technique to FTIR methods in the analysis of solid-bound organic molecules.

2. Sample requirement. KBr pellet, DRIFT, and photoacoustic methods generally require 5–10 mg of beads. Among these methods, DRIFT and the photoacoustic method are not destructive to the sample. The sample can be reused for synthesis after the analysis. This is clearly an advantage. However, under regular SPOS scale, the reaction has to be interrupted by analysis with these methods due to the demanded sample size. Macro FT Raman needs about 0.2 mg of resin beads for a 5-minute analysis (laser power 350 mW). Single-bead FT Raman can be performed using FT Raman microspectroscopy in 10–20 min and a 1000 mW laser output. The results obtained with both FT Raman methods are comparable (Yan et al., 1999). ATR and the beam condenser methods only require enough beads to cover the area of ATR crystal (Φ 0.5 mm) and the diamond window (Φ 1.5 mm) of the compression cell. Although oversample can ensure the coverage, 50–100 beads are the minimum amount of sample to be used. The most sensitive method is the single-bead FTIR method. It has been demonstrated in the past (Yan et al., 1996a) that the detection limit of the method is about 125 fmol of sample, which is about 1/3840 of a single bead loading. The amount of sample required by single-bead FTIR, single-bead FT Raman, beam condenser, and ATR analysis is so small that the reaction can be monitored without the need to interrupt the reaction (real-time reaction monitoring). This is clearly the advantage similar to what TLC offers in the monitoring of the solution synthesis.
3. Speed of analysis. No sample preparation is required for all methods except the KBr pellet method. All FTIR methods exhibit high speed of analysis (1–2 minutes), except that the KBr pellet method requires the making of pellet (~10 min). At a moderate laser power (~350 mW), the accumulation of an FT Raman spectrum from a 0.2-mg sample requires ~5 min. An increase in the laser power can reduce the accumulation time. However, it may cause photoreactions or photodecompositions of the sample. Single-bead FT Raman analysis requires 10–20 min and at least 1000 mW laser power. However, the measuring time highly depends on the Raman activity of the sample.
4. Automation. The single-bead FTIR and single-bead FT Raman methods can be automated by using a computer-controlled motorized stage (SpectraTech, Shelton, CT). Automated analysis using

DRIFT can be obtained commercially (Pike Technologies, 2919 Commerce Park Drive, Madison, WI 53719; Spectros Instruments, Inc., 175 North Street, Shrewsbury, MA 01545). Macro FT Raman and KBr methods can be automated by a rotating wheel assembly. Automation of ATR and photoacoustic spectroscopy are not known to our knowledge.

5. Instrumentation cost. FTIR is the most popular analytical instrument in organic chemistry laboratories. To use the above-mentioned techniques for analyzing polymer-bound organic molecules, accessories to a regular FTIR are required. They can be arranged in order of increasing cost in the following: KBr, beam condenser, ATR, DRIFT, photoacoustic, single-bead FTIR, macro FT Raman, and single-bead FT Raman.
6. DRIFT and photoacoustic methods. The comparison with these two methods was on the basis of the published papers assuming that authors had optimized their methods. In DRIFT measurement, several milligrams of resin was used. The analysis time was similar to other FTIR methods. The spectral quality can be seen in Figure 3.1. In general, the fingerprint region is subject to signal overlapping in DRIFT methods. The cost of the accessory is, however, low (similar to ATR).

   In photoacoustic measurement, less than 10 mg of resin was used. The measuring time is a few minutes (500 scans), which is fast enough for routine applications. The spectral quality is similar to DRIFT and ATR spectra. Signal overlapping in the fingerprint region is also observed (Figure 3.2). The cost of this accessory is higher than DRIFT, ATR, and beam condenser.

The above comparisons are summarized in Table 3.1 in terms of numeric scores from 1 to 5. Scores based on relative comparisons in each category are summed, and the final scores are compared. In general, the most appropriate technique for a laboratory is the technique that is conveniently available. However, the value/cost ratios of these methods are varied. For those chemists who are searching for a method to monitor SPOS, the comparisons offered in Table 3.1 are informative. Single-bead FTIR is an impressive method. It provides highest quality spectrum, fast analysis, and compatibility, particularly with one-bead-one-compound libraries. The beam condenser method offers the similar quality spectrum with a few more beads (50–100 beads). With the same speed of analysis, the operation is much simpler. Furthermore, the cost of a beam condenser is much lower than an IR microscope. The ratings of other methods are also listed.

3.1.4 Monitoring of Polymer-Supported Organic Reactions

The single-bead FTIR method has been extensively used in kinetics studies (Yan et al., 1995; Yan et al., 1996a, b; Yan & Li, 1997; Sun & Yan, 1998; Li & Yan, 1998). To establish the kinetics capability of the beam-condenser method, the kinetics of the reaction in Scheme 3.1 was studied. The reaction of succinimidyl 6-(N-(7-nitrobenz-2-oxa-1,3-diazo-4-yl)amino)hexanoate with Wang resin produced compound **7** (Figure 3.4). The disappearance of the starting material bands at 3577 and 3458 cm^{-1} with time coincides with the formation of bands at 1732, 1580, and 3352 cm^{-1} that are all attributable to the product. The completion of the reaction can be confirmed with confidence by the disappearance of the hydroxyl bands at 3577 and 3458 cm^{-1} (Figure 3.4). The area integrations of the emerging band at 1732 cm^{-1} were plotted against time, and the time course was fitted to a pseudo first-order rate equation with a rate constant of 3.8×10^{-4} 1/s. Both methods showed comparable quality.

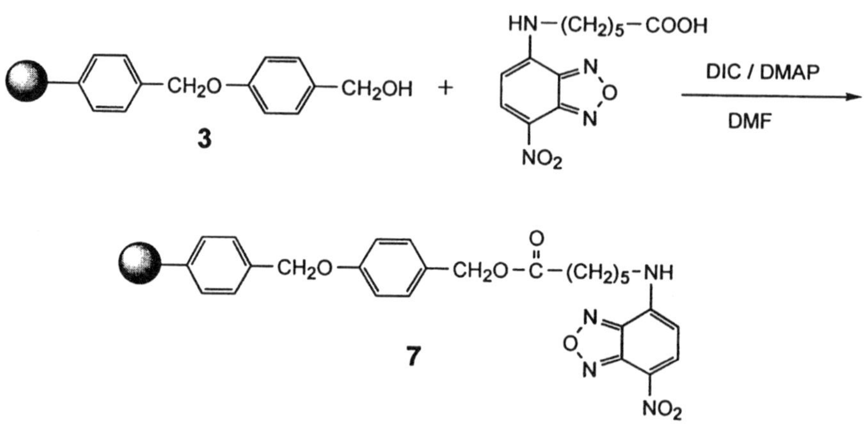

Scheme 3.1

Above results show that both single-bead FTIR and beam-condenser FTIR methods are superb methods for the purpose of solid-phase reaction monitoring. The beam-condenser method requires more beads (50–100 beads) compared to the single-bead FTIR method, but this disadvantage is well balanced by the low cost of the beam condenser and the ease of operation.

The single-bead and beam-condenser FTIR methods meet several requirements as a TLC equivalent for solid-phase synthesis:

1. Only 50–100 beads are needed for analysis. The removal and examination of this tiny amount of beads from the reaction suspension practically does not perturb the reaction so that the reaction is virtually monitored in real-time like TLC.
2. A high-quality spectrum can be recorded within a few minutes.
3. No sample preparation is required, reducing the whole analysis time even shorter than a TLC analysis.
4. These two FTIR methods provide qualitative, quantitative (the percentage of conversion), and kinetics information on organic reactions carried out on solid supports.
5. From the synthetic chemist's point of view, the low cost and the ease of operation are key factors to make the beam-condenser method the better method as a TLC equivalent in SPOS.

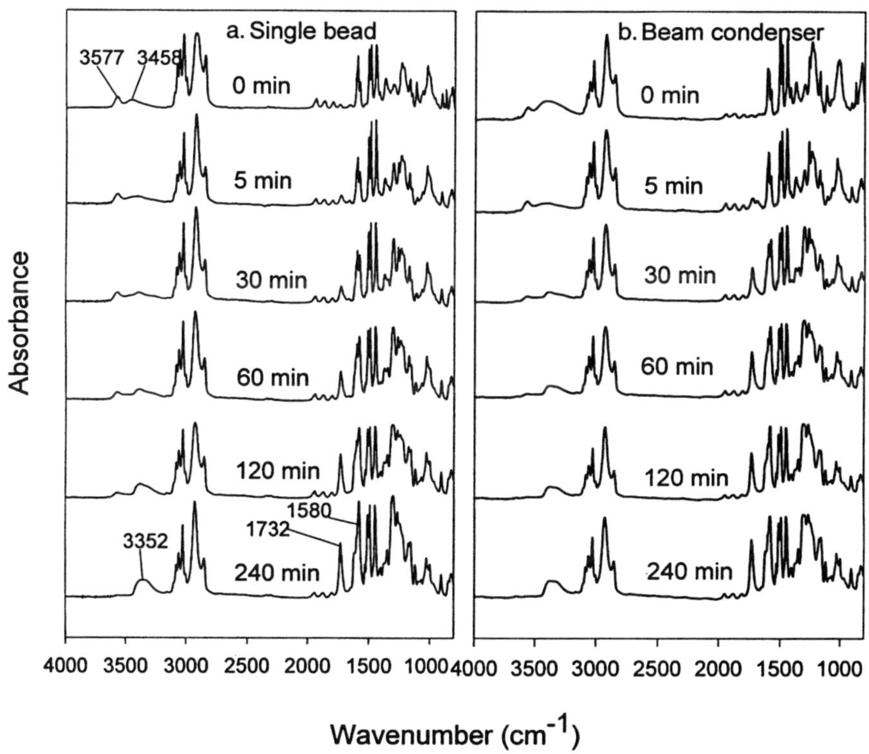

***Figure 3.4** IR Spectra taken at various times during the reaction in Scheme 3.1: (a) by single-bead FTIR method; (b) by beam-condenser FTIR method.*

The ATR method, although not the best method for resin analysis, is the method of choice for the surface-grafted solid supports, such as multi-pins (Gremlich & Berets, 1996) and "MicroTubes" (IRORI, San Diego, CA). More discussions on this technique will be given in section 3.4. In such analysis, it may be superior to other surface FTIR analysis methods. The FT Raman method scored low in our comparison due to low information content and higher cost. However, it provides complementary information to FTIR, especially for resin-bound compounds containing symmetric vibrations of organic functional groups, such as NO_2 and bands below 600 cm^{-1}, such as S-S and C-Cl stretching vibrations.

3.2 THE MONITORING OF REACTIONS ON POLYSTYRENE-BASED RESIN

A large portion of SPOS reactions have been carried out on PS resins. Most on-resin analysis by FTIR techniques has also been done on PS resins. This kind of analysis will be increasingly popular and powerful in the future because the resulting IR spectra are often the same or better quality than the spectra from small molecule samples.

3.2.1 Yes-and-No Information

Similar to the role of TLC in solution organic synthesis, FTIR provides a quick answer to the most important question any chemist would ask first: is my reaction working? An example of the monitoring of a multistep indazole synthesis is shown in Scheme 3.2 (Yan & Gstach, 1996). Shown in Figure 3.5, the coupling of benzophenone-4-carboxyphenylhydrazone to the Merrifield resin **8** was unambiguously identified by the appearance of many spectral features in **9.** The sharp peak at 3345 cm^{-1} is assignable to the N-H stretch of the hydrazone. A very strong carbonyl band also appears at 1710 cm,$^{-1}$ corresponding to the aromatic ester functionality introduced in this step. In addition, prominent C-O-C skeletal vibrations emerged at 1280 cm^{-1} and 1100 cm^{-1}. The quantitative oxidation of the hydrazone **9** to the α-arylazo-benzhydryl acetate **10** by $Pb(OAc)_4$ is indicated by the complete disappearance of the N-H band, due to oxidation of the hydrazone to the azo-functionality, and the appearance of a second well-resolved carbonyl stretch at 1759 cm^{-1}. The latter is assignable to the stretch of the carbonyl in the newly formed aliphatic ester. The cleavage of the N,O-acetal **10** by BF_3 and the successful formation of resin-bound indazole **11** is supported by the

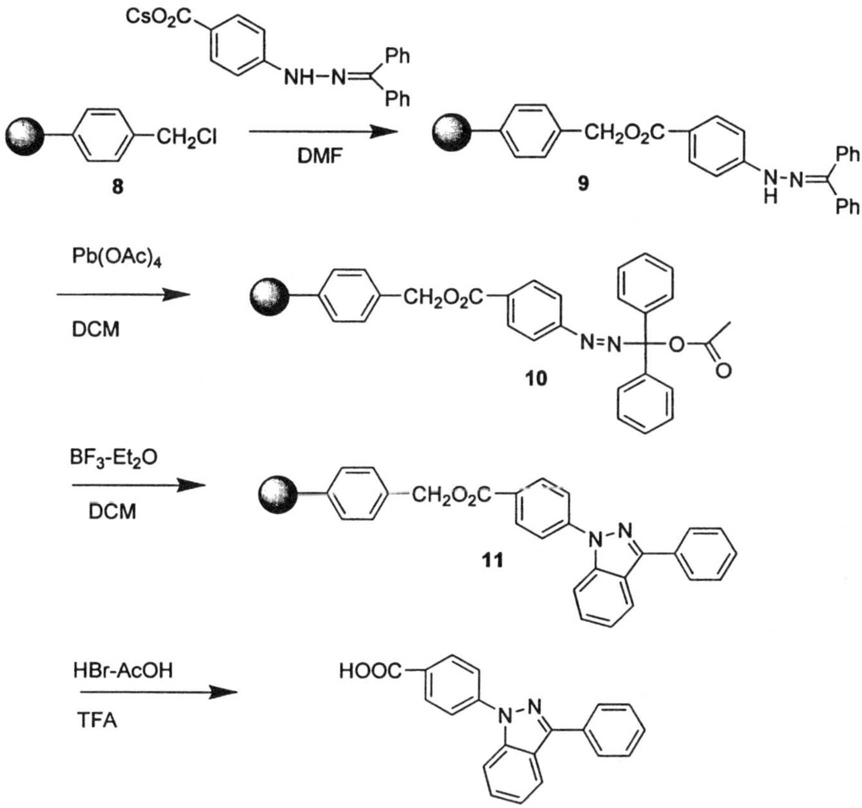

Scheme 3.2

disappearance of this carbonyl band. The isolated final product was confirmed by ^1H and ^{13}C NMR, and the overall yield is 79%.

Another example is the monitoring of the solid-phase synthesis via 5-oxazolidinones (Marti et al., 1997). As shown in Scheme 3.3, starting from the N-carbamate protected aspartic acid, the oxazolidinone **13** was first prepared in high yield (> 90%) and purity by known procedures (paraformaldehyde, p-TsOH, azeotropic removal of water). The Alloc protected oxazolidinone **13** was used due to the facile removal of the Alloc group by Pd^0. In the next step, the oxazolidinone **13** was attached to Knorr resin **12** using standard DIC/HOAt coupling methods. The analysis of the polymer-bound oxazolidinones **14** by single-bead FTIR microspectroscopy showed the characteristic carbonyl IR band at 1800 cm^{-1} (Figure 3.6). This unique band served as a useful

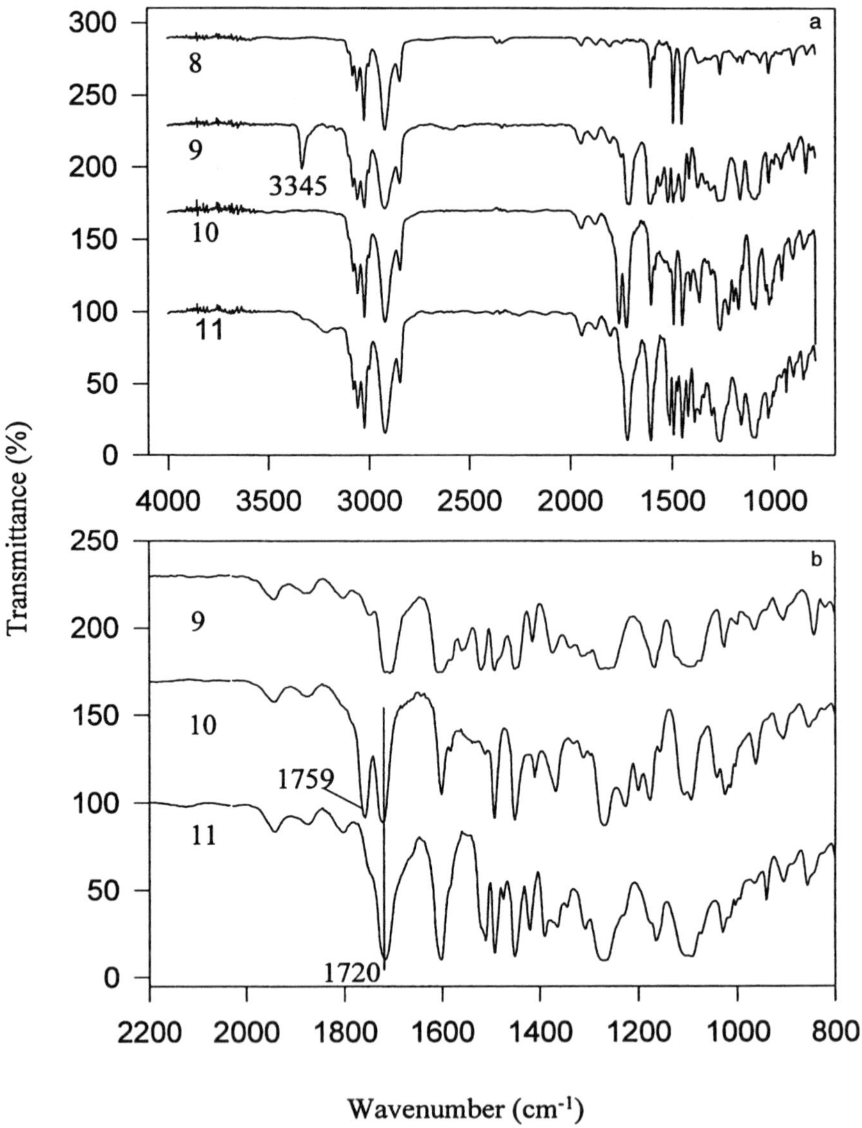

Figure 3.5 *The single-bead IR spectra of starting material, intermediates, and product in Scheme 3.2. This reaction series starts with Merrifield resin. The spectra labeled 1–4 are for compounds **8–11,** respectively. Panel b shows spectra of **9–11** on an expanded scale from 800 to 2200 cm^{-1}.*

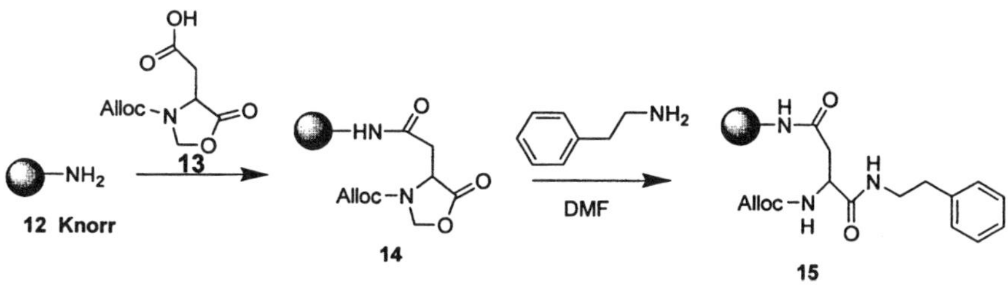

Scheme 3.3

and distinct indicator for product identification in a series of similar sythesis reactions.

Absorbance saturation in the fingerprint region is observed for spectra in this figure. Care should be taken to make sure the band that is used for analysis not be saturated. The cause for spectral saturation is the high local concentration in the bead. Bead flattening can effectively reduce the saturation effect and can improve the

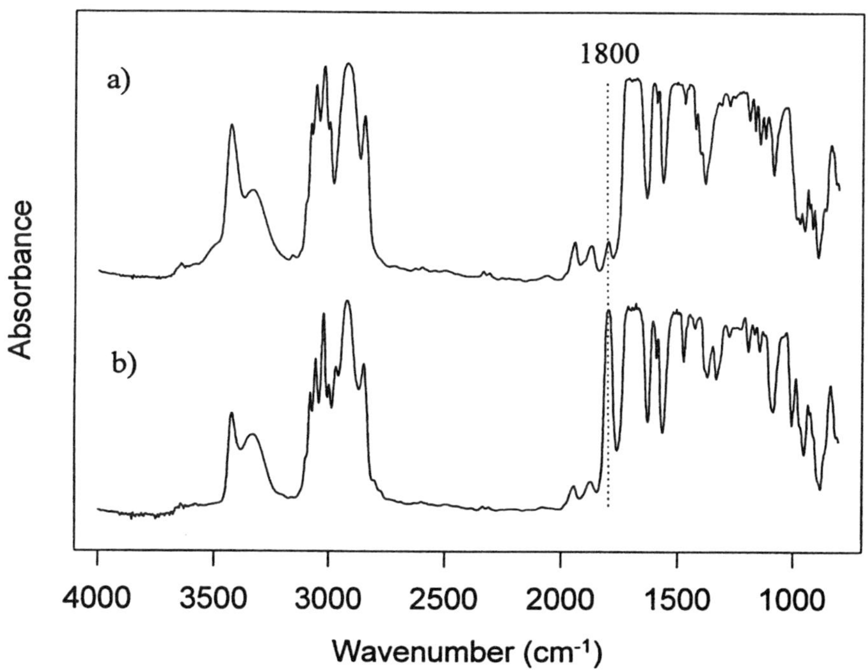

***Figure 3.6** The single-bead IR spectra for Scheme 3.3. (a) The IR spectrum of **12**; (b) The IR spectrum of **14**.*

spectrum (Yan & Kumaravel, 1996). However, after the resin is treated with aqueous solution or methanol (as in Figure 3.6), the bead is shrunken and hardened and, as a result, is not easily flattened.

3.2.2 Reaction Kinetics on Resin Support

Reactions used for the library synthesis should go to completion, preferably at a high rate. This is particularly important at intermediate steps in order to ensure a speedy and complete reaction and a relatively high purity in the final product. To fully realize the advantage of SPOS in multistep organic synthesis, the kinetics of solid-phase reactions need to be known. However, there is a serious lack of kinetic studies for polymer-supported reactions. As a consequence, solid-phase organic reactions are commonly carried out for 12–24 hours indiscriminately before cleaving compounds off for analysis. There are two reasons for this. First, the general lack of sensitive and rapid analytical tools for on-resin analysis in the past has limited one's capability to pursue such studies. Second, most available methods (such as gel-phase NMR and most FTIR methods, such as KBr pellet method) would require too much sample (> 10 mg) for a single time point. This is impractical for general use in kinetic studies considering the small-scale synthesis and the cost of resins. In contrast, the single-bead FTIR provides a convenient tool for kinetics studies (Yan et al., 1995; Yan et al., 1996a, b; Yan & Li, 1997; Pivonka et al., 1996; Pivonka & Simpson, 1997; Sun & Yan, 1998; Li & Yan, 1998).

To determine the reaction kinetics of the second step in Scheme 3.3, the progress of the reaction of resin **14** with a 10-fold excess of phenethylamine in DMF was monitored by taking spectrum of a single bead at various times during the course of this reaction, and results are shown in Figure 3.7. The intensity of the band at 1800 cm^{-1} decreases with time, indicating a rapid formation of the resin-bound product **15.** Integrations of areas under the oxazolidinone carbonyl band at 1800 cm^{-1} were plotted against time in Figure 3.8. Data were analyzed by curve-fitting to a first-order rate equation. The best fit is shown in Figure 3.8 as the solid line. The pseudo-first-order rate constant is 1.23×10^{-3} 1/s for this ring-opening reaction. The reaction time course is single exponential, suggesting the rate only depends on the available reactive sites on the resin when the other reagent is in excess. Solid-phase reactions have often been fitted to a pseudo-first-order reaction rate equation (Yan et al., 1996a). The lack of IR peaks of side products and the fit to a first-order reaction rate equation strongly

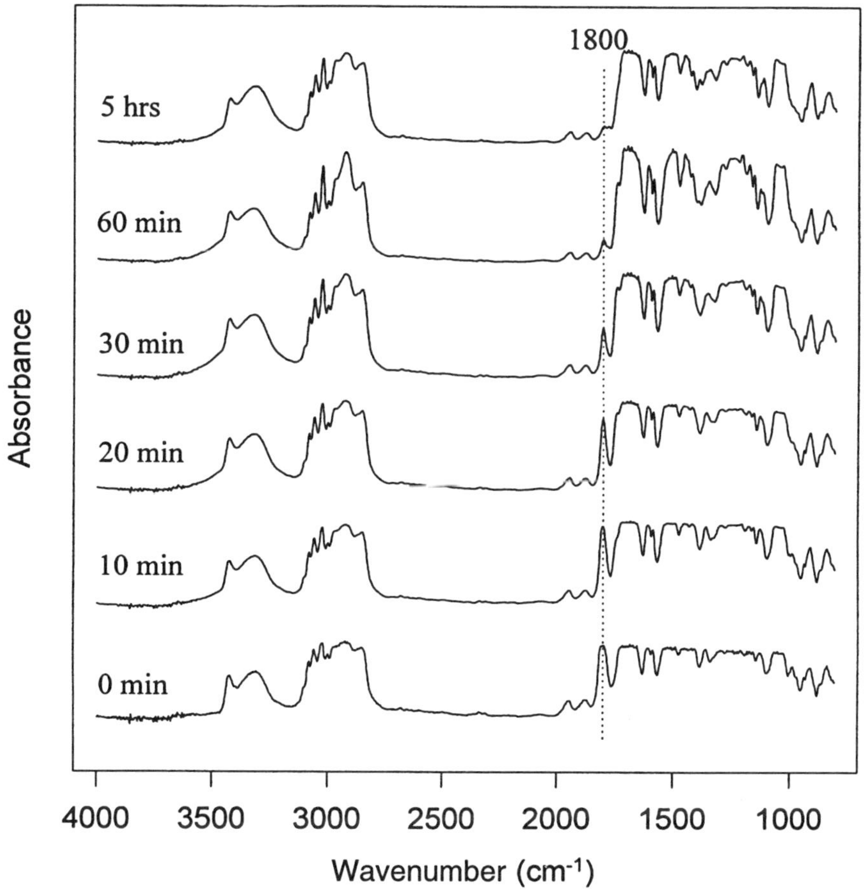

Figure 3.7 *Single-bead IR spectra at various times during the synthesis of **15**. The starting resin is resin **14**.*

suggest there are no significant side reactions. This was further confirmed by analyzing the cleaved product (TFA:H_2O, 95:5) by HPLC, MS, and NMR. The cleaved product was obtained in a "crude" yield of 80% in a purity of > 90% by RP-HPLC.

In Scheme 3.4, the catalytic oxidation of resin **3** to a resin-bound aldehyde **16** was monitored by single-bead IR (Figure 3.9). The areas of IR bands, which undergo changes, were integrated and plotted against time (Figure 3.10). The data are fit by a pseudo-first-order rate equation to obtain rate constants as shown. The rate depends on the

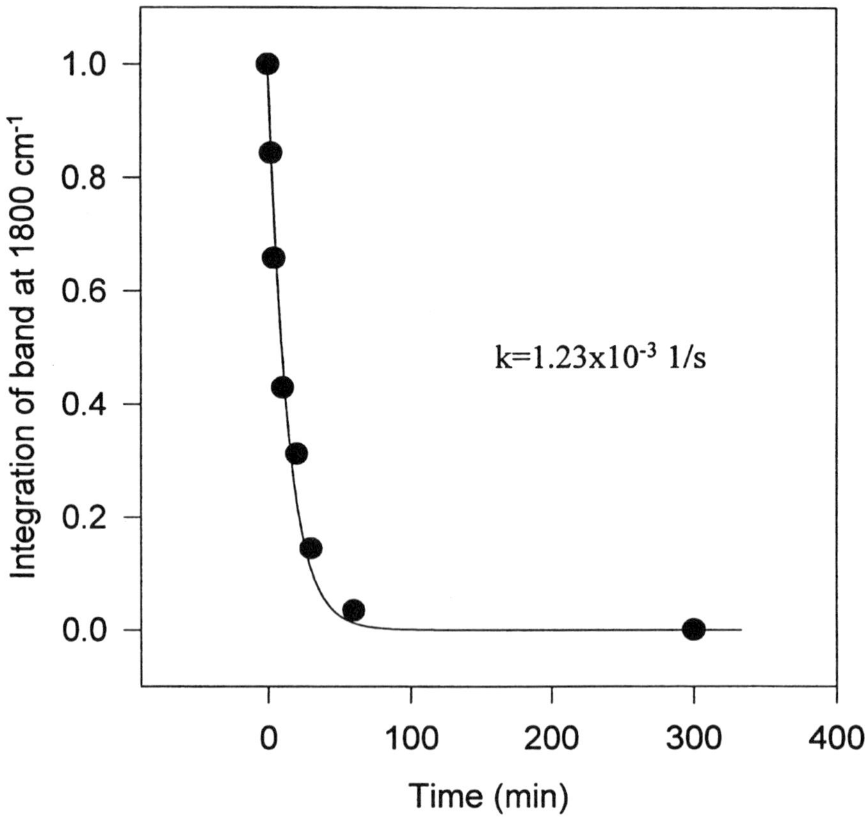

Figure 3.8 *The reaction time course for the second step reaction in Scheme 3.3. The integrations of peak area at 1800 cm⁻¹ for spectra in Figure 3.5 were plotted against time.*

Scheme 3.4

amount of catalyst tetra-n-propylammonium perruthenate (TPAP) used for the oxidation (Figure 3.10).

Attenuated total reflection (ATR) or internal reflection is a versatile, nondestructive technique for obtaining IR spectra from the sample surface. Using an ATR objective in the same IR microscope for recording

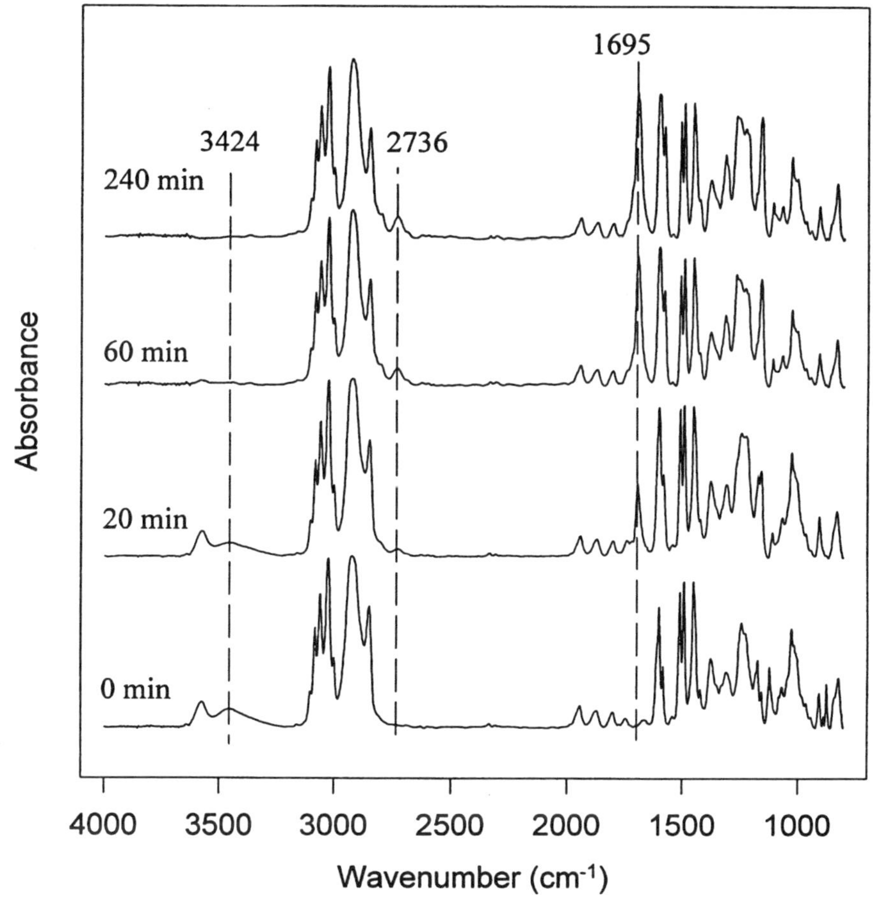

Figure 3.9 *IR spectra taken from a single bead at various times in the course of Scheme 3.4. Spectra were taken from a single flattened bead at 0, 20, 60, and 240 min after the initiation of the oxidation reaction. The hydrogen-bonded and unbonded hydroxyl stretch near 3400–3600 cm^{-1} disappears as the intensities of the bands for the aldehyde C-H (2736 cm^{-1}) and the aldehyde carbonyl (1695 cm^{-1}) increase.*

the single-bead transmission IR spectra, an IR spectrum from the partial surface of a single bead can be obtained in a few minutes. In order to investigate the problem of the possibly hindered diffusion of a reagent into the bead interior, the reaction rate on the bead surface and in the bead interior was monitored simultaneously using ATR mode and transmission mode, respectively (Yan et al., 1996a). Spectra taken from a single-bead surface and interior at various times during

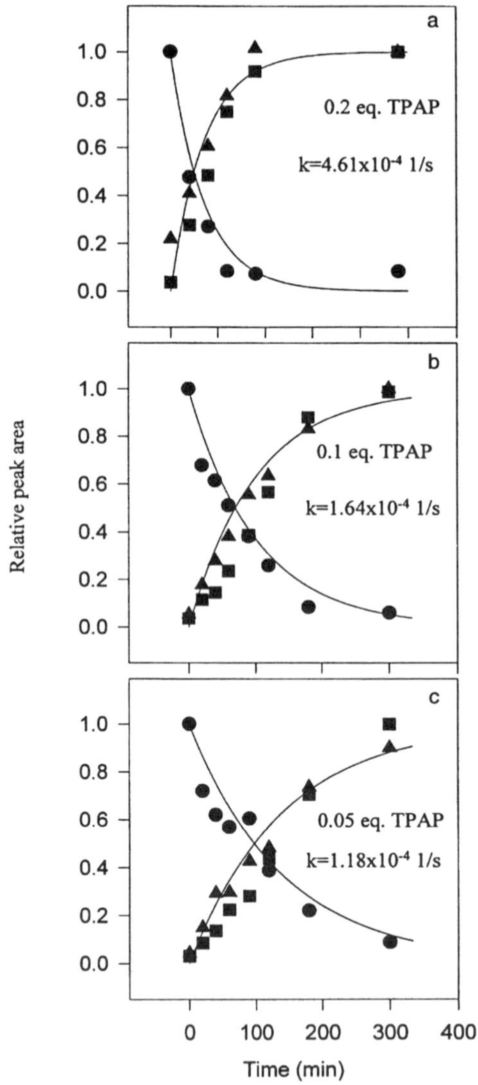

Figure 3.10 *The time course of the reaction in Scheme 3.4. All spectra were normalized by making the intensity of the polystyrene band at 1945 cm^{-1} equal. The area integration for the hydroxyl band from 3181 to 3637 cm^{-1} (circles), aldehyde C-H band from 2664 to 2766 cm^{-1} (squares), and aldehyde carbonyl band from 1641 to 1765 cm^{-1} (triangles) for spectra at various times were plotted against time. Lines were calculated from the best fit to a first-order reaction equation with a rate constant shown. The amount of catalyst TPAP was 0.2 eq. in (a) 0.1 eq. in (b), and 0.05 eq. in (c).*

the reaction in Scheme 3.5 are shown in Figure 3.11. Reaction time courses on the bead surface and in the bead interior for two other reactions (Schemes 3.6 and 3.7) were also studied (Yan et al., 1996a), and no difference was found in the reaction rate on the surface and in the interior of the bead for these reactions.

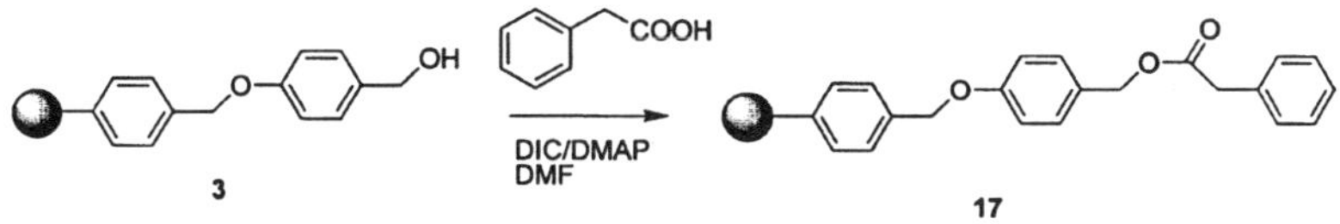

Scheme 3.5

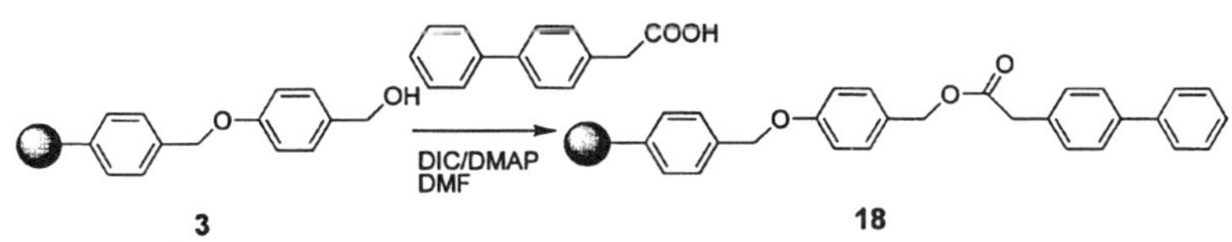

Scheme 3.6

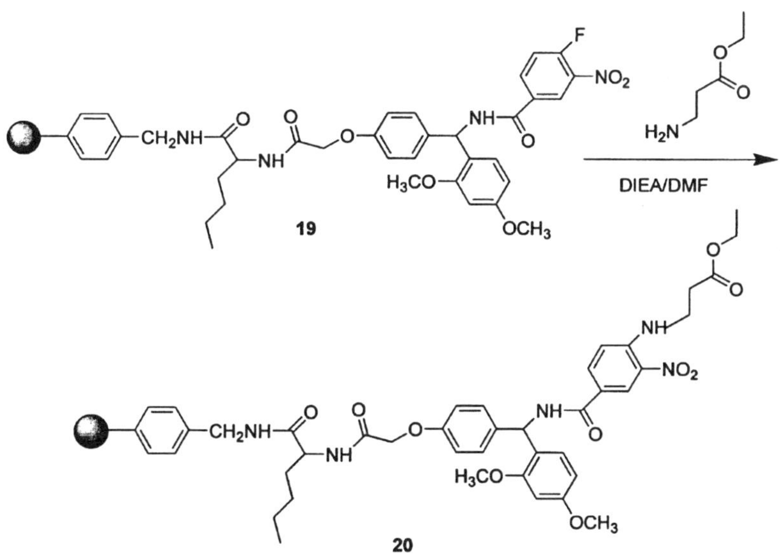

Scheme 3.7

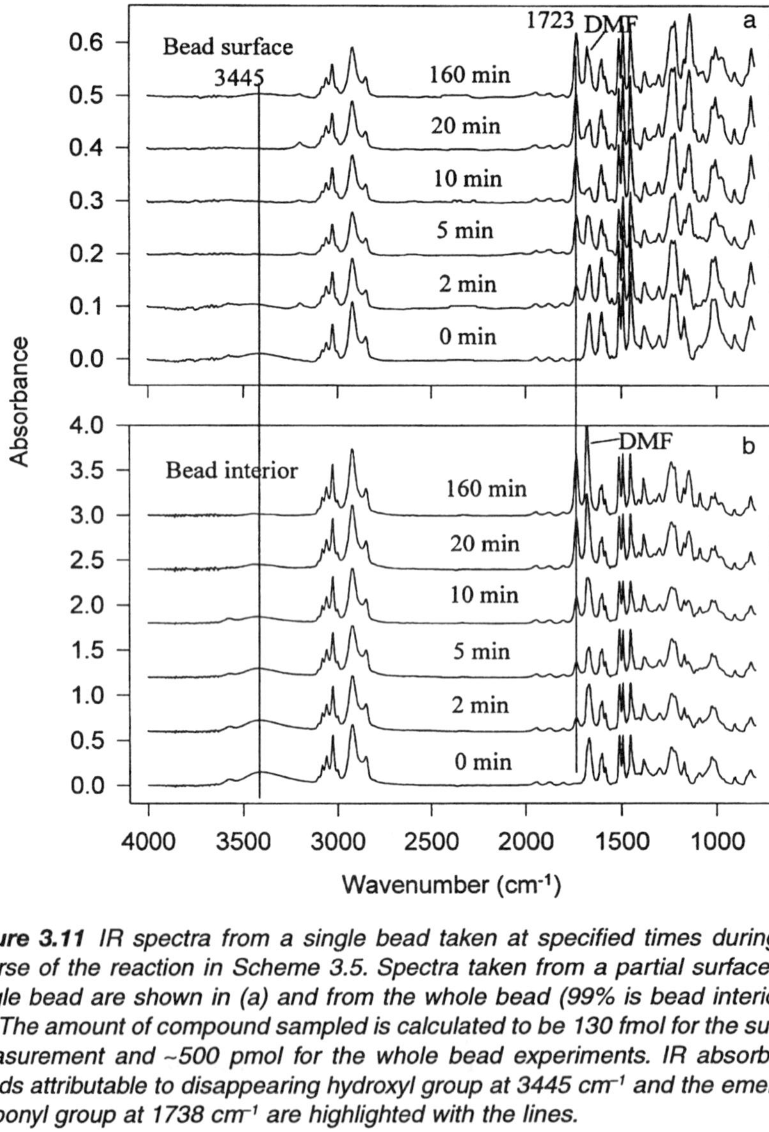

Figure 3.11 *IR spectra from a single bead taken at specified times during the course of the reaction in Scheme 3.5. Spectra taken from a partial surface of a single bead are shown in (a) and from the whole bead (99% is bead interior) in (b). The amount of compound sampled is calculated to be 130 fmol for the surface measurement and ~500 pmol for the whole bead experiments. IR absorbance bands attributable to disappearing hydroxyl group at 3445 cm^{-1} and the emerging carbonyl group at 1738 cm^{-1} are highlighted with the lines.*

Using the combination of FTIR microspectroscopy and a flow-through cell, organic reactions on a single flattened bead can be monitored *in situ* in the selected IR-friendly solvent (such as $CHCl_3$ or CH_2Cl_2). This instrument set-up has been used to monitor a benzoylation reaction of the aminomethylstyrene resin and a series of coupling reactions to the polystyrene tritylchloride resin (Pivonka et al., 1996).

The *in situ* synthesis and analysis of compounds prepared for solid-phase synthesis were documented. Pathlength stability of the *in situ* analysis combined with dynamic spectral processing using single-beam data collected internal to the run as spectral backgrounds have been shown to provide very high-quality spectra with enhanced spectral selectivity tailored to the problem at hand.

The feasibility of using infrared spectroscopy for the real-time, *in situ* analysis of resin-bound photocleavable linkers from a single bead is also illustrated using infrared microscopy (Pivonka & Simpson, 1997).

Figure 3.12a illustrated the nitrile band absorbance region of compound **21** for a time-resolved spectral series collected from the photo-cleavage reaction (Scheme 3.8). Figure 3.12b illustrates the first-derivative data calculated from the Figure 3.12a spectra. First-derivative data effectively eliminate baseline drift without the need for baseline correction.

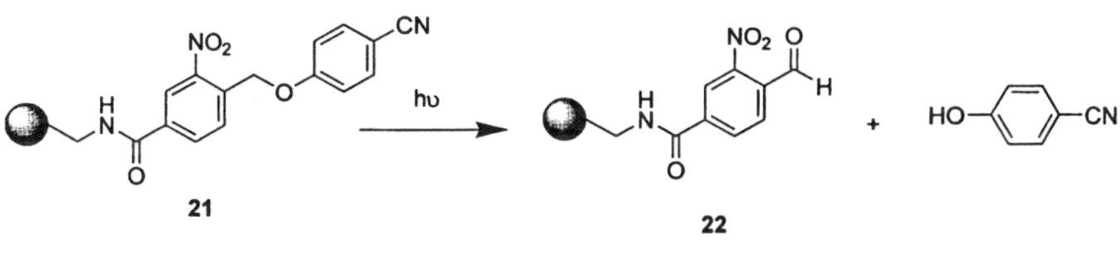

Scheme 3.8

This work showed the *in situ* infrared analysis of reaction on the solid-phase support involving the single-bead analysis of photolinker cleavage kinetics. The basic methodology presented by analysis of this reaction provides a pathway for the investigation and optimization of photo-cleavage parameters (solvent, temperature, wavelength, and intensity). This method also allows for a rapid measurement of relative $t_{1/2}$ values for various types of groups to be cleaved in preparation for combinatorial library synthesis. In addition, the intrabead reagent dynamics has also been studied using IR spectroscopic techniques (Pivonka & Palmer, 1999).

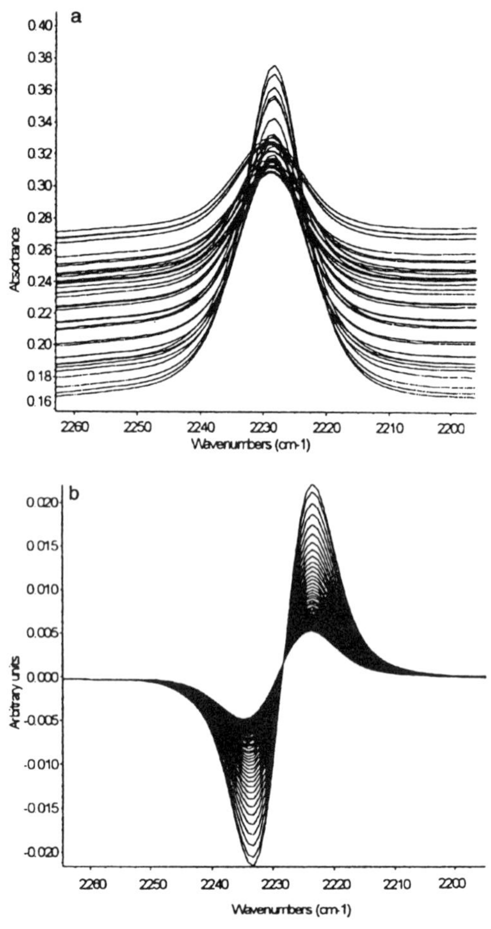

Figure 3.12 *(a) Time-resolved data of the CN region of resin* ***21;*** *(b) The first derivative data from (a) (Reproduced with permission from* Anal. Chem. *1997, 69, 3852. Copyright© 1997, American Chemical Society).*

3.2.3 Quantitative Estimation of Chemical Conversions (%) on Resin

In a typical TLC experiment, if the spot of a starting material disappears and the product spot is formed, the chemical transformation can be estimated as completed. In a similar fashion, chemical conversions (%) can be conveniently estimated by single-bead FTIR analysis by observing the disappearance of the starting material band quantitatively by peak area integration (see Figures 3.5 and 3.7 for examples). All the quantitative analysis, like the previously discussed kinetics studies, is pathlength and bead-size independent by using a polystyrene IR band as internal reference for quantitation.

The percentage conversion of a solid-phase benzimidazole synthesis (Scheme 3.9) was monitored by single-bead IR method (Sun & Yan, 1998). The IR spectra, taken at various times during the first step in Scheme 3.9, are shown in Figure 3.13a. Among other spectral features, two distinct IR bands in compound **23** are introduced: an ester carbonyl band at 1720 cm^{-1} and an aldehyde carbonyl band at 1703 cm^{-1}. At the

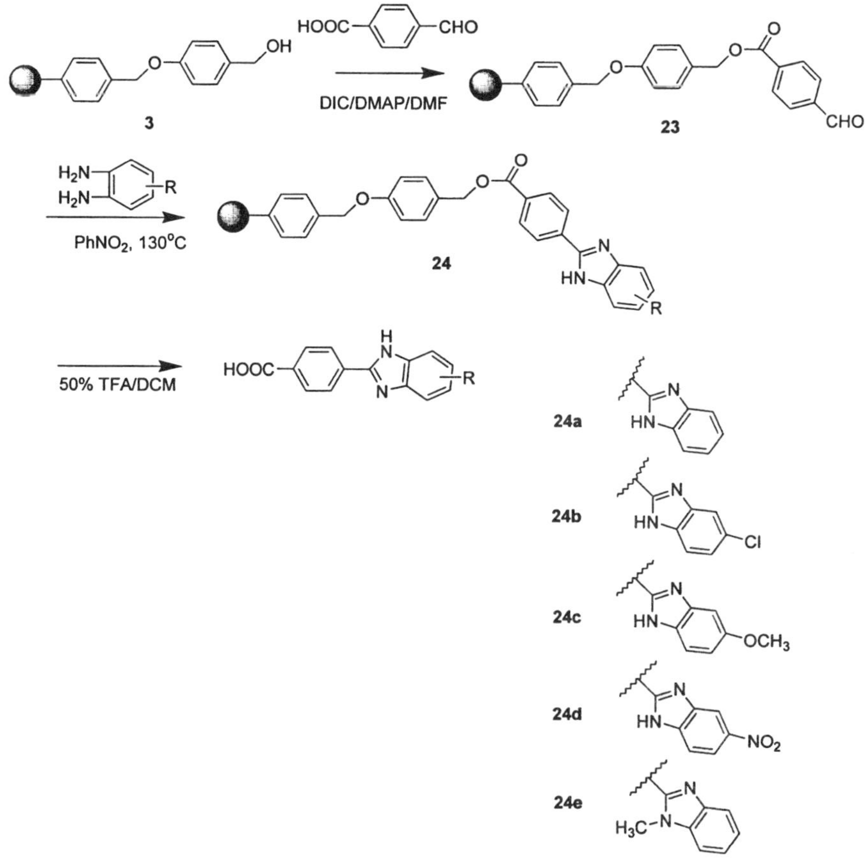

Scheme 3.9

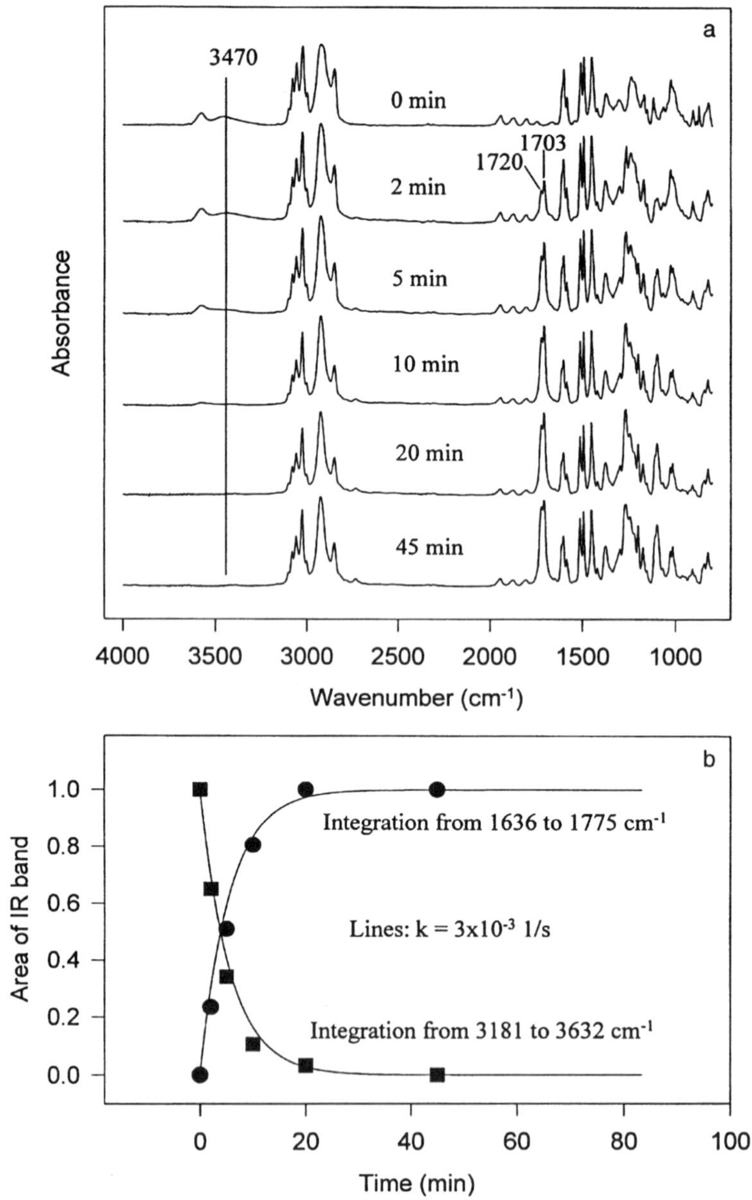

Figure 3.13 *(a) Single-bead IR spectra at various times during the synthesis of* ***23.*** *The starting material is Wang resin. (b) The time course of the reaction for the synthesis of* ***23.***

same time, the hydroxyl band at ~3500 is diminishing. In about 20 minutes, the intensity of the hydroxyl band almost reaches zero, and those of the carbonyl bands reach a constant maximum value suggesting a quantitative conversion. The time course of this reaction is shown in Figure 13b. The yield of 97% for this transformation was independently determined directly on the resin using a 6-mg resin sample and a novel fluorescence method (Yan & Li, 1997). A series of benzimidazoles **24a–24e** were then synthesized. The reaction was carried out at 130° C for eight hours using nitrobenzene as a solvent without stirring. The success of this reaction is confirmed by the disappearance of the aldehyde carbonyl band at 1703 cm^{-1} (Figure 3.14) and the formation of the N-H stretch for **24a–d** at ~3400 cm^{-1}. Due to the overlap between the ester and aldehyde carbonyl bands (Figure 3.14), the ratio of the two bands (1703/1720) in compounds **24a–e** was determined by peak-fit analysis (for example, see Figure 3.19 later) to obtain the on-resin purity and the yield of the product.

Another example is the catalytic oxidation of a secondary alcohol to a ketone. In solution, the oxidation of a secondary alcohol to a ketone by tetra-n-propylammonium perruthenate (TPAP) is generally slower than the oxidation of a primary alcohol, suggesting the catalyst is sterically demanding. The sterically hindered secondary alcohol **25** in Scheme 3.10 would be expected to be further hindered by the attachment to the polystyrene resin. It is important to know the rate of this reaction. The starting material Rink Acid was treated with 0.2 eq. TPAP in the presence of N-methylmorpholine N-oxide (NMO). IR spectra taken on a single bead from the reaction mixture at various time intervals after the initiation of the reaction are shown in Figure 3.15. The disappearance of a band at 3568 cm^{-1} attributed to the hydroxyl group and the increasing band at 1654 cm^{-1} attributed to the ketone carbonyl are evident. Areas under these two peaks, when plotted against time, fit with a first-order rate equation. The average rate constant is 1.9×10^{-4} 1/s. The same rate for the disappearance of the

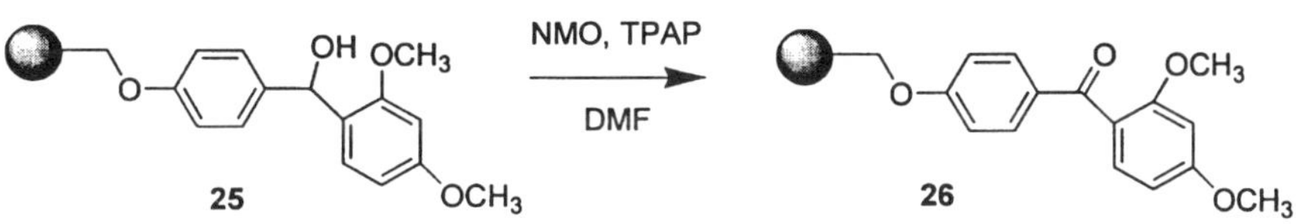

Scheme 3.10

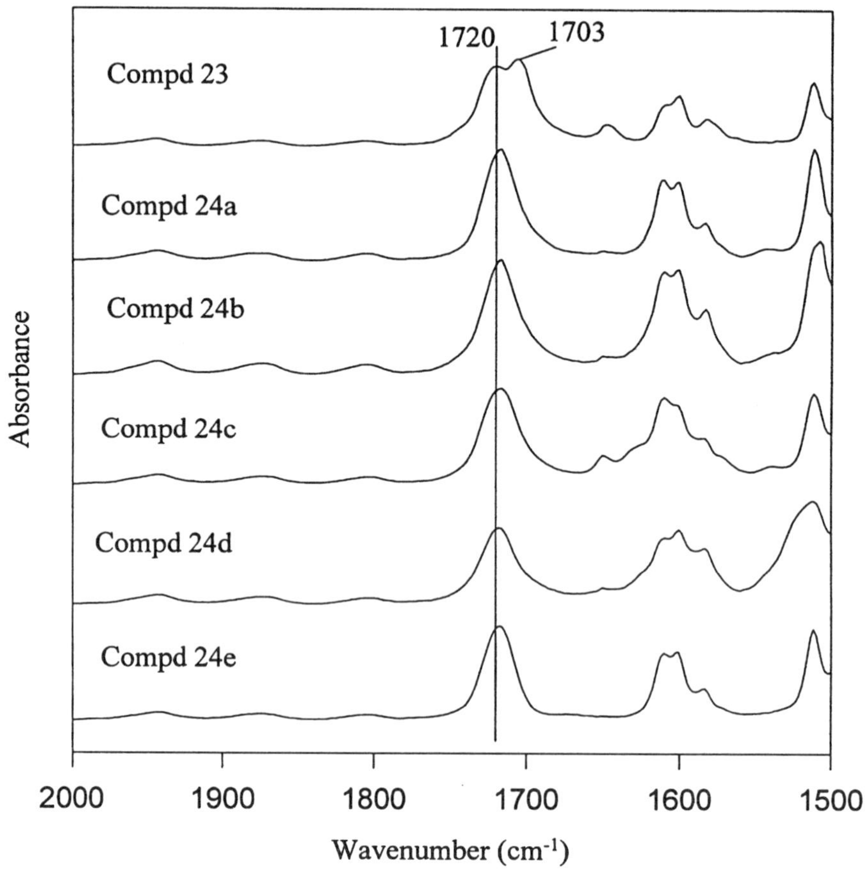

***Figure 3.14** Single-bead IR spectra for resin-bound product **23** and **24a–24e**. Five sealed tubes with dried resin (100 mg, 1.0mmol/g), one of five diamines a–e (1 mmol), and 3 mL nitrobenzene were heated at 130° C overnight. After they were cooled to room temperature, the resins were transferred to the filtration tubes and washed. Single-bead IR spectra were taken for each product.*

hydroxyl group and the appearance of the ketone carbonyl indicates that only one reaction (i.e., the oxidation of the alcohol **25** to the ketone **26**) is occurring. Based on the change in area integrations under the 3568 cm^{-1} band, a 97% conversion of **25** to **26** takes about four hours. More examples on the quantitative aspect of the single-bead FTIR can be found in related references (Yan et al., 1996a, b; Li & Yan, 1998; Yan & Gstach, 1996). Note that the estimation of the reaction yield with single-bead IR was obtained without the need to stop or interrupt the reaction or cleave the product from the support.

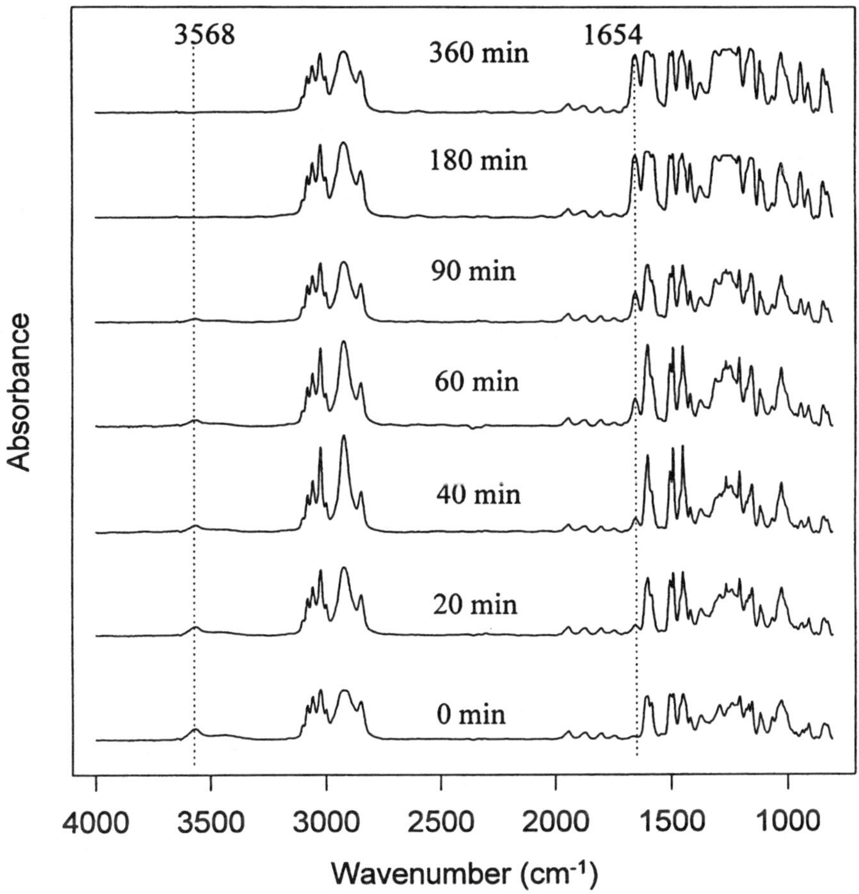

Figure 3.15 *IR spectra taken from a single bead at various times in the course of the reaction in Scheme 3.10. Spectra were taken from a single flattened bead at 0, 20, 40, 60, 90, 180, and 360 min after the initiation of the oxidation reaction. The hydrogen-bonded and unbonded hydroxyl stretching bands of resin **25** at 3420 and 3568 cm^{-1} disappear as the ketone carbonyl group at 1654 cm^{-1} of the product **26** increase.*

By resorting to the isotopic synthesis and a separately predetermined standard calibration curve, reactions using isotope-labeled protecting groups or substrates have been quantitated (Russell et al., 1996). The synthesis and/or application of deuterium isotope containing protecting groups, including acetyl-d_3 chlorides, benzoyl-d_5 chloride, and 2-[[[(tert-butyl-d_9)oxy]carbonyl]-oximino]-2-phenylacetonitrile (BOC-ON-d_9), are also introduced as highly selective infrared signatures amenable to quantitation by infrared spectroscopy. Calculation of first-derivative carbon-deuterium (C-D) stretching absorbance relative to resin backbone derivative absorbance provided pathlength-independent quantitation of deuterium content and allowed for the determination of chemical yields in solid-phase resin reactions involving changes in deuterium content. However, discrete calibrations were found to be necessary for each type of coupling, i.e., amide or ester. This technique facilitates analysis of solid-phase reactions where the number of C-D bonds changes throughout the course of a reaction. The use of deuterated protecting groups, such as Boc-d_9, provides for a simple method of labeling amino-containing substrates. The magnitude of the infrared absorbance from Boc-d_9 was found to be independent of environment; thus, the number of lysines in a multiple lysine solid-phase sample could be determined.

3.2.4 The Selection of Optimal Reaction Conditions Using Single-Bead IR

Because of the speed, sensitivity, and convenience of the single-bead FTIR method, the optimal solvent, resin, building block, linker, catalyst, mixing method, and other conditions have been routinely selected rapidly with the aid of this method. As an example, the selection of mixing method is shown below.

SPOS offers several advantages over solution phase techniques for achieving high-throughput synthesis. At the same time, SPOS also poses problems that do not exist in solution synthesis. One of the problems is that one has to select the most efficient reaction mixing method to maximize encounters between the solid-bound reactant and the soluble reagent. Diverse mixing techniques are currently used in various laboratories without a clear understanding of their relative efficiency. Bead suspensions are usually mixed by shaking the tubes, which are fixed at a 45° angle, horizontally by an orbit shaker, by angular shaking (20–30°) on a wrist shaker, by rotating tubes at an angle of 180° or 360° (whole turn) rotations at a certain rpm, by bubbling N_2 gas from the bottom of the reactor, and by the conventional magnetic stirring.

To compare the mixing efficiency, the formation of a resin-bound dansylhydrazone through the reaction of formylpolystyrene resin (NovaBiochem, 0. 58 mmol/g, 1% DVB, 100–200 mesh) and the dansylhydrazine was studied using six different mixing methods (Li & Yan, 1997). Single-bead IR and the fluorescence quantitation method (Yan & Li, 1997) were used to monitor the reaction completion. Figure 3.16 shows IR spectra from 40 and 30 individual beads after a 30-min reaction

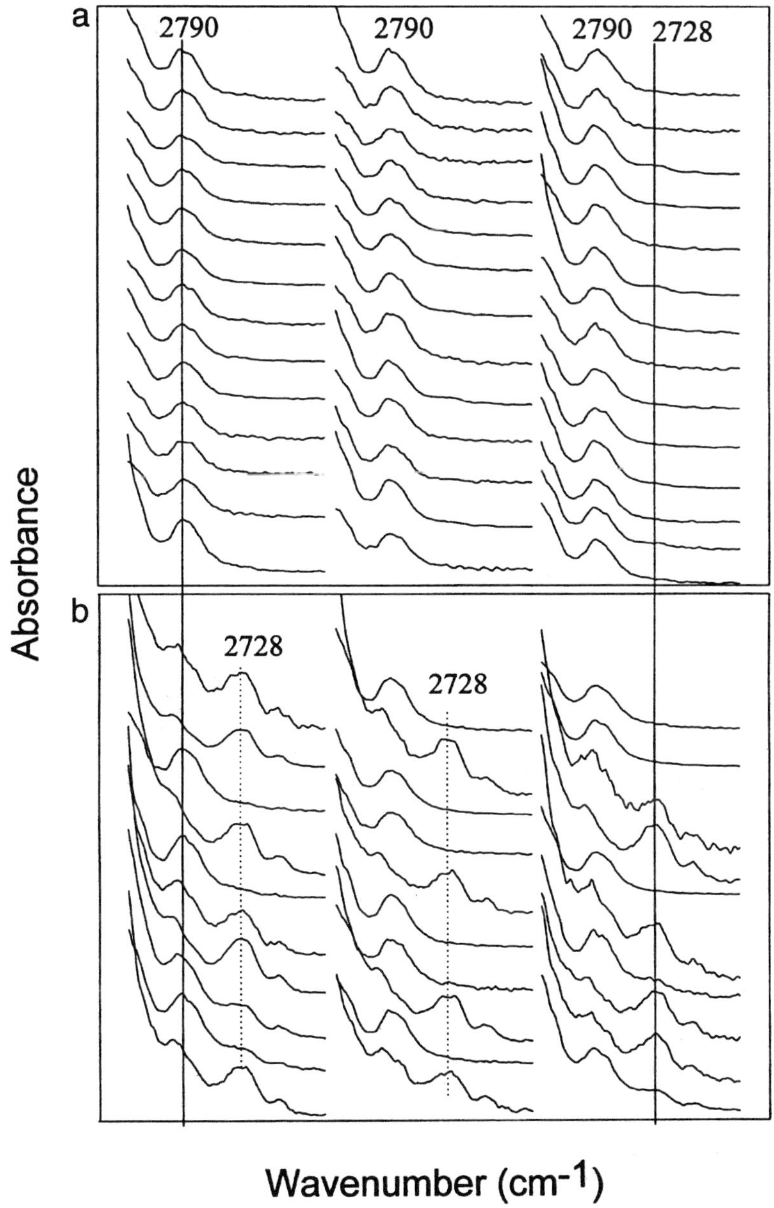

Figure 3.16 *IR spectra taken from individual formylpolystyrene beads after a 30-min reaction with a twofold excess of dansylhydrazine in DMF mixed with (a) a wrist shaker and (b) an orbit shaker with the reaction tube fixed at a 45° angle.*

(twofold excess dansylhydrazine in DMF), shaking with a wrist shaker or an orbit shaker. For the reaction studied, the 360° rotation and nitrogen bubbling methods provided highest mixing efficiency under mild conditions. Wrist shaking also showed high mixing efficiency with the maximum setting (vigorous shaking). Magnetic stirring showed the same high mixing efficiency, but this method often breaks beads. The method with 180° rotation mixing required longer reaction time and may not be preferable for slow reactions. Mixing with an orbit shaker did not give satisfactory reaction yields as compared with the other techniques. A larger excess of reagent (10-fold) and a doubled reaction time, however, can drive this reaction to completion.

3.3 THE MONITORING OF REACTIONS ON PS-PEG RESINS: COMPARING THE REACTION RATE ON PS AND PS-PEG RESINS

3.3.1 Background

IR spectra of PS-PEG resins are easily obtained with single-bead FTIR or other FTIR techniques. The main difference between IR spectra of PS resins and PS-PEG resins is the CH_2 vibration band of the latter is prominent. This does not interfere with the analysis in most cases. For PS beads, the flattening of beads usually improves the spectral quality (Yan & Kumaravel, 1996). PS beads can be flattened with a compression cell or simply by pressing the beads between two NaCl plates. However, the elasticity and mechanical stability of some PS-PEG resin beads are so low that a slight press will shatter the bead. A new PS-PEG resin, NovaGel, has an improved elastic property and can be flattened for FTIR studies. The single-bead FTIR measurement of PS-PEG resins is usually done with the globular bead when flattening is not possible. Spectra with reasonable quality can be generated and used for the monitoring of organic reactions on PS-PEG resins.

A common problem in SPOS practice is that reaction conditions cannot simply be transferred from one kind of support to another. A set of reaction conditions may work well for polystyrene resins, but may fail for pin-or PS-PEG resin-based synthesis. Reaction rate may also be drastically different on different supports. Understanding the effect of various supports on reactions is critical for a successful synthesis.

PS contains mainly 1% DVB crosslinked polystyrene backbones (~96–98% by weight), which is highly hydrophobic. Due to the short linker length, reactions on PS resin tend to be affected more by the hydrophobic PS backbones. The architecture of PS-PEG resin is based on a very small portion of crosslinked 1% DVB-PS backbones extensively grafted with long PEG linkers (40–60 ethylene oxide units). The PEG content is up to 70% of the resin weight. Therefore, the PEG chain determines the mechanical, physico-chemical behavior of the resin. Because the reactive sites in PS-PEG resins are located at the end of long and flexible spacers and are totally separated from the PS backbone, they are not affected by PS matrix. Additionally, the dominant PEG linkers exhibit greater miscibility with most solvents, including water. PS-PEG is generally regarded as providing a quasi-homogeneous and "solution-like" reaction support. It is commonly assumed that reactions carried out on PS-PEG resins proceed faster than those carried out on PS resins. Peptide synthesis reactions would be more favored on PS-PEG resins, because it is more polar than PS resins. However, peptide bond formation is only a special reaction. In SPOS, a different set of rules may apply.

3.3.2 Examples

To compare the reaction rate on PS-and PS-PEG-resins, several reactions (Schemes 3.4, 3.11, 3.12, and 3.13) were carried out on these two resins under identical conditions. An example shown in Figure 3.17 is the synthesis of a dansyl hydrazone (Scheme 3.12). Resin-bound **27** reacted with dansylhydrazine under identical conditions on PS-PEG and PS resins. Single-bead IR spectra for the transformation from **27** to **28** is monitored. The ester carbonyl bands at 1734 cm^{-1} remain, while the ketone carbonyl band at 1716 cm^{-1} disappears accompanying the formation of dansylhydrazone. The formation of the band for hydrazone N-H stretch at 3230 cm^{-1} and the band for $N(CH_3)_2$ stretch on the dansyl group at ~2790 cm^{-1} on PS resin product proved the product formation. Although the product band at 2790 is not observed due to the overlaps between the strong CH_2 band and the hydrazone signal in PS-PEG resin, the appearance of the band at 3230 and the disappearance of the band at 1715 cm^{-1} are evident.

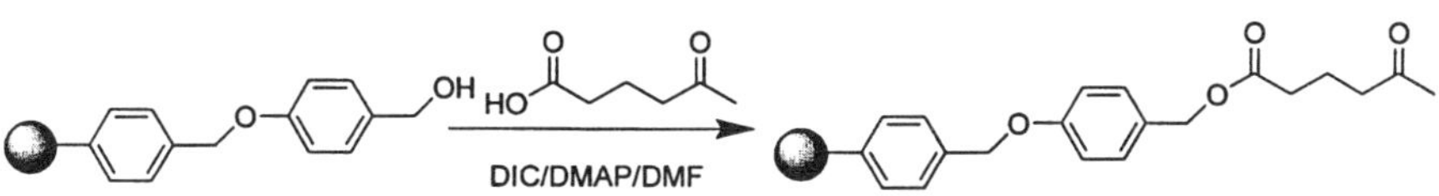

Scheme 3.11

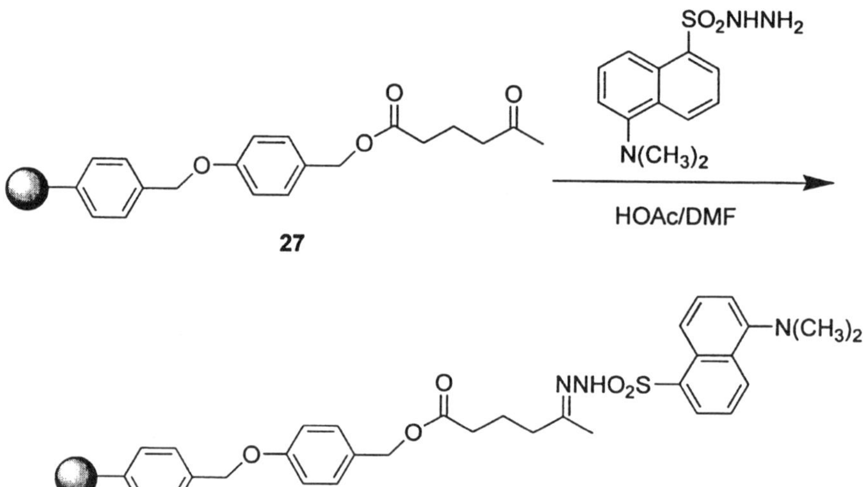

Scheme 3.12

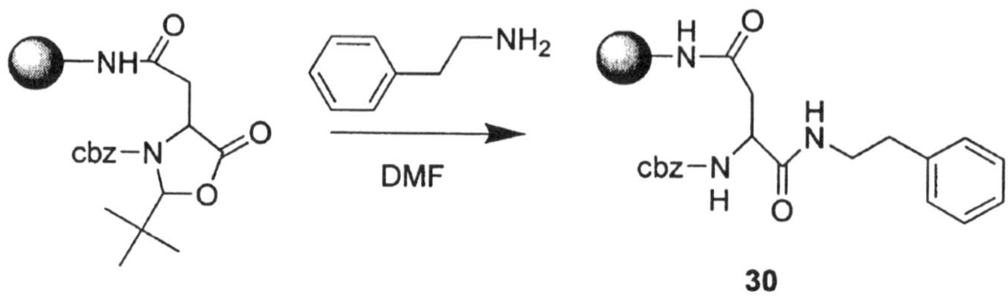

Scheme 3.13

Due to the overlap between the ester and the ketone carbonyl bands, a peak deconvolution analysis using PeakFit (Jandel Scientific, San Rafael, CA) was carried out for a clear-cut quantitation. As an example for the fitting method, the fitting of the data for PS reaction at 0 and 60 min (taken from Figure 3.17) is shown in Figure 3.18. It demonstrates that overlapping bands can be resolved to aid a quantitative analysis. The areas of the resolved ketone carbonyl band after PeakFit analysis were used to obtain the time course for these reactions. Results of area integrations are plotted in Figure 3.19. In this case, the reaction on PS resin is 2.2 times faster than that on PS-PEG

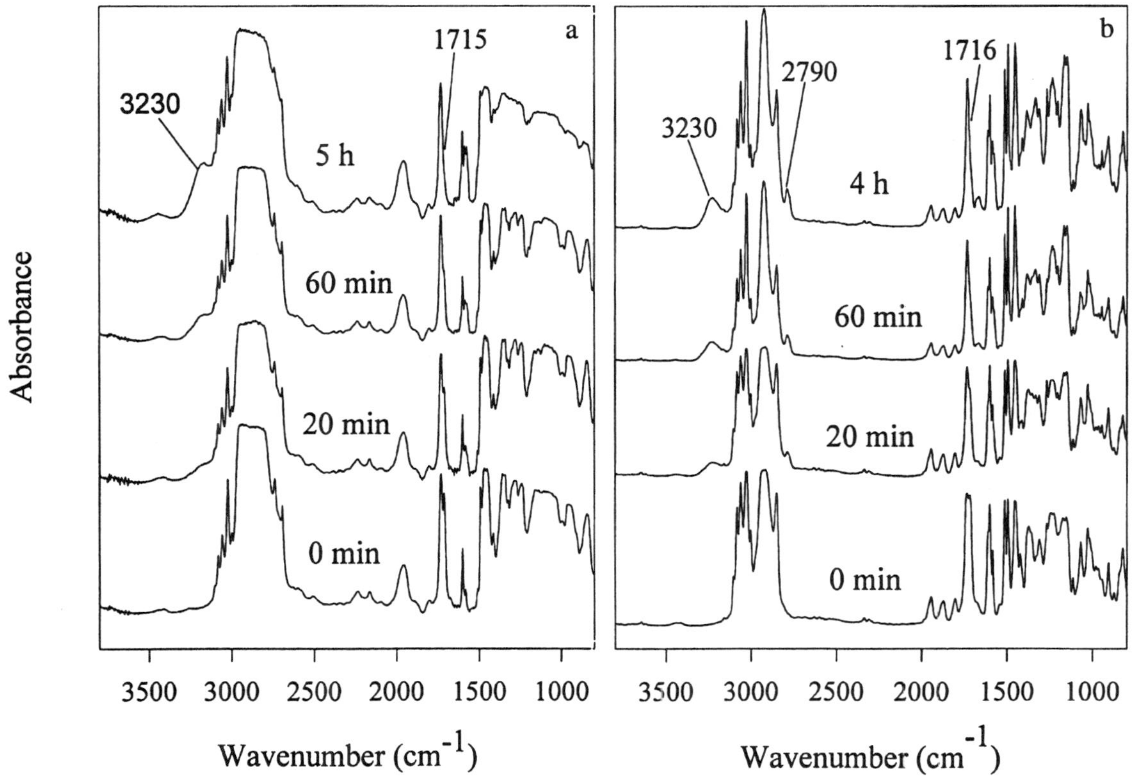

Figure 3.17 *IR spectra taken from a single PS-PEG-based bead (a) or a single flattened PS-based bead (b) at various times during the reaction in Scheme 3.12.*

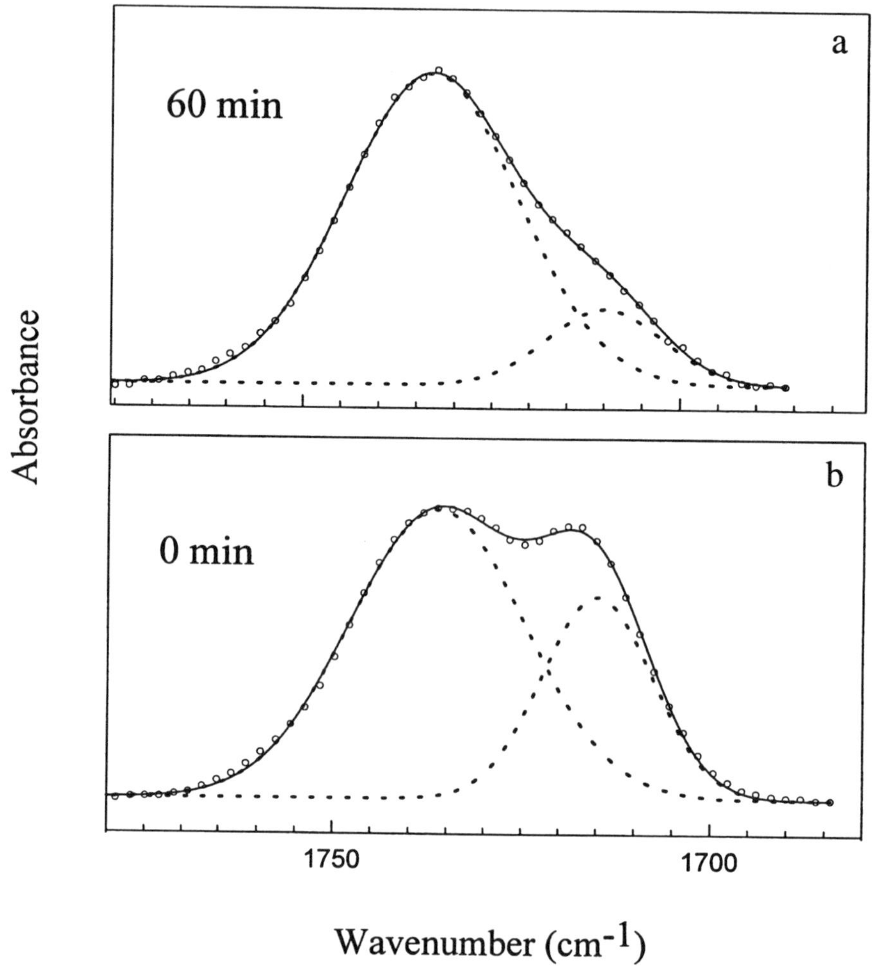

Figure 3.18 *PeakFit analysis of overlapping IR bands. Overlapping IR bands at ~1716 cm⁻¹ from Figure 3.18(b) were analyzed with the PeakFit program (Jandel Scientific, San Rafael, CA).*

resin. Reaction in Scheme 3.13 was carried out on PS and PS-PEG resins. The conversion of **29** to **30** was shown in spectra taken at various times (Figure 3.20), and the time courses for reactions on PS and PS-PEG resins are shown in Figure 3.21.

The results from studies on other reactions (Schemes 3.4 and 3.11) are listed in Table 2.1 in Chapter 2. The effect of polymer matrix on the reaction rate is similar to the effect of solvent on a solution-phase

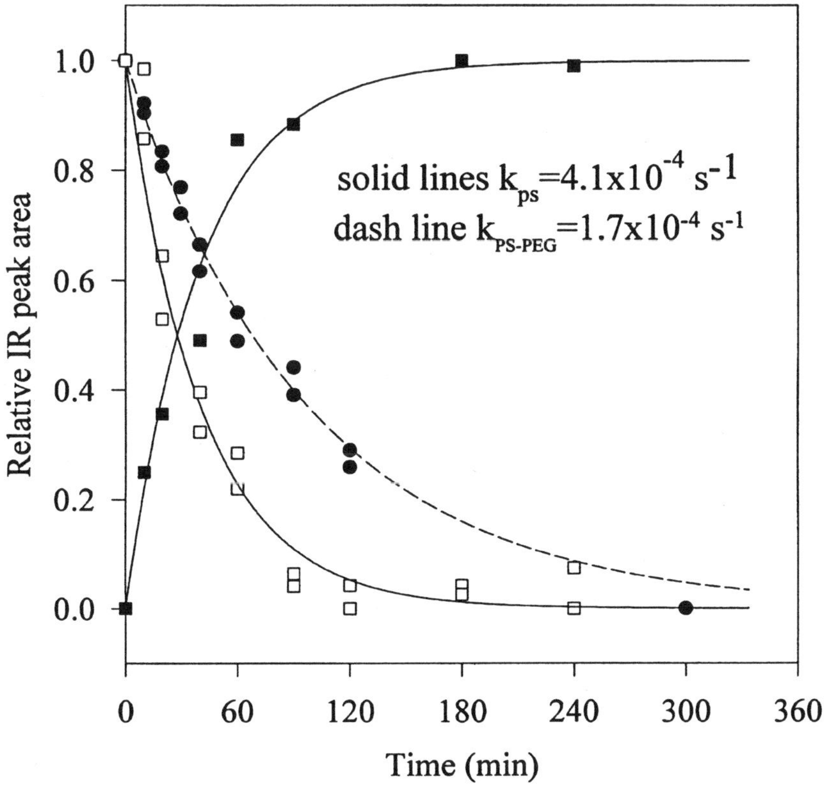

Figure 3.19 *Integrations of peak areas for band at 3230 cm^{-1} for PS-based resin (solid squares) and PeakFit resolved peak areas at 1715/1716 cm^{-1} (open squares for PS- and solid circles for PS-PEG resins). Solid lines are a theoretical time course with a rate constant of 4.1 × 10^{-4} 1/s, and the dashed line is a theoretical time course with a rate constant of 1.7 × 10^{-4}.*

reaction rate. These data are consistent with the conclusion reached by Czarnik (1998) that solid supports are like solvents. The reaction rate on resin supports should be proportional to the compatibility of the reagent with the solvent and the resin, and the adsorption property of the reagent into the resin. The selective adsorption of a compound into the resin depends on its similarity to the resin in terms of molecular interactions, such as hydrogen bonding, hydrophobicity, and polarity. There is no single polymer support that favors all reactions. Depending on whether the SPOS requires polar or non-polar medium, PS-PEG-or PS-based resins, respectively, should be chosen.

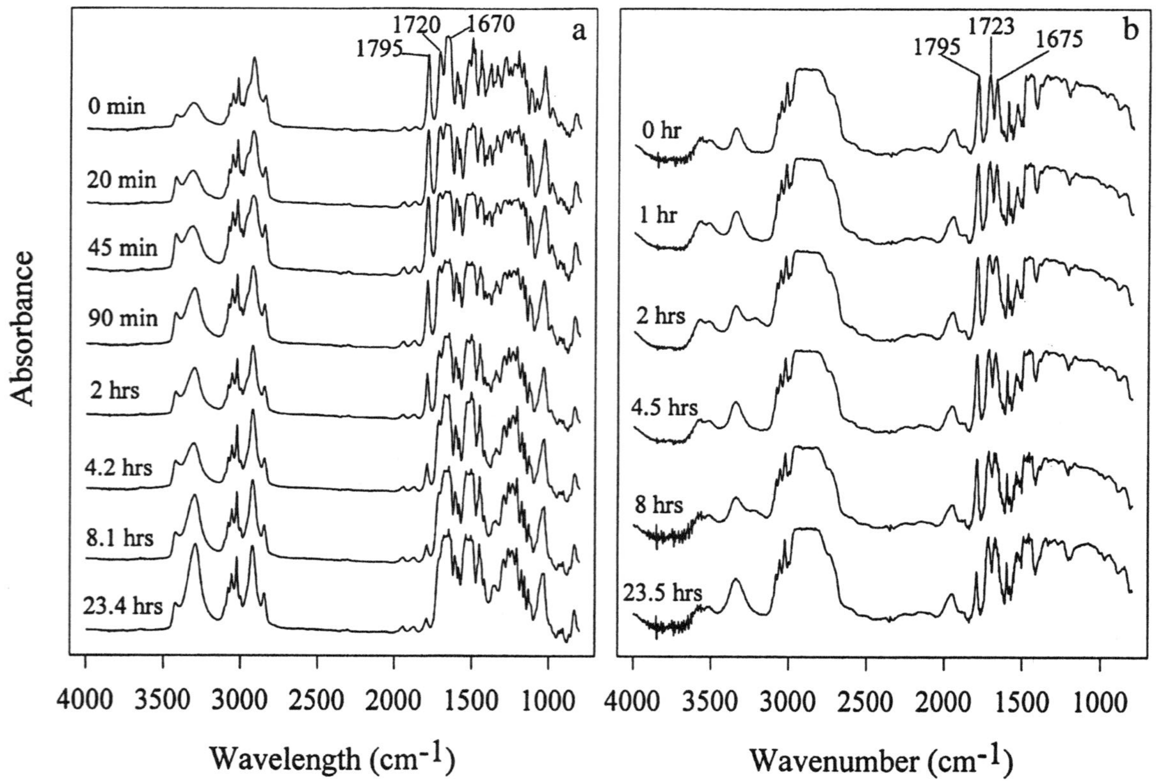

Figure 3.20 *IR spectra taken from a single flattened PS bead (a) or a single PEG-PS bead (b) at various times during the reaction in Scheme 3.13.*

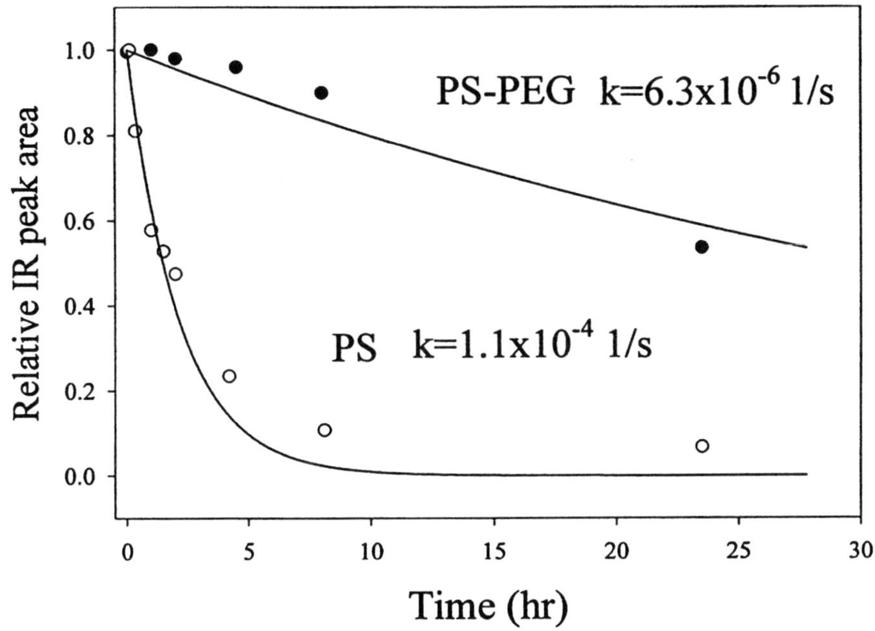

Figure 3.21 *Integration of peak areas for band at 1795 cm⁻¹ for PS-based resin (open circles) and for PS-PEG resins (filled circles). Solid lines are theoretical time courses with rate constants shown.*

3.4 THE MONITORING OF REACTIONS ON SURFACE-FUNCTIONALIZED POLYMERS

Chiron "Crowns" (Maeji et al., 1995) and "MicroTubes" (IRORI, San Diego, CA) are surface-coated polymers with various dimensions. Reactive groups coupled with linkers are grafted onto the polymer. In this section, the monitoring of reactions directly on crowns and Micro Tubes using macro- and micro-FTIR internal reflection spectroscopy is described.

3.4.1 Techniques

Internal reflection spectroscopy, also known as attenuated total reflectance (ATR), is a versatile, non-destructive technique for obtaining the infrared spectrum of the surface of a material or the spectrum of materials either too thick or too strongly absorbing to be analyzed by standard transmission spectroscopy.

Direct observation of the chemical synthesis on crowns is possible with FTIR accessories like the SplitPea™ (from Harrick Scientific Corporation, Ossining, NY; see Harrick et al., 1991), the Golden Gate (Graseby Specac Inc., Fairfield, CT), and the DuraSamplIR (ASI SensiIR Tech., Norwalk, CT), which combine internal reflection with an FTIR spectrophotometer. Micro-ATR has also been used to obtain IR spectra for multipin crowns and microtubes. In internal reflection spectroscopy, no sample preparation is needed—the crown is simply clamped onto the internal reflection element. This makes it a straightforward, extremely convenient, and efficient method for the direct spectral analysis of crowns.

3.4.2 Example

Synthesis on crowns sometimes starts with the removal of the 9-Fluorenylmethoxycarbonyl (Fmoc) protecting group from the linker. This reaction was monitored qualitatively by macro-ATR method (Gremlich & Berets, 1996).

The reaction kinetics on the surface of MicroTube reactors were also monitored by micro-ATR FTIR method. Reaction in Scheme 3.5 was carried out on the surface of MicroTube reactors, and FTIR spectra were recorded at various times during the reaction (Figure 3.22). Micro ATR studies allow a quantitative examination of reaction kinetics on polymer surface. This reaction was found slightly faster on this support than PS resins (Yan et al., 1999).

Very limited analytical methods are available for analyzing these surface-modified polymer samples. The above example shows that internal reflection measurements are a convenient, quick, and reliable means for direct analysis of the surface of polymers. It is particularly advantageous for the monitoring of combinatorial chemistry reactions, which take place in a thin stratum above the "crown" or "tube" polymer. The resulting spectra, therefore, have a high information content regarding the synthesis rather than the underlying crown polymer.

With these advantages, micro-and macro-internal reflection spectroscopy are excellent tools for directly checking successive steps of a combinatorial synthesis on polymer surface. It is extremely helpful for the elaboration of a synthesis when reaction conditions are to be optimized.

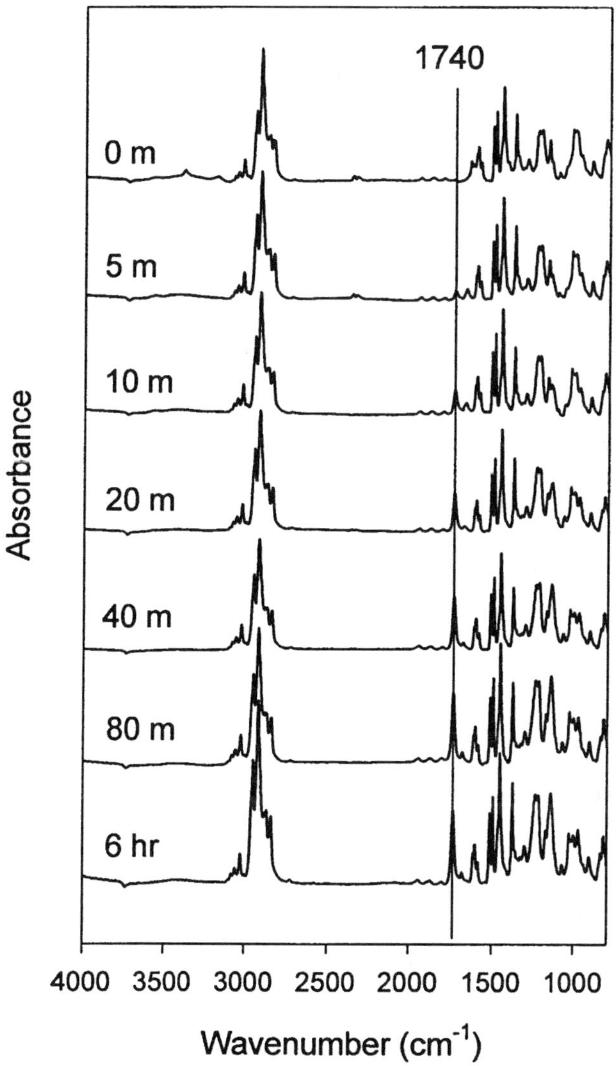

Figure 3.22 *IR spectra taken from the surface of MicroTube reactors using micro ATR FTIR at various times during the reaction in Scheme 3.5.*

3.5 SUMMARY

FTIR in general has been one of the most useful methods for monitoring organic reactions on solid supports. Various FTIR methods can all be used, although the amount of sample and the cost of the instrument need to be considered. As a TLC-equivalent method, single-bead FTIR is a simple, sensitive, fast, and convenient analytical method to monitor SPOS without stopping the reaction or cleaving the product. As with TLC, single-bead FTIR provides a wide range of information, such as qualitative assessment, quantitative determination, and reaction kinetics. Studies with the single-bead FTIR have not only provided a tool for daily monitoring of the solid-phase reactions, but a way to understand the properties of polymer-bound substrate and the nature of polymer supports and polymer-supported organic reactions. It has assisted in the selection of a wide range of reaction conditions for SPOS in the rehearsal phase of combinatorial chemistry. Due to its convenience and efficiency, FTIR internal reflection spectroscopy has evolved as a useful analytical method for monitoring combinatorial chemistry reactions directly on polymer surface.

3.6 REFERENCES

Balkenhohl, C., von dem Bussche-Hunnefeld, F., Lansky, A., Zechel, C. Combinatorial chemistry. *Fresenius J. Anal. Chem.* 1997, 359, 3.

Blanton, J. R., Salley, J. M. Reducing alkyl halides using a polymer-bound crown ether/tin hydride cocatalyst. *J. Org. Chem.* 1991, 56, 490.

Chan, T. Y., Chen, R., Sofia, M. J., Smith, B. C., Glennon, D. High throughput on-bead monitoring of solid phase reactions by diffuse reflectance infrared Fourier transform spectroscopy (DRIFTS). *Tetrahedron. Lett.* 1997, 38, 2821.

Chen, C., Randall, L. A. A., Miller, R. B., Jones, A. D., Kurth, M. J. "Analogous" organic synthesis of small-compound libraries: Validation of combinatorial chemistry in small-molecule synthesis. *J. Am. Chem. Soc.* 1994, 116, 2661.

Crowley, J. I., Rapoport, H. Unidirectional Dieckmann cyclizations on a solid phase and in solution. *J. Org. Chem.* 1980, 45, 3215.

Czarnik, A. W. Solid-phase synthesis supports are like solvents. *Biotech. Bioeng. (Combi. Chem.),* 1998, 61, 77.

Deben, I. E. A., Goorden, J., van Doornum, E., Ovaa, H., Kellenbach, E. Micro-DRIFT: Rapid and efficient IR analysis of solid phase organic chemistry reactions. *Eur. J. Org. Chem.* 1998, 697.

Deben, I. E. A., van Wijk, T. H. M., van Doornum, E., Hermkens, P. H. H. Characterization of reaction products on solid supports by FT-IR. *Mikrochim. Acta. [Suppl.]*, 1997, 14, 245.

Egner, B. J., Bradley, M. Analytical techniques for solid-phase organic and combinatorial synthesis. *Drug Discovery Today.* 1997, 2, 102.

Fitch, W. Analytical methods for quality control of combinatorial libraries. *Annual Reports in Combinatorial Chemistry and Molecular Diversity.* 1998, 1, 59.

Frechet, J. M. J., Schuerch, C. Solid-phase synthesis of oligosaccharides. I. Preparation of the solid support. Poly[p-(propen-3-ol-1-yl)styrene]. *J. Am. Chem. Soc.* 1971, 93, 492.

Gallop, M. A., Fitch, W. L. New methods for analyzing compounds on polymeric supports. *Curr. Opin. Chem. Biol.* 1997, 1, 94.

Gosselin, F., Renzo, M. D., Ellis, T. H., Lubell, W. D. Photoacoustic FTIR spectroscopy, a nondestructive method for sensitive analysis of solid-phase organic chemistry. *J. Org. Chem.* 1996, 61, 7980.

Gremlich, H.-U., Berets, S. L. Use of FT-IR internal reflection spectroscopy in combinatorial chemistry. *Appl. Spectr.* 1996, 50, 532.

Harrick, N. J., Milosevic, M., Berets, S. L. Advances in optical spectroscopy: The ultra-small sample analyzer. *Appl. Spectr.* 1991, 45, 944.

Larsen, B. D., Christensen, D. H., Holm, A., Zillmer, R., Nielsen, O. F. The Merrifield peptide synthesis studied by Near-Infrared Fourier-Transform Raman spectroscopy. *J. Am. Chem. Soc.* 1993, 115, 6247.

Letsinger, R. L., Kornet, M. J., Mahadevan, V., Jerina, D. M. Organic and biological chemistry. *J. Am. Chem. Soc.* 1964, 86, 5163.

Li, W., Yan, B. A direct comparison of the mixing efficiency in solid-phase organic synthesis by single bead IR and fluorescence spectroscopy. *Tetrahedron Lett.* 1997, 38, 6485.

Li, W., Yan, B. Effects of polymer supports on the kinetics of solid-pahse organic reactions: A comparison of polystyrene and TentaGel-based resins. *J. Org. Chem.* 1998, 63, 4092.

Maeji, N. J., Bray, A. M., Valerio, R. M., Wang, W. Larger scale multipin peptide synthesis. *Peptide Res.* 1995, 8, 33.

Marti, R. E., Yan, B., Jarosinski, M. A. Solid-phase synthesis via 5-oxazolidinones. Ring-opening reactions with amines and reaction monitoring by single-bead FT-IR microspectroscopy. *J. Org. Chem.* 1997, 62, 5615.

Moon, H.-S., Schore, N. E., Kurth, M. J. A polymer-supported C_2-symmetric chiral auxiliary: Prepatation of non-racemic 3,5-disubstituted-γ-butyrolactones. *Tetrahedron Lett.* 1994, 35, 8915.

Narita, M., Honda, S., Umeyama, H., Ogura, T. Infrared spectroscopic conformational analysis of polystyrene resin-bound human proinsulin C-peptide fragment. β-sheet aggregation of peptide chains during solid-phase peptide synthesis. *Bull. Chem. Soc. Jpn.* 1988, 61, 1201.

Narita, M., Isokawa, S., Honda, S., Umeyama, H., Kakei, H., Obana, S. Individuality of amino acid residues in protected peptides. Conformational and β-sheet structure-disrupted behaviors of resin-bound peptides. *Bull. Chem. Soc. Jpn.* 1989, 773.

Narita, M., Tomotake, Y., Isokawa, S., Matsuzawa, T., Miyauchi, T. Synthese and properties of resin-bound oligopeptides. Infrared spectroscopic conformational analysis of cross-linked polystyrene resin bound oligoleucines in the swollen state. *Marcromolecules,* 1984, 17, 1903.

Pivonka, D. E., Palmer, D. L. Direct infrared spectroscopic analysis of reagent postitioning in polystyrene bead supported solid phase reaction chemistry. *J. Combin. Chem.* 1999, 1, 294.

Pivonka, D. E., Russell, K., Gero, T. Tools for combinatorial chemistry: *In situ* infrared analysis of solid-phase organic reactions. *Appl. Spectr.* 1996, 50, 1471.

Pivonka, D. E., Simpson, T. R. Tools for combinatorial chemistry: Real-time single-bead infrared analysis of a resin-bound photocleavage reaction. *Anal. Chem.* 1997, 69, 3851.

Reggelin, M., Brenig, V. Towards polyketide libraries: Iterative, asymmetric aldol reactions on a solid support. *Tetrahedron Lett.* 1996, 37, 6851.

Russell, K., Cole, D. C., McLaren, F. M., Pivonka, D. E. Analytical techniques for combinatorial chemistry: quantitative infrared spectroscopic measurements of deuterium-labeled protecting groups. *J. Am. Chem. Soc.* 1996, 118, 7941.

Shapiro, M., Lin, M., Yan, B. On-resin analysis in combinatorial chemistry. In (ACS Book) *A Practical Guide to Combinatorial Chemistry,* Czarnik, A. W., DeWitt, S. H., Eds., 1997, 123.

Sun, Q., Yan, B. Single bead IR monitoring of a novel benzimidazole synthesis. *Bioorg. Med. Chem. Lett.* 1998, 8, 361.

Yan, B. Monitoring the progress and the yield of solid phase organic reactions directly on resin supports. *Acc. Chem. Res.* 1998, 31, 621.

Yan, B., Fell, J. B., Kumaravel, G. Progression of organic reactions on resin supports monitored by single bead FTIR microspectroscopy. *J. Org. Chem.* 1996a, 61, 7467.

Yan, B., Gremlich, H.-U. Role of Fourier transform infrared spectroscopy in the rehearsal phase of combinatorial chemistry: a thin-layer chromatography equivalent for on-support manitoring of solid-phase organic synthesis. J. Chromatogr. B. 1999, 725, 91–102.

Yan, B., Gremlich, H.-U., Moss, S., Coppola, G. M., Sun, Q., Liu, L. A comparison of various FTIR and FT Raman methods: applications in the reaction optimization stage of combinatorial chemistry. *J. Combin. Chem.* 1999, 1, 46—54.

Yan, B., Gstach, H. An indazole synthesis on solid support monitored by single bead FTIR microspectroscopy. *Tetrahedron Lett.* 1996, 37, 8325.

Yan, B., Kumaravel, G. Probing solid-phase reactions by monitoring the IR bands of compounds on a single "Flattened" resin bead. *Tetrahedron,* 1996, 52, 843.

Yan, B., Kumaravel, G., Anjaria, H., Wu, A., Petter, R., Jewell, Jr. C. F., Wareing, J. R. Infrared spectrum of a single resin bead for real-time monitoring of solid-phase reactions. *J. Org. Chem.* 1995, 60, 5736.

Yan, B., Li, W. Rapid fluorescence determination of the absolute amount of aldehyde and ketone groups on resin supports. *J. Org. Chem.* 1997, 62, 9354.

Yan, B., Sun, Q. Crucial factors regulating site interactions in resin supports determined by single-bead IR. *J. Org. Chem.* 1998, 63, 55–58.

Yan, B., Sun, Q., Wareing, J. R., Jewell Jr., C. F. Real-time monitoring of the catalytic oxidation of alcohols to aldehydes and ketones on resin support by single-bead Fourier transform infrared miscrospectroscopy. *J. Org. Chem.* 1996b, 61, 8765.

Yan, B., Tang, Q., Marepalli, H. Effects of polymer supports on the kinetics of solid-phase organic reactions: A comparison of four different solid supports. 1999, submitted.

4

Reaction Optimization Using MS and NMR Methods

IN THE REACTION optimization stage of combinatorial chemistry, it is essential to know if the desired product is formed. This information would be ideally obtained directly on solid support because the cleavage of the product from the support is more time and labor consuming. The requirement for cleavage reaction also presents a significant limitation for any chemist who needs to either monitor a reaction in progress or characterize an intermediate in a multistep synthesis. However, the "cleave-and-analyze" method has frequently been used whenever on-support techniques are not available. This chapter will give an overview of on-support MS and NMR methods that have been used in the reaction optimization stage.

4.1 MS METHODS

In the solid-phase reaction optimization process, MS is a useful tool for monitoring solid-phase organic synthesis (SPOS) due to its high sensitivity and short analysis time. Compounds from a fraction of a single-bead loading can be detected easily by electrospray ionization (ESI) or matrix-assisted laser desorption ionization (MALDI) MS. A drawback of these methods is that the compound has to be cleaved before analysis. Approaches such as using photolabile linkers or using the partial cleavage by TFA vapor alleviate the problem in some cases. TFA vapor partial cleavage and MALDI-MS analysis were used to monitor compounds attached to Wang, Rink, and HMPB linkers on a single bead or 10–100 beads (Egner et al., 1995 a, b).

When synthesized on a photolabile linker (such as α-methyl-phenacyl-ester linker and 3-amino-3-(2-nitrophenyl) propionyl linker), a compound can be partially cleaved with a UV laser beam (e.g., ~337 nm) at any time during the synthesis. The released fraction of the compound can be analyzed by any conventional methods. Photocleavage combined with the MALDI-MS analysis using 1–50 beads has been used to monitor reactions (Fitzgerald et al., 1996).

Carrasco (Carrasco et al., 1997) has reported a construct of a photolabile linker, an ionization moiety (basic group to facilitate protonation), and a chemically cleavable linker on a support to facilitate both analysis and synthesis. Photocleavage can produce a compound that contains an ionization moiety for MS characterization on 50–100 beads, and the final chemical cleavage releases the desired product.

Chu and coworkers (Yu et al., 1998) reported MALDI-MS determination of cyclic peptidomimetic compounds on single beads. The quality control of reactions was achieved by the identification of the desired and side products. The semi-quantitation of the cyclization efficiency was found to be both ring size- and sequence-dependent. MALDI-TOF-MS has been used to analyze the results of SPOS on a single bead using TFA vapor. The potential of this technique to monitor low levels of bead surface-associated compound was demonstrated. A single component from a five-member combinatorial library was positively identified using the method (Haskins et al., 1995). The molecular weights of femtomole quantities of small peptides attached to PS beads have been determined with imaging time-of-flight secondary ion MS. The analysis is made possible by the selective clipping of the bond linking the peptide to bead with TFA vapor before the secondary ion MS assay (Brummel et al., 1994).

4.2 NMR METHODS

NMR is the primary tool for structure elucidation in low-throughput solution synthesis. However, much less information can be obtained from NMR spectra of solid-phase samples due to broadened spectral lines and other interference. If the purpose is only to confirm the formation of the product, NMR sometimes can be used as an analytical tool in combinatorial chemistry. The capability of NMR in structural elucidation on solid support is also under investigation in many laboratories.

4.2.1 Gel-Phase NMR

Advancements in the development of NMR techniques for on-resin analysis have recently been reviewed (Egner & Bradley, 1997; Keifer, 1997, 1998; Gallop & Fitch, 1997; Shapiro & Wareing, 1998).

Gel-phase NMR has been a well-established method to acquire NMR data on solid-phase samples for decades. Because the solvent-swollen resin slurries are neither fully solid nor fully liquid, they are referred to as gel-phase samples. The term gel-phase NMR now refers to the use of a conventional liquids probe (a probe in which the sample-spinning axis is aligned along the magnet bore [z] axis) to acquire the NMR data.

It is assumable that a solvent that causes more swelling will produce narrower linewidths due to increased motion within the swollen resin. Systematic studies of the influence of different solvents upon the ^{1}H NMR spectra of different resins have demonstrated, however, that this generalization is far too simplistic to account for the observed linewidths.

All substances exhibit a physical property called "magnetic susceptibility," that provides a measure of how that substance affects a surrounding magnetic field. When solution samples are analyzed, the uniform (homogeneous) substance will cause only a uniform change of the applied (static) magnetic field. This is typically corrected by "shimming" the static field of the NMR magnet to make the sample magnetically homogeneous. A resin sample is composed of discrete beads, which can be described as having an organic moiety attached via a variable length linker to a more rigid polymer matrix. Because the susceptibility of the resin and the solvent are different, solid-phase synthesis resins are heterogeneous samples. They have regions of differing magnetic susceptibility throughout the sample. This difference caused localized variations (distortions) in the magnetic field that is not correctable by "shimming" the field. The magnetic susceptibility mismatches cause additional linebroadenings that scale up with the NMR observation frequency.

Reduced molecular motion in solid samples can also broaden the NMR linewidth. ^{1}H or ^{13}C NMR spectra of the dry beads consist only of broad, unresolved resonance, as is typical of solid polymers. If the resin beads are swollen to form slurry, many lines will narrow due to an increase in the molecular motions that allow the chemical shifts to average to their isotropic values (Rapp, 1996, see Figure 4.1). While the relative rigid polymer matrix would still be expected to generate broad NMR resonance, the attached organic moieties should have

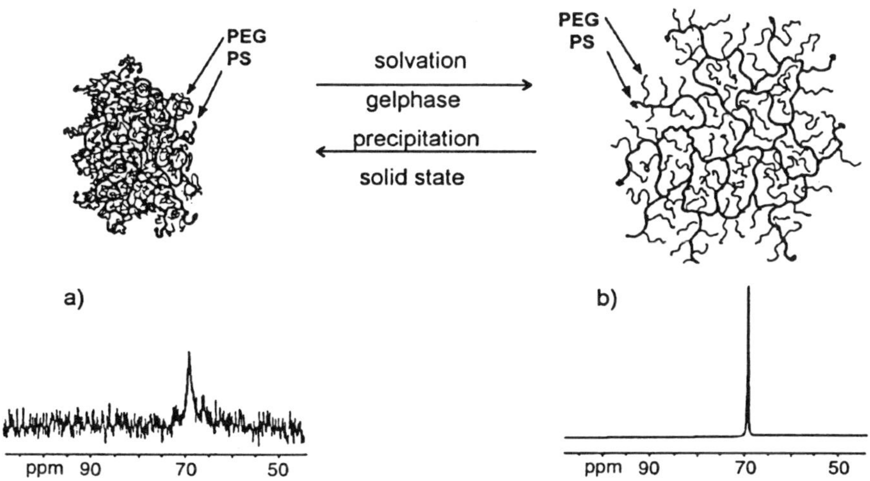

Figure 4.1 *Top: resin in shrunken and swollen condition. The bottom shows the corresponding gel-phase ^{13}C NMR of the PEG resin: in swollen state, the tentacles show high T_1 values (highly flexible, well solvated) resulting in sharp ^{13}C resonance, whereas in the nonswollen state, the ^{13}C resonance shows broadening. (Reproduced with permission from Rapp, W., in* Combinatorial Peptide and Nonpeptide Libraries *edited by Jung, G., 1996, 828. Copyright© 1996, Wiley-VCH, Weinheim).*

sufficient molecular motion to generate narrow linewidths. Gel-phase NMR was born this way. Bayer and coworkers (1990) studied the spin-lattice relaxation times of the tripeptide Boc-Gly-Pro-Pro-OMe in solution, on soluble supports, and bound to various solid supports. The investigation revealed that the T_1 value and, hence, the mobility of the carbon atoms of the peptide esters decrease in the sequence methyl ester>soluble polyethyleneglyco (PEG)-bound ester≥PEG-PS ester>PS ester. The results suggest that the grafted copolymer PS-PEG allows more internal molecular motion than PS resins (Figure 4.2).

Due to linebroadening in gel-phase NMR spectroscopy, the application of gel-phase NMR is usually for lower frequency nuclei that have a wide chemical shift dispersion. For example, although the linewidth for ^{13}C spectra may be 25–75 Hz or more, the resonance is still resolved well enough to allow structural assignments. Gel-phase NMR spectroscopy has found applications for nuclei like ^{13}C, ^{19}F, ^{2}H, ^{31}P, and ^{15}N. ^{13}C and ^{19}F NMR are most useful. Incorporation of isotopes, such as ^{13}C and ^{15}N, has been investigated and found very useful.

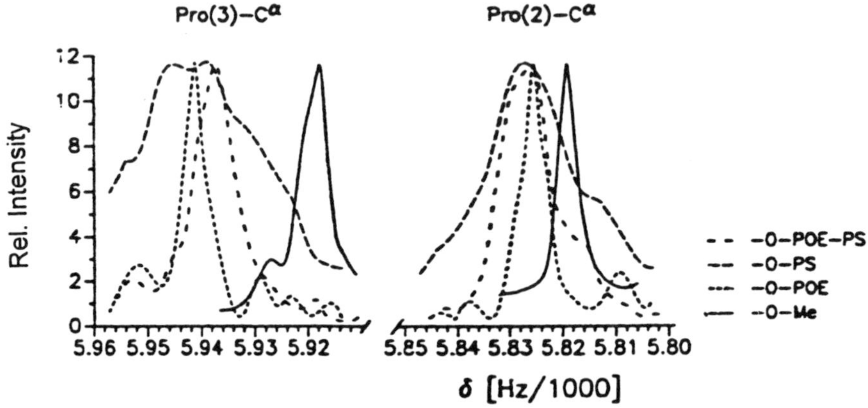

Figure 4.2 *Line width of the signals of α-carbon atoms of Pro(2) (57.8–57.9 ppm) and Pro (3) (58.8–59.1 ppm) of the tripeptide (Boc-Gly-Pro-Pro) esters (at 100.6 MHz). (Reproduced with permission from* Macromolecules, *1990, 23, 1937–1940. Copyright© 1990, American Chemical Society).*

4.2.1.1 ^{13}C GEL-PHASE NMR

In a reaction monitoring of solid-phase peptide synthesis (Giralt et al., 1984), gel-phase sample was prepared by mixing the resin sample with $CDCl_3$ for 30 min after agitation with a vortex shaker or, sometimes, ultrasonic agitation. ^{13}C NMR spectra were obtained at 50.31 MHz on a Varian XL-200 spectrometer. An array of starting resins were analyzed from a quantitative and a semi-quantitative point of view. These resins include the commonly used ones, such as Merrifield, benzhydrylamino, and Wang resins. All the carbons were assigned. A synthesis of a tripeptide, H-Asn-(N-Me)Ala-Thr-NH_2, was monitored step by step. There is a general tendency for the amino acid carbons to move to a lower field when the Boc group is removed and to a higher field when the next amino acid is attached. The effect suggests the possibility of steric compression. In this monitoring procedure, the appearance of side reactions during the growing of the peptide chain can be detected.

Measurements of relaxation times were used to obtain an estimate of the mobility of the different parts of the peptide molecule at each stage, providing a quantitative basis to improve the synthetic experimental conditions. It was found that the Boc methyl groups of Boc-Thr-(Bzl)-OH in solution and of Boc-tripeptide-resin show similar relaxation rates. This gives a direct verification of the fact that solvation in gels in which polymer-supported peptide synthesis is carried out can be almost as efficient as solvation in solution and, therefore, accounts for the possible similar reactivity observed in solid phase and in solution.

In peptide synthesis, the coupling step can be monitored to completion by a sensitive amine test. Unfortunately, there are not many corresponding tests to check for quantitative removal of the matrix-bound peptide N-terminal protecting groups. It is inevitable that any residual protected amino groups will become exposed in a subsequent chain elongation cycle. Erroneous peptide sequences will then result. Epton et al. (1980) used gel-phase ^{13}C NMR spectroscopy in a study of polymer-supported peptide synthesis on a phenolic crosslinked poly-(acryloyl-morpholine)-based matrix. The ^{13}C NMR spectra were obtained using a Bruker NH90FT NMR spectrometer operating at 22.63 MHz, with broad-band irradiation to decouple the protons. Satisfactory signal-to-noise ratio was achieved using 30,000–150,000 pulses at 0.7 s intervals with 0.15-g of resin swollen in 2 mL DMSO-d_6.

Peaks due to almost all the carbons in groups pendant on the polymer backbone are readily discerned and have been assigned by comparison with the spectra of the monomers. The backbone carbons cannot be seen because (a) they fall in the 35–42 ppm region, which is obscured by DMSO-d_6; (b) the intensity will be spread over several lines resulting from tacticity effects; and (c) the line will be broadened by the relatively low mobility of the backbone.

The ^{13}C NMR spectrum of the DMSO-d_6 swollen phenolic poly (acryloylmorpholine)-based matrix after substitution with Boc leucine (~0.5 mmol/g by amino acid analysis) and re-O-acetylation was recorded. In addition to the peaks due to the parent matrix, the expected leucine peaks and Boc peaks are all well resolved and readily assigned. The fact that the major sample peak in the spectrum is due to the three magnetically equivalent methyl carbons, $-C(CH_3)_3$ of the Boc group is most important because Boc is one of the widely used protecting group in both SPPS and SPOS.

A step-by-step synthesis of a peptide enkephalin on polymer support was studied. As the peptide chain is lengthened, the signals from the carbons nearer the support tend to diminish in height and to broaden. The peptide unit at the end of the chain, next to Boc, always gives the sharpest lines, which is understandable in terms of decreased mobility of the nuclei nearer the support. When the next amino acid unit is added, the α-carbon signal from the unit previously adjacent to Boc tends to move upfield. This is a normal end-group effect as mentioned previously. This example showed that the application of gel-phase ^{13}C NMR spectroscopy to SPPS yields directly hitherto inaccessible information on the peptide chain. Almost certainly, the removal of other protecting groups, such as Fmoc, could also be readily checked.

The application of gel-phase NMR in monitoring solid-phase organic synthesis was reported by Leznoff and coworkers (Jones et al., 1982). They reported the ^{13}C NMR spectra of a wide variety of organic substrate bound to 2% crosslinked polystyrenes. Their ^{13}C NMR spectra were determined in the Fourier mode using a Bruker HFX-270 spectrometer at 67.89 MHz. In general, spectral width of 15 kHz was observed in 16 K data points using an approximate pulse angle of 60° and averaging 2–4 K acquisitions. Delays of 10–15 s between pulses were used when integration was desired. Samples were in 10-mm tubes. Solvent-induced swelling of 0.75 g of polymers in 2–2.5 cm of $CDCl_3$ generally produced an adequate sample volume of the appropriate slurry consistency.

The results of the study, like results from SPPS synthesis, were consistent with the notion that segmental motion of side-chain substituents attached to the backbone polymer will result in sharply defined ^{13}C NMR signals for the more mobile portion of the substituent molecules. They showed that resonance of the carbon atoms close to the polymer-binding site are broadened relative to those associated with portions of the side chains considered to be more freely rotating. These indicated increasing T_1 and T_2 values for the carbon atoms of the side-chain substituents with distance from the polymer backbone. Most significant from an assignment point of view are the obvious broad lines (relative shorter T_2 values) generally associated with the first three carbon atoms of the side-chain substituents from the polymer.

Figure 4.3a shows ^{13}C NMR spectrum of the 2% crosslinked polystyrene. The quaternary aromatic carbons (145.8 ppm), the ortho and meta carbons (128.0 ppm), the para carbons (125.7 ppm), and the backbone methylene and methine carbons (40.4 and 43.0 ppm) are observed. The spectrum of polymer-bound trityl alcohol is shown in Figure 4.3b. The characteristic broad line at 81.9 ppm is attributed to the quaternary carbon atom in the triphenylcarbinol group and that at 147.0 ppm to the α-carbons in the phenyl ring.

The spectrum at the bottom of Figure 4.3 shows the ^{13}C spectrum of a polymer-bound C12 bromoalkyne polymer-Tr-O-$(CH_2)_2C{\equiv}C(CH_2)_8Br$ derived using an alternative synthetic precursor, the monotrityl ether of ethylene glycol. The broad lines for resin backbone and carbon atoms closer to the backbone (C1–C5) and the sharper lines for carbon atoms remote from the backbone are evident in the spectrum. These results remind us of spectra of resin-bound peptides in previous discussions.

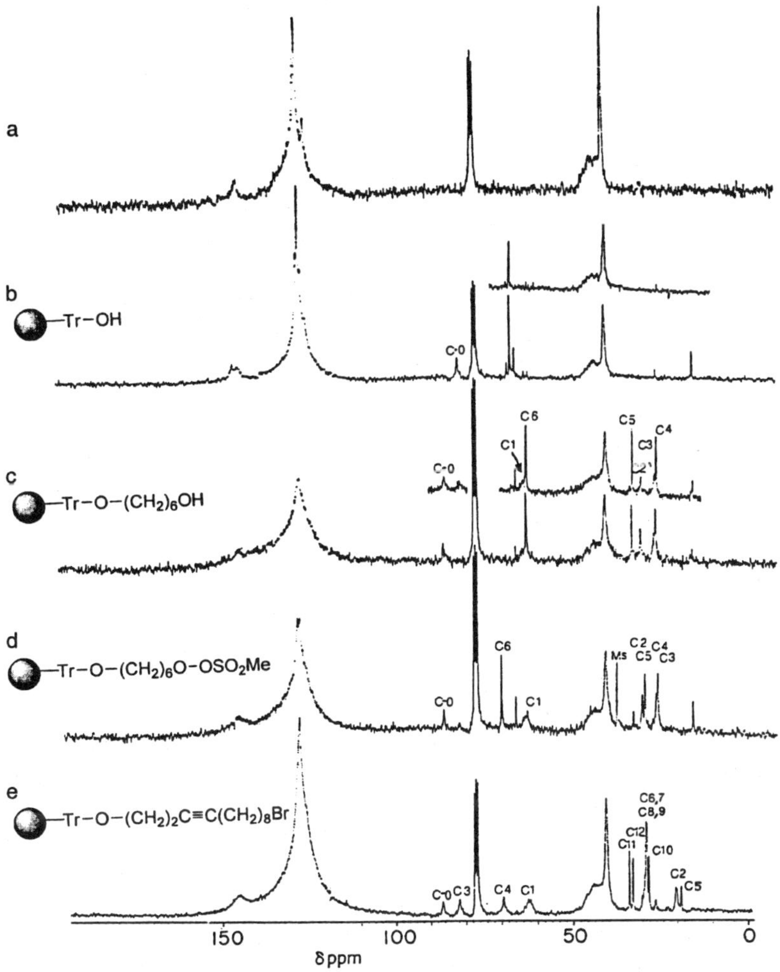

Figure 4.3 *The* [13]*C NMR spectra, determined at 67.89 MHz, of solvent ($CDCl_3$) swollen slurries of (a) Eastman 2% divinylbenzene cross-linked polystyrene; (b) polystyrene-bound trityl alcohol and the inset shows the same spectrum after solvent evaporation and resuspension; (c) the polymer-bound monotrityl ether of hexanediol and the inset is the spectrum obtained using a pulse delay of 10 s; (d) the polymer-bound methylsulponate derivative and (e) the polymer-bound C_{12} bromoalkyne. The side-chain carbon atoms are numbered from the trityl ether anchoring site, C-O refers to the trityl carbinol carbon, and Ms refers to the methylsulphonate methyl carbon. (Reproduced with permission from Jones et al.,* Org. Magnetic Resonance, *1982, 18, 237. Copyright© 1982, John Wiley & Sons Limited).*

Blossey et al. (1990) reported that the reaction monitoring of synthetic transformations of the bound steroids containing carbonyl and hydroxyl groups and esterification of hydroxyl functionality was monitored by ^{13}C NMR spectra of solvent swollen polymer gels. The ^{13}C NMR spectra of the bound dehydrocholic acid (3,7,12-trioxo-5β-cholanoic acid) and cholic acid (3α,7α,12α-trihydroxyl-5β-cholanoic acid) permit structural assignments to be made, on a qualitative basis, of the intact polymer without resorting to cleavage reactions of the bound materials.

There is good agreement between the literature values for solution spectrum of methyl cholate and polymer-bound cholate recorded in this study. The mono-, di-, and triesterification of the compound are determined from the ^{13}C NMR spectrum, with separation of chemical shift values for each of the carbons bearing the ester groups at three, seven, and 12 positions. Oxidation of the polymer cholate with DMSO-oxalyl chloride (Swern oxidation) gave a quantitative yield of the triketo derivative, with a disappearance of the hydroxyl-bearing carbons at 67–70 ppm and the appearance of small signals for carbonyl carbons in the 208–212 ppm range.

By using the ^{13}C-enriched building blocks, ^{13}C NMR experiments were used to monitor SPOS on as little as 20 mg of resin (Look et al., 1994). Although labeled compounds can be quite costly, the application of this method is still possible in small-scale synthesis for reaction optimization studies. In this work, the PS-PEG resin is suspended in the NMR solvent benzene-d_6 and is administered to the tube insert for immediate observation. The use of ^{13}C-enriched building blocks enables one to obtain high-quality spectra at the center of interest without interference from unenriched peaks and solvent. Meaningful data can be obtained in as few as 64 transients. Typical turnaround times from preparation of the resin to the recovery of the beads for subsequent reactions are from 15 to 30 min. The enriched center need not be directly involved in the reaction that is monitored because changes in hybridization as well as electron density can be easily detected.

4.2.1.2 GEL-PHASE ^{19}F NMR

The first application of gel-phase ^{19}F NMR in monitoring of SPPS was reported by Manatt et al. in 1980. Proton decoupled ^{19}F 94.1 MHz NMR spectra of fluorobenzyloxycarbonyl and 2-(4-fluorophenyl)-2-propyloxycarbonyl derivatives of amino acids and peptides bonded to 1% crosslinked polystyrene were studied (starting resin “Bio Beads” S-X-1, 200–400 mesh, 1.25 mmol/g in CH_2Cl_2, 5-mm sample tubes of slurries of

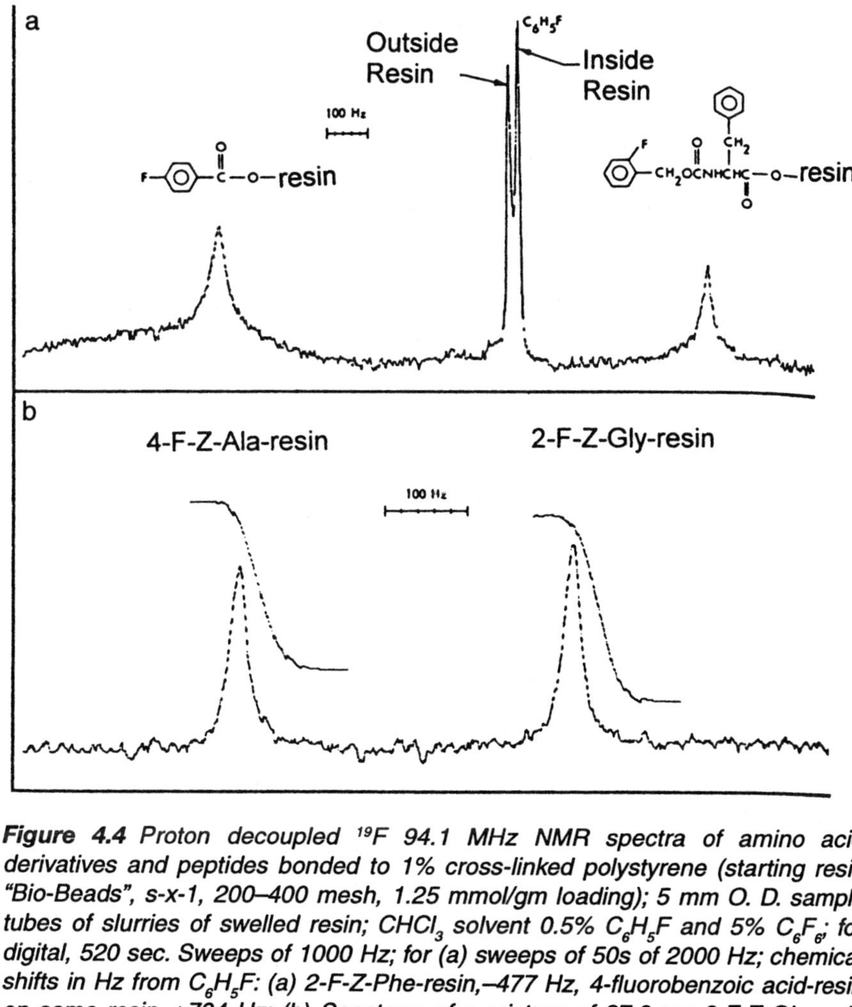

Figure 4.4 *Proton decoupled* ^{19}F *94.1 MHz NMR spectra of amino acid derivatives and peptides bonded to 1% cross-linked polystyrene (starting resin "Bio-Beads", s-x-1, 200–400 mesh, 1.25 mmol/gm loading); 5 mm O. D. sample tubes of slurries of swelled resin;* $CHCl_3$ *solvent 0.5%* C_6H_5F *and 5%* C_6F_6*; for digital, 520 sec. Sweeps of 1000 Hz; for (a) sweeps of 50s of 2000 Hz; chemical shifts in Hz from* C_6H_5F*: (a) 2-F-Z-Phe-resin,–477 Hz, 4-fluorobenzoic acid-resin on same resin, +704 Hz; (b) Spectrum of a mixture of 27.0 mg 2-F-Z-Gly-resin beads (0.98 mmol/g) with 22.5 mg 4-F-Z-Ala-resin beads (0.91 mmol/g) with integration from 16 20-second sweeps. (Reproduced with permission from* Tetrahedron Lett. *1980, 21, 1397–1400. Copyright© 1980, Elsevier Science Ltd.).*

swollen resin, $CHCl_3$ solvent 0.5% C_6H_5F, and 5% C_6F_6). Relative narrow spectral lines were obtained (Figure 4.4). This offered the potential for studying and monitoring the SPPS. Internal concentration references are easily derived by several approaches. A ^{19}F NMR probe with a chemical shift different from that of the shift in the protecting group and inert to SPPS conditions may be coupled to a portion of the resin functional groups (Figure 4.4a). Alternately, resin beads possessing an inert ^{19}F NMR probe may be added to the resin beads on which the peptide will be synthesized (Figure 4.4b).

The time-dependent change of a nucleophilic aromatic substitution (S_NAr) reaction on Rink amide resin was monitored by gel-phase ^{19}F NMR using about 1 g of resin (Scheme 4.1). Two time points at 15 and 30 min were shown in Figure 4.5, and the reaction was finished within 2 hr (Shapiro et al., 1996).

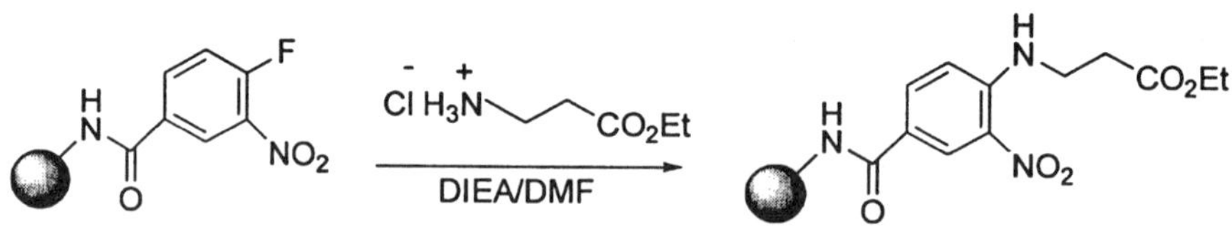

1 2

Scheme 4.1

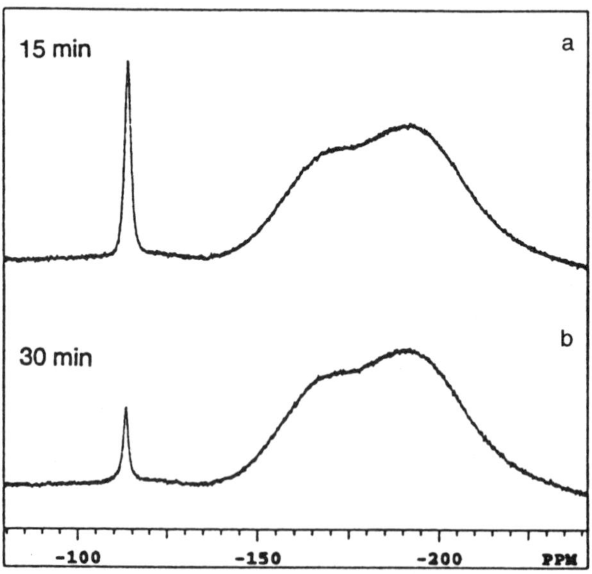

Figure 4.5 *Gel-phase ^{19}F NMR spectra of benzene-d_6-swollen resin 1 at (a) 15 min and (b) 30 min after exposure to ethyl 3-aminopropanoate HCl/DIEA. δ(1) = –115.3 ppm. (Reproduced with permission from* Tetrahedron Lett. *1996, 37, 4671–4674. Copyright© Elsevier Science Ltd.).*

Gel-phase ^{19}F NMR spectroscopy has also been used to characterize products from a variety of reactions of fluorinated aromatic compounds linked to a PS-PEG resin (Svensson et al., 1996). Spectra of high quality were obtained in less than 10 min on 50–200 mg of resin. Linewidths for the compounds linked to the solid support approached those in solution and indicated that shift difference of ≤0.5 ppm can be detected. Substantial chemical shift differences were obtained for almost all of the synthetic transformations illustrating the potential of ^{19}F NMR for rapid monitoring of reactions in SPOS.

4.2.1.3 GEL-PHASE ^{31}P NMR

Gel-phase ^{31}P NMR was used to monitor solid-phase oligonucleotide synthesis by Giralt and coworkers (Bardella et al., 1993). This method has several advantages. First, different phosphorous derivatives are involved in the key steps of the synthetic cycle in oligonucleotide synthesis. Second, its NMR sensitivity is much higher than that of ^{13}C (383/1 considering the natural abundance of the nuclei). Finally, the signals of the polymer do not interfere with the observed ^{31}P NMR resonance.

The half-height width of the ^{31}P NMR signal of resin-bound DMT-C^{bz}-P(Ar)-G^{ib} is 26 Hz. The ^{31}P NMR spin-lattice relaxation time (T_1) of this resin-bound compound measured by using the inversion recovery method is 1.006±0.013 s. Considered together, linewidth and T_1 values provide a clear indication that the considerable degree of swelling of the dinucleotide-resin in $CDCl_3$ provides a highly flexible microenvironment for the phosphorous atoms and, therefore, of the reactive sites. By showing several reaction examples, authors pointed out that ^{31}P NMR is a useful technique for monitoring solid-phase oligonucleotide synthesis and is particularly interesting in methodological studies in order to ensure quantitative yields or to determine the extent of a side reaction on the resin. The use of ^{31}P nuclei for NMR gives great sensitivity in the analysis, especially considering the size of this kind of molecule, and takes advantage of the fact that the solid support is transparent. In addition, the great difference between the chemical shifts of P^{III} and P^{V} makes gel-phase ^{31}P NMR especially useful as a non-destructive method for monitoring the synthesis of oligonucleotide using the phosphorous-triester approach.

Gel-phase ^{31}P NMR was used to monitor a synthesis of the solid-supported α,β-unsaturated amide using the reaction of phosphonoacetamide esters as Horner-Wadsworth-Emmons (HWE) reagents with a variety of aldehydes (Johnson & Zhang, 1995). The reaction progress was monitored by gel-phase ^{31}P NMR of the resin suspended in acetonitrile.

^{31}P gel-phase NMR shows the resonance of the starting resin-bound diethylphosphonoacetamide as a narrow multiplet at δ22. After the reaction is complete, the δ22 peaks have disappeared, and a new broad resonance for diethylphosphate has appeared near 80.

4.2.1.4 GEL-PHASE ^{15}N NMR

Incorporation of an isotopically enriched ^{15}N-Alanine into the synthetic scheme and the application of ^{15}N gel-phase NMR were combined to monitor a synthetic pathway, which provides functionalized piperazin-2-one ring systems (Swayze, 1997). This technique provided a convenient label for following the progress of the reaction sequence, as the amino acid-derived nitrogen is involved in the key ring forming reactions and, therefore, exhibits significant differences in chemical shift. The sensitivity obtained was sufficient to gauge the success or failure of reaction, as byproducts were evident by the presence of multiple peaks when inappropriate reaction conditions were chosen.

4.2.1.5 GEL-PHASE DEUTERIUM NMR

Gel-phase deuterium NMR spectra were reported for a series of swollen Merrifield resins containing protected glycine oligomers (Ludwick et al., 1986), with the goal of delineating molecular-level interactions that can affect desired reactivity of these materials. Figure 4.6 shows that the static signal increases as the chain length increases. Heating samples (gly-d_2)$_7$ and (gly-d_2)$_9$ to 111° C resulted in significant decreases in the static portions of the spectra (Figure 4.7). Results were consistent with a model in which partial aggregation of the glycine oligomers occurs after the pendant chain reaches a critical length ($n > 5$). Lengths greater than this correspond to the overlap necessary to form at least one helix repeat of the polyglycine II structure. The polystyrene matrix is, therefore, concomitantly immobilized, presumably due to additional effective crosslinks caused by the aggregation. The mechanism show in this work agrees with other works presented in Section 2.2.4 and explains some difficult and inefficient synthesis of particular peptides.

4.2.2 Magic Angle Spinning (MAS) NMR

Solvent swelling is needed to make the active functional groups accessible to reagent in synthesis. The solvent motion in the resin network leads to chain expansion observable on a macroscopic scale as the swelling of bead and is accompanied by an enhanced local mobility of the attached molecules at the microscopic level. As a consequence of

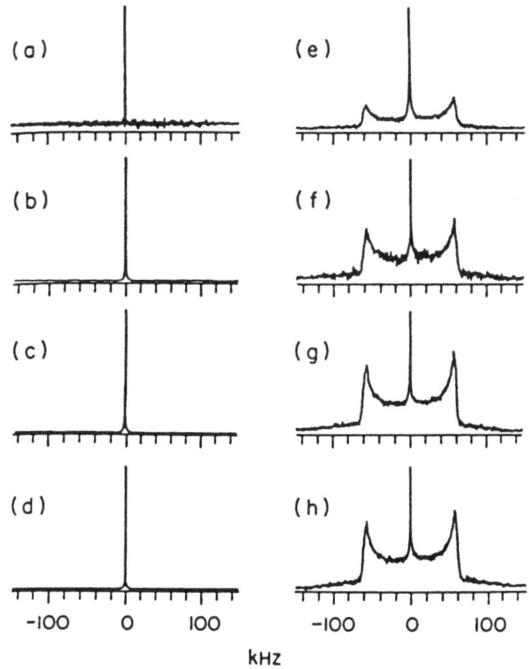

Figure 4.6 *Quadrupole echo deuterium NMR spectra of DMF swollen materials for (a) the polystyrene matrix and for solid-phase peptide samples (b) I (gly-d_2), (c) II((gly-d_2)$_3$), (d) III ((gly-d_2)$_5$), (e) IV((gly-d_2)$_7$), (f) V ((gly-d_2)$_8$), (g) VI ((gly-d_2)$_8$)gly, and (h) VII ((gly-d_2)$_9$). The spectra were obtained with a recycle delay time of 2 s. (Reproduced with permission from* J. Am. Chem. Soc., *1986, 6493–6496. Copyright© 1986, American Chemical Society).*

the enhanced molecular mobility, the various factors that govern the linewidth of the NMR spectrum will be motionally averaged. These factors include the homonuclear dipolar interaction that can broaden proton signals in a solid-state sample. However, this motional averaging is incomplete because of the anisotropy environment of the solvent molecules in the resin. In addition to these magnetic-susceptibility interface differences within the sample, magnetic-susceptibility interfaces inherent in the construction of the probe are another factor for spectral linebroadening. It has long been recognized that the magnetic-susceptibility-induced linebroadenings in heterogeneous samples can be eliminated by spinning the sample at the magic angle (54.7° from Z) (Drain, 1962; Doskocilova & Schneider, 1970) at a spinning rate higher than the non-spinning residual linewidth. MAS has provided

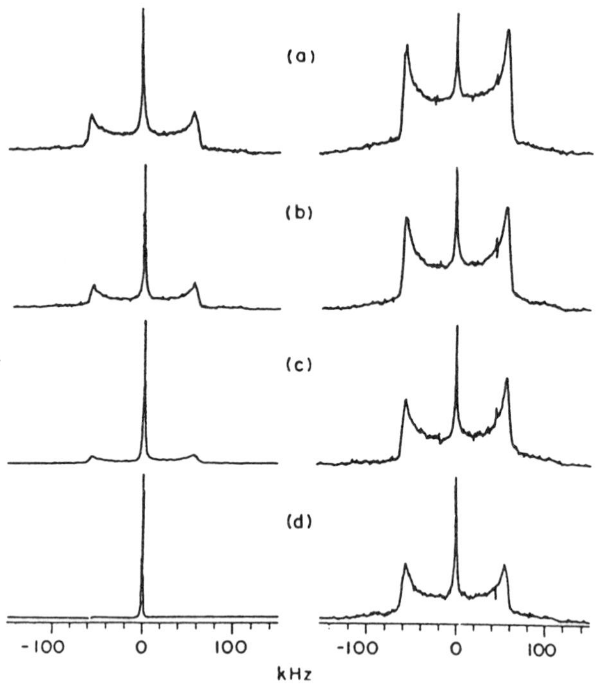

Figure 4.7 *Quadrupole echo deuterium NMR spectra of DMF swollen sample IV (left column) and DMF swollen sample VII (right column) obtained at (a) 23, (b) 50, (c) 82, and (d) 111° C. (Reproduced with permission from* J. Am. Chem. Soc., *1986, 6493–6496. Copyright© 1986, American Chemical Society).*

substantial line narrowing in one-dimensional ^{13}C and ^{1}H NMR spectrum of solvent-swollen polymer gels with conventional MAS probes. MAS removes the magnetic-susceptibility broadening caused by the difference in magnetic susceptibility between the resin and the pure solvent, combined with the irregular shape of the solvent/resin interface. MAS is also advantageous for suppressing the lines of the more rigid supporting resin while obtaining more liquid-like linewidth for the more mobile attached compounds. However, the contribution of magnetic-susceptibility discontinuities within the materials used for the probe construction will not be eliminated by MAS and can only be removed by a careful probe design. Very narrow ^{1}H linewidths (1–8 Hz) have been obtained with a probe called Nanonmr™ probe, which combined the use of both MAS and "magnetic-susceptibility-matched" materials.

4.2.2.1 1D NMR SPECTROSCOPY OF RESIN SAMPLES

The quality of NMR spectra obtained for resin-bound N-Fmoc-L-aspartic acid using conventional high-resolution, MAS, and high-resolution MAS (Nanonmr™) probes was compared by Keifer and coworkers (1996). Because the primary criterion of quality in NMR spectra for resin samples is the resonance linewidth, all possible factors that may affect linewidth were investigated and discussed. Factors that affect the achievable linewidth include the swelling of the resin with solvent, which increases molecular mobility, the microscopic discontinuities of magnetic susceptibility within the resin, which can be removed by MAS, the mobility of the nuclei at the end of the tether, which is controlled by both the length of the tether and the solvation, the spectral-resolution limitation of the probe, and within the more rigid region of the resin polymer, possibly solid-state effects. Magnetic susceptibility differences in the sample are the dominant line-broadening mechanism for the resonance of the mobile nuclei within the resin. Magnetic-susceptibility mismatches within the sample are removed by spinning the sample about the magic angle, while magnetic susceptibility mismatches within the probe are not. The largest difference was seen in ^{1}H spectra, which became significantly narrower with MAS and using the ^{1}H Nanoprobe. The three main findings from this study are: a) ^{1}H spectra must be obtained in a Nanoprobe, where as ^{13}C spectra could also be obtained in either a Nanoprobe or a CP/MAS probe; b) a Nanoprobe will always generate higher resolution spectra than a CP/MAS probe; c) the Nanoprobe will be the most efficient choice for small quantities of sample while a CP/MAS probe may provide higher S/N spectra if large quantities of sample are available.

Fitch and coworkers (1994) reported the application of the high-resolution ^{1}H NMR in SPOS. The conventional 500 MHz NMR spectrum of a reaction product (Diagram 4.1) in DMSO-d6 was obtained

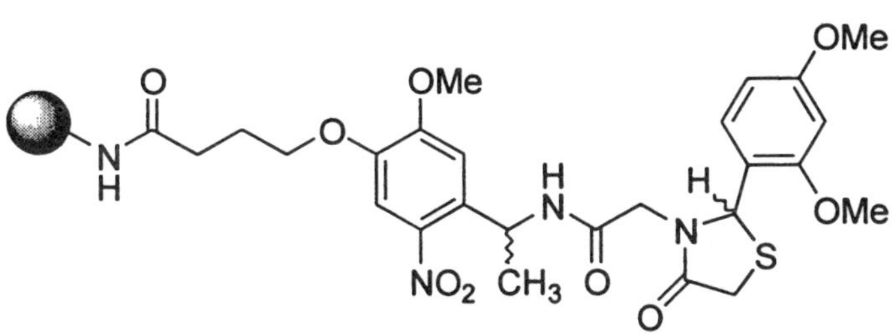

3

Diagram 4.1

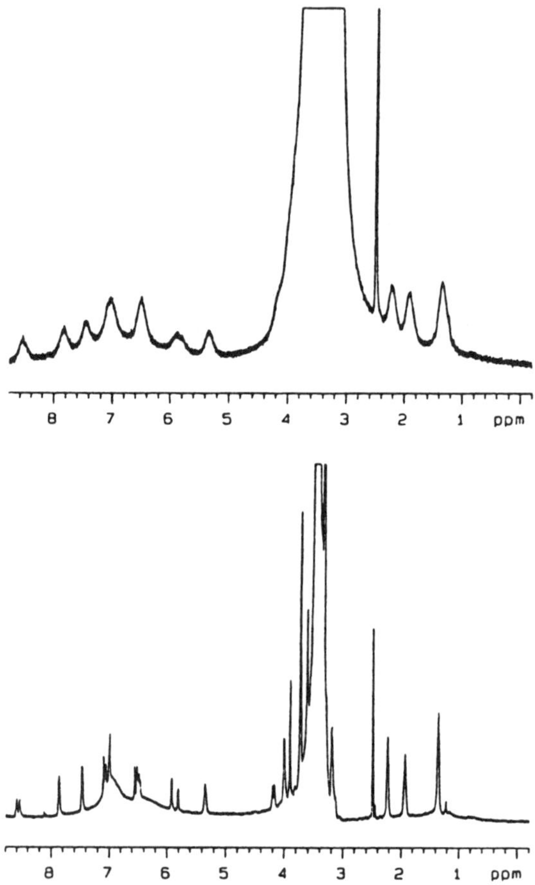

Figure 4.8 *(Top) A 16-scan 500 MHz NMR spectrum of **3** (100 mg suspended in 600 μL of DMSO-d_6). The spectrum was obtained using a conventional 5-mm liquids probe. (Bottom) A 16-scan 500 MHz NMR spectrum of **3** (10 mg suspended in 30 μL of DMSO-d_6). The spectrum was obtained by spinning at 2000 Hz about the magic angle in a Varian Nano-NMR probe. (Reproduced with permission from* J. Org. Chem., *1994, 59, 7955–7956. Copyright© 1994, American Chemical Society).*

using either conventional 5 mm liquid probe (top spectrum in Figure 4.8) or by spinning a 2000 Hz about the magic angle in a Varian Nanonmr probe (bottom spectrum in Figure 4.8).

Although the PS-PEG resin has been proven to possess the best NMR properties with respect to linewidth and spectral purity, high quality MAS NMR spectra of polystyrene-based resins were obtained (Wehler & Westman, 1996). The author showed that PS-based resin are well suitable for NMR analysis with a Nanoprobe.

The utility of the spin echo sequence (90°-τ-180°-τ) with a selected τ values, along with MAS at a speed of 20 kHz in suppressing signals from the PS matrix of the resin has been reported (Garigipati et al., 1996). It was shown that by the appropriate choice of τ value one can significantly minimize the effect due to polystyrene peaks.

4.2.2.2 2D NMR SPECTROSCOPY OF RESIN SAMPLES

MAS ^1H NMR spectra of resin-bound compounds, with the exception of PS-PEG resins, are relatively broad and featureless. The additional complication of polymer peaks can be attenuated by using a spin echo sequence, although the residual peaks can still be problematic. The loss of coupling information relegates the interpretations to chemical shift information only, making peak assignment difficult. Coupling information can be reintroduced by performing a 2D-J-resolved experiment. Usually, the 2D-J-resolved spectrum is "tilted" by rotating the data to align the chemical shift axis so that, upon projection of the data, a "proton decoupled" proton NMR spectrum can be obtained. However, the untilted projection can yield 1D proton NMR spectra of resin-bound compounds that are of high quality. The top spectrum in Figure 4.9 is the spin-echo MAS NMR spectrum for DMF-swollen crude product of isoleucine on Wang resin obtained by Fmoc chemistry (Shapiro et al., 1997). The quality of the spectrum is typical of this derivatized 1% DVB crosslinked PS resin. In contrast, the bottom spectrum in Figure 4.9b shows the spectrum obtained for the same compound by projection onto the chemical shift axis of the untilted 2D-J-resolved data. The resonance patterns for the two methyl groups of isoleucine (inset) serve as a comparison of the difference in the resolution.

Reduction of inhomogeneous line broadening in 2D MAS NMR spectra of molecules attached to swollen resin and the extraction of crucial J coupling constants information otherwise buried in the linewidth were achieved by using a pulse sequence for multiple-quantum filtered COSY or E.COSY with scaling in the evolution period (Meissner et al., 1997). By swelling the resin and spinning the resin at the magic angle, line broadening caused by the restricted molecular motions within a sample and the magnetic-susceptibility are reduced. The resulting linewidth is still not narrow enough to reveal J coupling constant. Because the inhomogeneity is associated with the chemical shift terms of Hamiltonian, a solution is to scale down these terms relative to the J coupling terms. This is done by inserting the element $kt_{1/2}$-π-$kt_{1/2}$ into

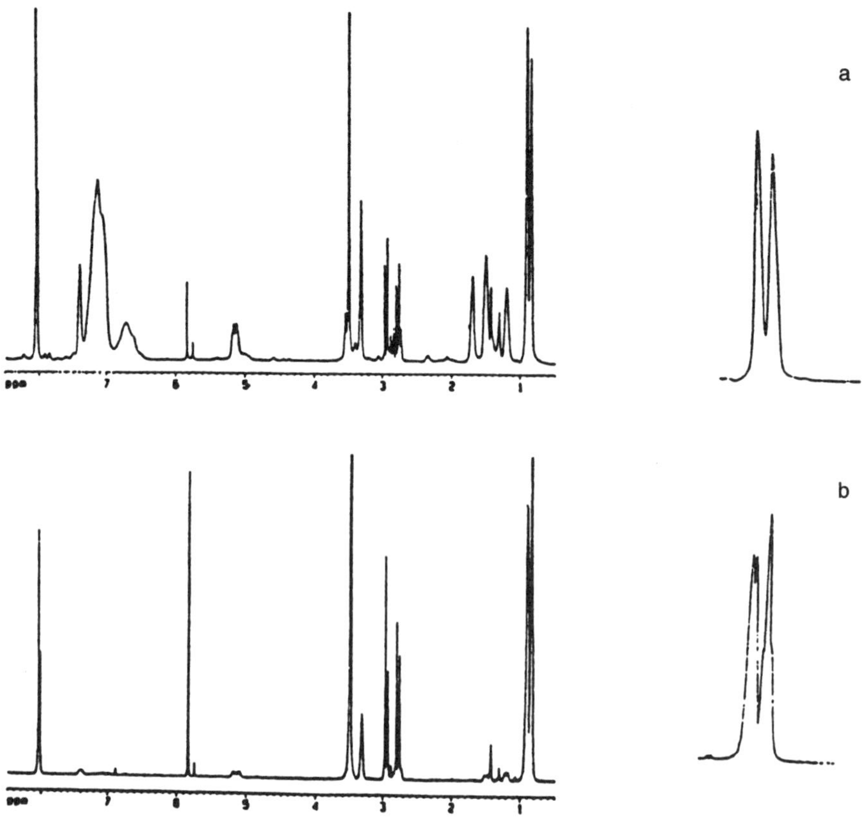

Figure 4.9 (a) Spin echo MAS ¹H NMR spectrum for DMF swollen isoleucine Wang resin. Inset are the two methyl resonances at δ = 0.95 and 0.89. (b) Projection of the homonuclear 2D J-resolved MAS of DMF swelled isoleucine on Wang resin. Inset are the methyl resonances at δ = 0.95 and 0.89. (Reproduced with permission from Tetrahedron Lett. *1997, 38, 1333–1336. Copyright© 1997, Elsevier Science Ltd.).*

the evolution period of any existing 2D correlation sequence as indicated for multiple quantum-filtered COSY (Piantini et al., 1982). In this way, the chemical shifts are scaled by a factor of $(k + 1)^{-1}$, while the J coupling is unaffected.

The acquisition of the MAS ^{13}C-^{1}H-correlated NMR of resin-bound compounds with a standard solid-state probe was reported (Anderson et al., 1995a). The MAS ^{13}C NMR spectrum of a norbornane-2-carboxylic acid bound to Wang resin showed carbonyl resonance at

174.9, 173.8 ppm, which was not observed in gel-phase NMR spectrum of the same sample. The ratio of exo/endo epimers was semi-quantitatively determined as 60:40, based on the peak intensities in the MAS ^{13}C spectrum. The resonance of tether carbon atoms was also observed, and these signals provide a straightforward means of determining the extent of reaction because an internal reference is observed in the resonance of the tether. The linewidth of 26 Hz obtained in these experiments allows structural information to be obtained by evaluation of CH-correlated data. The MAS CH-correlated spectrum of these compounds (shown in Figure 4.10a) and the solution-state spectrum of norbornane-2-carboxylic acid (Figure 4.10b) are shown for comparison. This technique promises to be useful not only in the determination of the success or failure of a chemical transformation, but in the structural verification of molecules produced by SPOS, as well as detection of diastereomers and side products.

The total correlation spectroscopy (TOCSY) NMR spectrum and the heteronuclear multiple quantum correlation (HMQC) spectrum of Wang-Fmoc-lysine-Boc were obtained. Correlation of the lysine portion of the molecule are readily observed. The connectivity through the chain can be traced from the NH group to the α-proton through the ε protons and in the reverse direction. The chemical shift assignment follows directly from consideration of the TOCSY data and from expected proton and carbon 13 chemical shift relationships (Anderson et al., 1995b).

It would be advantageous if the enhanced resolution afforded by the projection of the 2D-J-resolved data can be retained, while spin system connectivity is also obtained. Although correlation spectroscopy (COSY) and TOCSY data allow for the connectivity to be obtained, the projection onto the chemical shift axis does not lead to enhanced resolution. In spin-echo correlated spectroscopy (SECSY), both spin connectivity and resolution enhancement can be obtained (Chin et al., 1998). The apparent increase in resolution as well as the ability to make spin-spin connectivity were shown by Wang resin-bound compounds. A major problem associated with 2D NMR data obtained using MAS is that artifacts and T_1 noise are abundant, leading to potential confusion in data interpretation. By using the gradient MAS probe, this problem is alleviated.

As in high-resolution liquid-state studies, the gradient is essential for removing T2 noise and is extremely helpful in coherence pathway selection. A series of heteronuclear correlation experiment, heteronuclear multiple quantum correlation (HMQC), and heteronuclear multiple bond correlation (HMBC) was carried out, and the level of in MAS

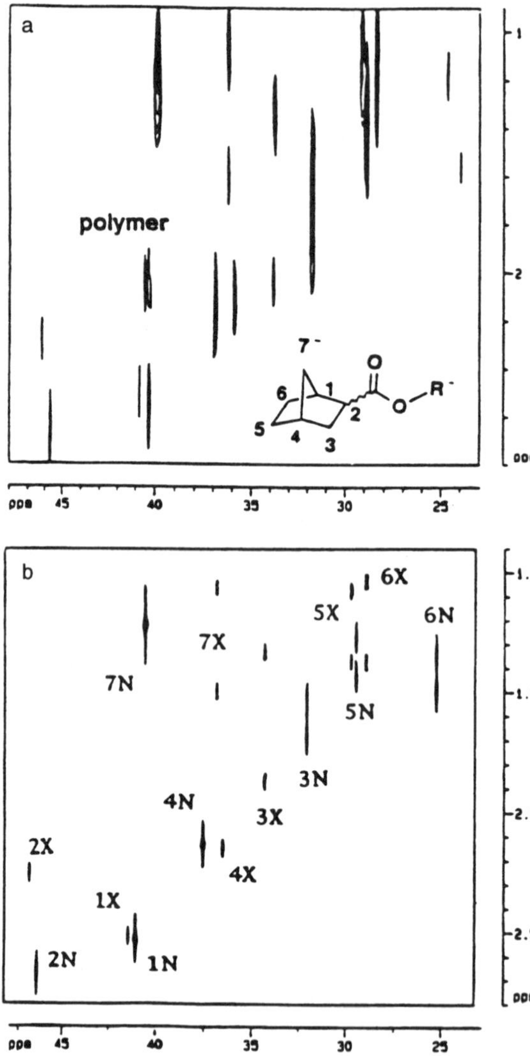

Figure 4.10 *CH-correlated MAS NMR spectrum of the resin-bound compound (a) swollen in benzene, where R=Wang resin. (b) Solution CH-correlated spectrum for a mixture of epimers of norbornane-2-carboxylic acid. Assignments are indicated for endo (N) and exo (X) epimers. (Reproduced with permission from J.* Org. Chem., *1995, 60, 2650–2651. Copyright© 1994, American Chemical Society).*

spectra of Wang-bound-N-Fmoc-N-Boc-L-Lysine swollen with CDCl3 was suppressed by using gradients (Maas et al., 1996). It is anticipated that gradient methods will become as common in MAS as they are in high-resolution liquid-state measurement. In fact, the need for gradient is more pronounced in the MAS case because the probe geometry and sample spinning build into the measurement an unavoidable modulation of the coupling of the spins to the receiver and, hence, pronounced artifacts.

2D ^{1}H NOE spectrum of a hexapeptide GPFRAF bound to amino PS-PEG resin was obtained (Jelinek et al., 1997). Short peptides in solution usually do not yield any NOE cross peaks because of their fast molecular motion and conformational flexibility. In this 2D ^{1}H NOE spectroscopic study, a relatively large number of NOE cross peaks was found. This suggests that a significant part of the population of the peptide on beads is folded into a defined stable conformation with less molecular motion than for the same peptide in solution.

A three-step SPOS involving a Heck reaction was monitored by a series of 2D MAS NMR measurements (Pop et al., 1996). All intermediate products were characterized on resin. The work demonstrated that the goal of reaction characterization and optimization *in situ* can be obtained in an entirely satisfying manner using conventional MAS NMR, although each 2D NMR experiment required 3 hours to complete.

4.2.2.3 NMR ON MULTIPIN CROWNS

SPOS uses resin beads as the primary format for synthesis. An alternate format is the multipin crown system where polystyrene is grafted onto a base polymer unit called a "crown," and arrays of these units are held on the ends of pins in a 96-array format for parallel synthesis. The first MAS NMR spectrum of crowns was reported (Chin et al., 1997). Spectra from a product of the Wittig reaction was recorded and clearly showed the disappearance of the peak from the starting material (aldehyde δ=10.17) and the formation of a peak from the product (olefin, δ=6.78, 7.72). The procedure is similar to that for resin beads. This work opened another possibility to effectively monitor reactions carried out on multipin crowns.

4.3. SUMMARY

Although most MS and NMR applications in the reaction optimization stage are still following the "cleave-and-analyze" protocol, on-resin applications have greatly increased. MS analysis always requires the cleavage before analysis. Two methods have accelerated the cleavage speed. One uses light and a photocleavable linker to release the bound compound. The other method uses TFA vapor with an acid labile linker. These methods at least have gotten rid of the time-consuming problem of the "cleave-and-analyze" protocol.

There is a long history of using gel-phase NMR for monitoring reactions on resin. On-resin NMR can be routinely obtained, and the most useful nuclei are ^{13}C, ^{19}F, ^{31}P, ^{2}H, and ^{15}N. However, the resolution and the spectral quality have not achieved the same level as those of solution samples. A question is whether we need the same resolution to accomplish our goal. For most SPOS optimization projects, the goal is to confirm the formation of a desired product, not elucidate an unknown structure. In many instances, this can be achieved without a solution-like quality spectrum. Gel-phase sampling method is a convenient way to record an NMR spectrum. Although tens to a hundred milligrams of resin are required for analysis, the method is still useful because the sample amount is not limited in reaction optimization work and the NMR sample can be reused for synthesis after analysis. Finally, the most important aspect is that the probe and instrumentation required for acquiring gel-phase NMR are widely available to nearly every chemistry laboratory.

MAS NMR and the high-sensitivity probes have greatly improved the resolution for ^{1}H spectra with a much smaller amount of a sample. 2D NMR spectra are also becoming available. Although there is a long way to go before any unknown structure can be resolved on-resin routinely, MAS technique can help solve various problems in the optimization efforts.

4.4. REFERENCES

Anderson, R. C., Jarema, M. A., Shapiro, M. J., Stokes, J. P., Ziliox, M. Analytical techniques in combinatorial chemistry: MAS CH correlation in solvent-swollen resin. *J. Org. Chem.* 1995a, 60, 2650.

Anderson, R. C., Stokes, J. P., Shapiro, M. J. Structure determination in combinatorial chemistry: Utilization of magic angle spinning HMQC and TOCSY NMR spectra in structure determination of Wang-bound Lysine. *Tetrahedron Lett.* 1995b, 36, 5311.

Bardella, F., Eritja, R., Pedroso, E., Giralt, E. Gel-phase ^{31}P-NMR. A new analytical tool to evaluate solid phase oligonucleotide4 synthesis. *Bioorg. Med. Chem. Lett.* 1993, 3, 2793.

Bayer, E., Albert, K., Willisch, H., Rapp, W., Hemmasi, B. ^{13}C NMR relaxation times of a tripeptide methyl ester and its polymer-bound analogues. *Macromolecules,* 1990, 23, 1937.

Blossey, E. C., Cannon, R. G. Synthesis, reactions, and ^{13}C FT NMR spectroscopy of polymer-bound steroids. *J. Org. Chem.* 1990, 55, 4664.

Brummel, C. L., Lee, I. N. W., Zhou, Y., Benkovic, S. J., Winograd, N. A mass spectrometric solution to address the problem of combinatorial libraries. *Science.* 1994, 264, 399.

Carrasco, M. R., Fitzgerald, M. C., Oda, Y., Kent, S. B. H. Direct monitoring of organic reactions on polymeric supports. *Tetrahedron Lett.* 1997, 38, 6331.

Chin, J., Fell, B., Pochapsky, S., Shapiro, M. J., Wareing, J. R. 2D SECSY NMR for combinatorial chemistry. High-resolution MAS spectra for resin-bound molecules. *J. Org. Chem.* 1998, 63, 1309.

Chin, J., Fell, B., Shapiro, M. J., Tomesch, J., Wareing, J. R., Bray, A. M. Magic angle spinning NMR for reaction monitoring and structure determination of molecules attached to multipin. *J. Org. Chem.* 1997, 531.

Doskocilova, D., Schneider, B. Narrowing of proton NMR lines by magic angle rotation. *Chem. Phys. Lett.* 1970, 6, 381.

Drain, I. F. The broadening of magnetic resonance lines due to field inhomogeneities in powdered sample. *Proc. Phys. Soc.* 1962, 80, 1380.

Egner, B. J., Bradley, M. Analytical techniques for solid-phase organic and combinatorial synthesis. *Drug Discovery Today.* 1997, 2, 102.

Egner, B. J., Cardno, M., Bradley, M. Linkers for combinatorial chemistry and reaction analysis using solid phase *in situ* mass spectrometry. *J. Chem. Soc., Chem. Commun.* 1995a, 2163.

Egner, B. J., Langley, G. J., Bradley, M. Solid phase chemistry: Direct monitoring by matrix-assisted laser desorption/ionization time of flight mass spectrometry. A tool for combinatorial chemistry. *J. Org. Chem.* 1995b, 60, 2652.

Epton, R., Goddard, P., Ivin, K. J. Gel phase ^{13}C n. m. r. spectroscopy as an analytical method in solid (gel) phase peptide synthesis. *Polymer.* 1980, 21, 1367.

Fitch, W. L., Detre, G., Holmes, C. P. High-resolution ^{1}H NMR in solid-phase organic synthesis. *J. Org. Chem.* 1994, 59, 7955.

Fitzgerald, M. C., Harris, K., Shevlin, C. G., Siuzdak, G. Direct characterization of solid phase resin-bound molecules by mass spectrometry. *Bioorg. & Med. Chem. Lett.* 1996, 6, 979.

Gallop, M. A., Fitch, W. L. New methods for analyzing compounds on polymeric supports. *Curr. Opin. Chem. Biol.* 1997, 1, 94.

Garigipati, R. S., Adams, B., Adams, J. L., Sarkar, S. K. Use of spin echo magic angle spinning ^{1}H NMR in reaction monitoring in combinatorial organic synthesis. *J. Org. Chem.* 1996, 61, 2911.

Giralt, E., Rizo, J., Pedroso, E. Application of gel-phase ^{13}C-NMR to monitor solid-phase peptide synthesis. *Tetrahedron.* 1984, 40, 4141.

Haskins, N., Hunter, D., Organ, A. J., Rahman, S. S., Thom, C. Combinatorial chemistry: Direct analysis of bead surface associated materials. *Rapid Commun. Mass Spectr.* 1995, 9, 1437.

Jelinek, R., Valente, A. P., Valentine, K. G., Opella, S. J. Two-dimensional NMR spectroscopy of peptides on beads. *J. Magn. Reson.* 1997, 125, 185.

Johnson, C. R., Zhang, B. Solid phase synthesis of aldenes using the Horner-Wadsworth-Emmons reaction and monitoring by gel phase ^{31}P NMR. *Tetrahedron Lett.* 1995, 36, 9253.

Jones, A. J., Leznoff, C. C., Svirskaya, P. I. Characterization of organic substrates bound to cross-linked polystyrenes by ^{13}C NMR spectroscopy. *Org. Magn. Reson.* 1982, 18, 236.

Keifer, P. A. High-resolution NMR techniques for solid-phase synthesis and combinatorial chemistry. *Drug Discovery Today.* 1997, 2, 468.

Keifer, P. A. New methods for obtaining high-resolution NMR spectra of solid-phase synthesis resins, natural products, and solution-state combinatorial chemistry libraries. *Drugs of the Future.* 1998, 23, 301.

Keifer, P. A., Baltusis, L., Rice, D. M., Tymiak, A. A., Shoolery, J. N. A comparison of NMR spectra obtained for solid-phase-synthesis resins using conventional high-resolution, magic-angle-spinning, and high-resolution magic-angle-spinning probes. *J. Magn. Reson. A.* 1996, 119, 65.

Look, G. C., Holmes, C. P., Chinn, J. P., Gallop, M. A. Methods for combinatorial organic synthesis: The use of fast ^{13}C NMR analysis for gel phase reaction monitoring. *J. Org. Chem.* 1994, 59, 7588.

Ludwick, A. G., Jelinski, L. W., Live, D., Kintanar, A., Dumais, J. J. Association of peptide chains during Merrifield solid-phase peptide synthesis. A deuterium NMR study. *J. Am. Chem. Soc.* 1986, 108, 6493.

Maas, W. E., Laukien, F. H., Cory, D. G. Gradient, high resolution, magic angle sample spining NMR. *J. Am. Chem. Soc.* 1996, 118, 13085.

Manatt, S. L., Amsden, C. F., Bettison, C. A., Frazer, W. T., Gudman, J. T., Lenk, B. E., Lubetich, J. F., McNelly, E. A., Smith, S. C., Templeton, D J., Pinnell, R. P. A fluorine-19 NMR approach for studying Merrifield solid-phase peptide syntheses. *Tetrahedron Lett.* 1980, 21, 1397.

Meissner, A., Bloch, P., Humpfer, E., Spraul, M., Sorensen, O. W. L. Reduction of inhomogeneous line broadening in two-dimensional high-resolution MAS NMR spectra of molecules attached to swelled resins in solid-phase synthesis. *J. Am. Chem. Soc.* 1997, 119, 1787.

Piantini, U., Sørensen, O. W., Ernst, R. R. Multiple quantum filters for elucidating NMR coupling networks. *J. Am. Chem. Soc.* 1982, 104, 6800.

Pop, I. E., Dhalluin, C. F., Déprez, B. P., Melnyk, P. C., Lippens, G. M., Tartar, A. L. Monitoring of a three-step solid phase synthesis involving a Heck reaction using magic angle spinning NMR spectroscopy. *Tetrahedron* 1996, 52, 12209.

Rapp, W. PEG grafted polystyrene tentacle polymers: Physico-chemical properties and application in chemical synthesis. In G. Jung (Ed.), *Combinatorial Peptide and Nonpeptide Libraries: A Handbook.* Wiley-VCH, Weinheim, 1996, 425–464.

Shapiro, M. J., Chin, J., Marti, R E., Jarosinski, M. A. Enhanced resolution in MAS NMR for combinatorial chemistry. *Tetrahedron Lett.* 1997, 38, 1333.

Shapiro, M. J., Kumaravel, G., Petter, R. C., Beveridge, R. ^{19}F NMR monitoring of a S_NAr reaction on solid support. *Tetrahedron Lett.* 1996, 37, 4671.

Shapiro, M. J., Wareing, J. R. NMR methods in combinatorial chemistry. *Curr. Opin. Chem. Biol.* 1998, 2, 372.

Svensson, A., Fex, T., Kihlberg, J. Use of ^{19}F NMR spectroscopy to evaluate reactions in solid phase organic synthesis. *Tetrahedron Lett.* 1996, 37, 7649.

Swayze, E. E. The solid phase synthesis of trisubstituted 1, 4-diazabicyclo[4.3.0]nonan-2-one scaffolds: On bead monitoring of heterocycle forming reactions using ^{15}N NMR. *Tetrahedron Lett.* 1997, 38, 8643.

Wehler, T., Westman, J. Magic angle spinning NMR: A valuable tool for monitoring the progress of reactions in solid phase synthesis. *Tetrahedron Lett.* 1996, 37, 4771.

Yu, Z., Yu, X. C., Chu, Y.-H. MALDI-MS determination of cyclic peptidomimetic sequences on single beads directed toward the generation of libraries. *Tetrahedron Lett.* 1998, 39, 1.

5

Reaction Optimization Using Spectrophotometric and Other Methods

ANALYTICAL METHODS BASED on spectrophotometric and fluorometric methods, in addition to qualitative analysis, provide a quantitative means to analyze organic compounds bound to solid supports. Traditional "cleaving, purifying, and weighing" method cannot be applied to combinatorial chemistry because there is not always enough cleaved compound for weighing. On the solid support, quantitation based on the weight gain of the support after synthesis is not accurate because this weight increase is usually very small and the entrapped reagent and solvent will cause errors. Only if the weight gain is significant relative to the polymer backbones may this method be used to estimate the yield. If the "cleave and analyze" method has to be adopted, there are more appropriate choices of methods for quantitation of minute amounts of samples, such as quantitative detectors (nitrogen-specific detector, evaporative light-scattering detector), NMR, or other analytical methods combined with a standard.

An important advantage of solid-phase organic synthesis (SPOS) is that it circumvents tedious intermediate isolation and purification procedures, such as recrystallization, distillation, and chromatography. Ideally, a solid-phase reaction can be driven to completion by excess reagent without causing problems in the purification. In reality, most organic reactions cannot easily go to completion. Unlike solution synthesis, the resin-bound synthetic intermediate in SPOS cannot be purified. The unreacted portion will accumulate until the end of the synthetic reactions and cause purity problems for the final product. Therefore, on-resin quantitative analysis of solid-supported compounds at intermediate stages is very advantageous to guide the multistep synthesis.

5.1 QUALITATIVE ANALYSIS OF AMINES

Resin lacking free amino groups exposed to the solution of bromophenol blue (3′, 3″, 5′, 5″-tetrabromophenolsulfophthalein, see Diagram 5.1) becomes yellowish to yellow-orange ($_{\lambda max}$429 nm). When free amino groups are present, the color turns to deep blue (91,800 $M^{-1}cm^{-1}$ at 600 nm). Note that most acid-base indicators do not change their color in the presence of free amino groups on resin. Bromocresol green and bromocresol purple also displayed satisfactory properties. This simple method (Krchnak et al., 1988a, b) gave comparable results to those using either picric acid (Diagram 5.1) or the ninhydrin method (Scheme 5.1).

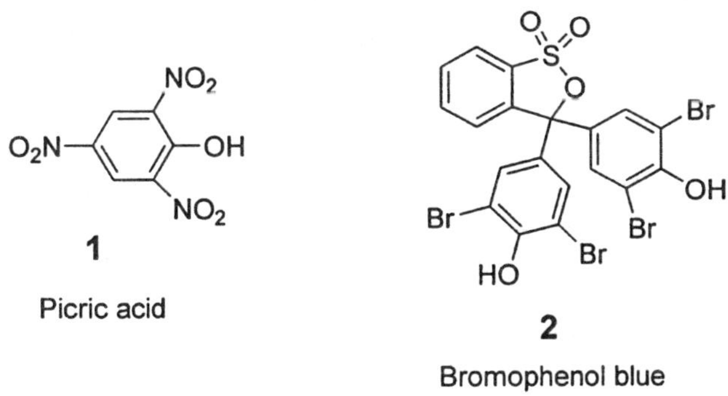

Diagram 5.1

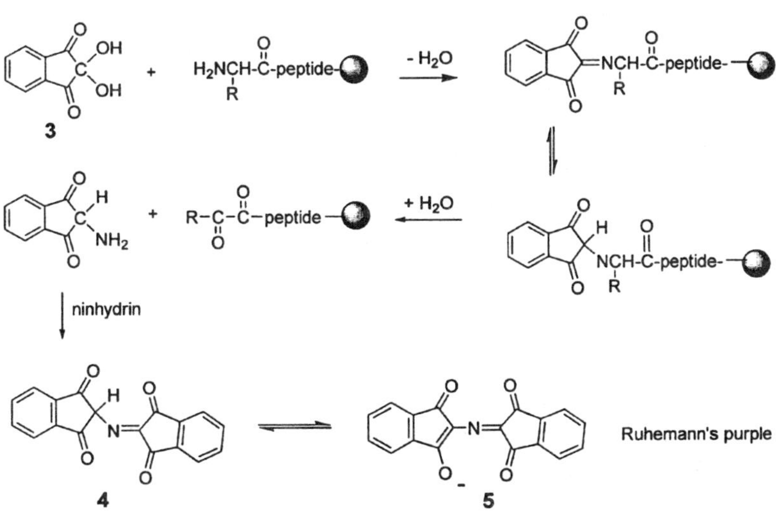

Scheme 5.1

The sensitivity of these monitoring methods brings about their own limitations:

1. The quality of the protected amino acids must be high. This high purity, in some cases, is very hard to achieve.
2. DMF used for the synthesis must be dimethylamine-free, and no base can be added during the coupling step.
3. The nature of the base used in the neutralization step is very important if the classical Merrifield's resin is used. The residual chloromethyl groups form quaternary ammonium salts with the tertiary amine, and this leads to blue coloring of the resin after the addition of the bromophenol blue.
4. Only a very small amount of monitoring dye can be used because it has the ability to decolorize itself by its leuko-form. Therefore, the amount of the indicator applied to the resin should be lower than the quantity of the free amino group, which one wants to monitor.

A rapid micro-test using 2,4,6-trinitrobenzenesulphonic acid (TNBS) has been developed (Hancock & Battersby, 1976) to detect incomplete coupling reactions in solid-phase peptide synthesis. This new test detects amino groups at a level of 0.003-mmol/g.

One gram of resin reacts with one drop of 1% (w/v) solution of TNBS in DMF for 10 min. After centrifuging the mixture for five minutes to remove the supernatant, the resin appears as red-orange color under the microscope.

A qualitative test (Christensen, 1979) is described that can be carried out, as some other reagent, in the presence of the reaction mixture. At intervals, a resin sample is removed from the reactor, and the chloranil test is carried out directly (Scheme 5.2). If free amino groups are present, a green to blue color develops on the resin beads in less than five minutes. The test can also be used when the N-terminal amino acid residue is proline. The sensitivity for detection of N-terminal proline is in the range of 0.002–0.005 mmol/g resin, and for primary amino

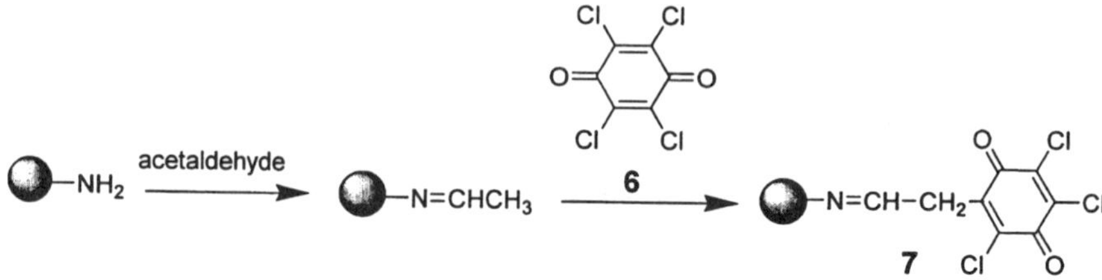

Scheme 5.2

groups, it is in the range of 0.005–0.008 mmol/g when approximately 1 mg of peptide resin is assayed. Several peptide syntheses have been carried out and monitored by the chloranil test, indicating that the method is generally applicable. However, when histidine is present, the test result may be blurred. In a few cases, discrepancy between the chloranil test and the ninhydrin test was observed.

A procedure (Vojkovsky, 1995) uses chloranil in the presence of acetaldehyde to detect secondary amines. Except for N (p-tolyl) glycine, all secondary amino groups that were examined were detected reliably in a substitution range down to 0.0028 mmol/g. Protected imidazole in His (Trt) did not interfere with the detection method.

Proline is known to form a blue color with isatin (Scheme 5.3). The isatin test (Kaiser et al., 1980) made it possible to detect incomplete couplings to proline in cases when the ninhydrin test was so weak that it could be interpreted as negative. The positive isatin test became negative when the coupling reaction was repeated or when the reaction time prolonged. Ala, Arg, Asp, Asn, Cys, Glu, Gln, His, Leu, Met, Thr, Phe, and Val all tested negative. Only with terminal prolyl residues did the color of the beads become blue. In addition to isatin, 4-N, N′-dimethylaminoazobenzene-4′-isothiocyanate (DABITC), and malachite green isothiocyanate (MGI) have also been used in qualitative analysis of resin-bound amines (Shah et al., 1997).

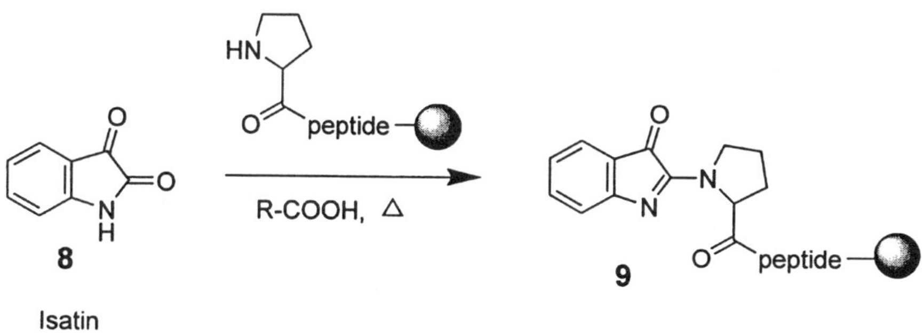

Scheme 5.3

The fluorescamine test (Scheme 5.4) (Felix & Jimenez, 1973) for the rapid detection of trace amounts of uncoupled products from solid phase peptide synthesis was reported. Because the test is carried out under mild conditions, certain side reactions are circumvented. The fluorophore-resins are easily viewed under long wavelength ultraviolet light, are stable at room temperature, and may be used for quantitative evaluation.

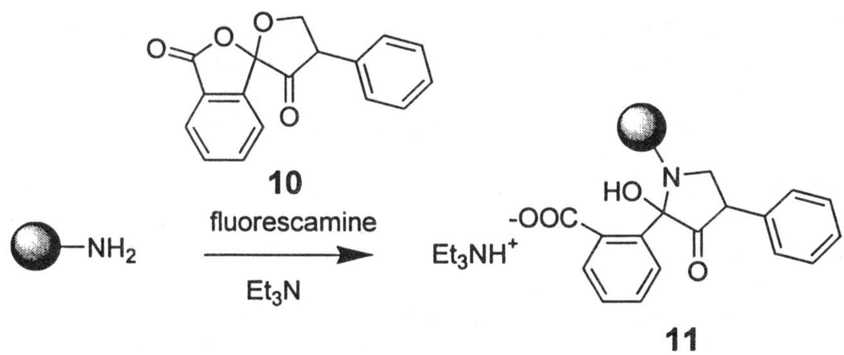

Scheme 5.4

Another highly sensitive fluorometric method (Garden & Tometsko, 1972) for determining free amine groups on the Merrifield peptide resin is based on the reaction with dansylchloride (Scheme 5.5). The method is capable of quantitatively analyzing amine groups at a 0.0002 mmol/g level, about 10-fold more sensitive than color test methods. The method was used to follow the kinetics of amino acid incorporation into resin-bound peptides.

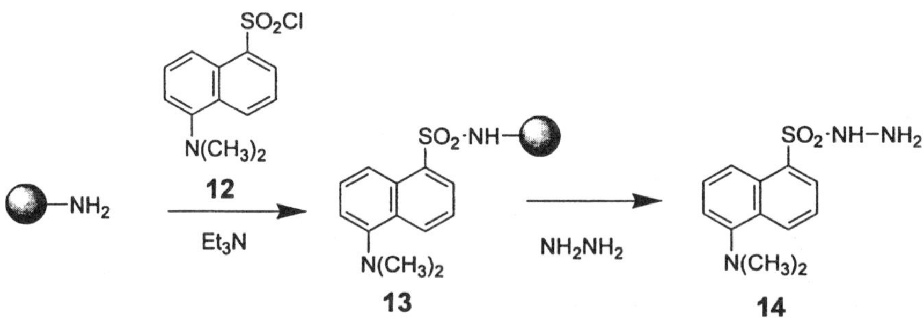

Scheme 5.5

5.2 QUANTITATIVE ANALYSIS OF AMINES

"Kaiser Test" (Kaiser et al., 1970, Scheme 5.1) is a rapid and sensitive method that determines if the formation of a peptide bond can reach 99.9% completion. This test is based on the reaction of ninhydrin reagents with resin-bound amines. The ninhydrin reagent described by Troll and Canan (1953) was found to be the most suitable for this purpose. In this reaction, a blue chromophore is formed. The absorbance of the violet-colored product, known as Ruhemann's purple, was observed at 570 nm with an average extinction coefficient of 1.57×10^4 $M^{-1}cm^{-1}$ (Sarin et al., 1981). The ninhydrin test was routinely used in SPPS.

After deprotection, a peptide resin sample is taken and tested in organic buffer with the ninhydrin reagent. In two to five minutes, an intense blue color develops on the beads and in the solution unless the terminal amino acid is proline or benzyl aspartate. With these two amino acids, there is brown or reddish-brown color. After each coupling step, a sample is taken and heated for five to 10 minutes with the ninhydrin reagent. If any blue color develops on the beads or in the solution, an incomplete coupling is indicated. The rate and extent of the color development are very sensitive to heating time, temperature, and individual peptide. The blue color is not stable and decays within 10–15 minutes. Pyridine, which is one of the reagents used in this quantitative test, partially deprotects Fmoc amino acid resins. The deprotected amino acid resin may then react with the ninhydrin reagent, resulting in a higher Ruhemann's purple concentration than the true coupling efficiency warranted. Premature removal of the Fmoc group can be minimized by heating the reaction mixture for five minutes instead of 10 minutes and by adding two to three drops (0.02–0.04 mL) of glacial acetic acid to each resin sample.

In aqueous buffers, a blue chromophore is formed and is stable at 100° C for two hours, and the quantitative analysis can easily be carried out. Because the curves did not show a stable plateau in organic medium, the rate and extent of color development were very sensitive to heating time, temperature, and individual peptide.

For example, ε'_{570} for HGly-resin was 1.01×10^4, and for HGlu (OBzl)-resin, it was 1.25×10^4. The accuracy of the calculation of free amino groups by this method depends on the quality of the secondary method used to establish the extinction coefficient, e.g., amino acid analysis, picrate titration, Edman end-group analysis. The precision of the assay on replicate samples was about ±4% over most of the analytical range, and the response curve was shown to be linear up to at least 1 mmol of amine per gram of resin.

For the routine monitoring of a synthesis, where an estimate of the extent of the coupling reaction is needed and where continued coupling to a constant endpoint is planned, the use of an average ε' is usually adequate. Where an exact value for the percent-uncoupled amine is required, it becomes necessary to establish the extinction coefficient for the particular peptide-resin under study. The rapid ninhydrin color test, as applied to samples of solid-phase resin peptides, makes it possible to detect incomplete couplings not discernible by amino acid analysis, to show diminishing amounts of free terminal amino groups by decreasing color intensities, and to obtain a positive test even though more than 99% of the terminal amino groups are reacted.

Picric acid (Diagram 5.1) was initially used for noninvasive monitoring of free amines (Gisin, 1972). The resin is treated with picric acid to form a salt, is rinsed to remove unbound acid, and is then reacted with diisopropylethylamine (DIEA). The picric acid/DIEA chromophore is measured quantitatively at 362 nm (extinction coefficient 14,500 $M^{-1}cm^{-1}$) to determine the amine content in the resin (Sarin et al., 1981). This method is used in the amine concentration range between $1–20 \times 10^{-5}$ M. Dilute acid labile groups are susceptible to removal by picric acid, making this technique incompatible with some Fmoc amino acid resins.

A method based on the condensation reaction of 2-hydroxy-1-naphthaldehyde with the free amino groups in the resin was reported (Esko et al., 1968). A quite stable aldimine (Schiff's base) absorbing at 420 nm is formed. The chromophore is then displaced from the polymer by an amine, e.g., benzylamine, and the amount of the soluble aldimine thus formed is determined. A 50-mg resin sample is required for an analysis.

Dorman (1969) described a simple, non-destructive procedure for quantitatively determining the completeness of solid-phase coupling reactions in conjunction with the use of N-t-butyloxycarbonyl (BOC) protected amino acids and active esters (Scheme 5.6). Following the coupling reaction and its subsequent washing steps, the resin peptide is treated with pyridine hydrochloride (Pyr.HCl) in dry methylene chloride to form the hydrochloride of any uncoupled resin and is then treated with triethylamine in DMF for three minutes to neutralize any hydrochloride formed. The filtrates are combined, chilled, acidified with excess dilute nitric acid, and titrated potentiometrically with standard silver nitrate. The equivalents of silver nitrate required represent the amount of uncoupled resin amine.

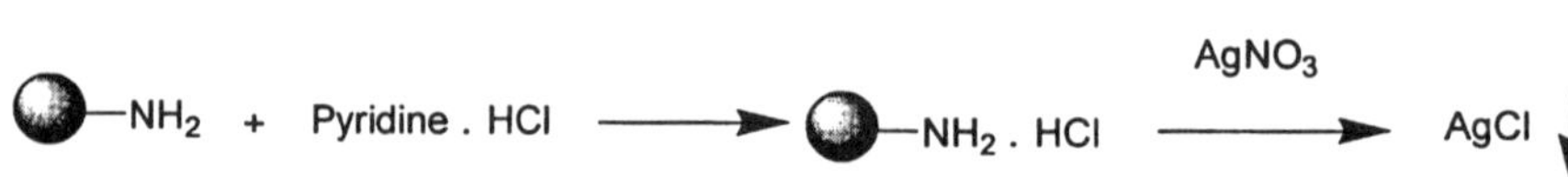

Dorman's test

Scheme 5.6

The success of this method rests on the ability of pyridine hydrochloride to form quantitatively the hydrochloride of uncoupled resin amine without causing premature removal of the t-butyloxycarbonyl protective group of the coupled amino acid. Unfortunately, Pyr.HCl/CH_2Cl_2 cleaves the o-nitrophenylsulfenyl (NPS) protective group at a substantial rate, thereby precluding the use of this reagent with NPS protective group.

A method (Reddy & Voelker, 1988), which essentially quantitates the unreactive N-terminal amino groups, involves tritylation of the uncoupled amino groups with 4,4′-dimethoxytrityl chloride (DMTCl) in the presence of tetra-*n*-butylammonium perchlorate and subsequent detritylation by a mild acid (Scheme 5.7). Measurement of the color of the released trityl cation determines the coupling efficiency.

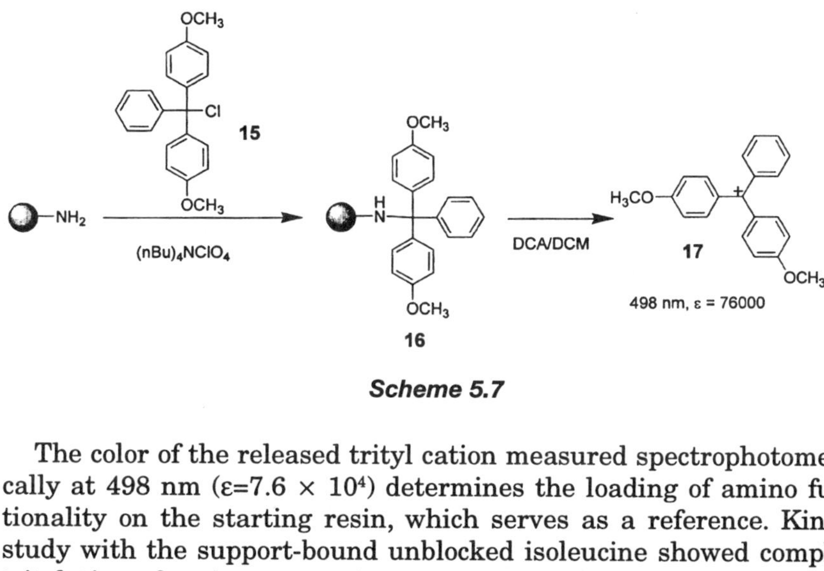

Scheme 5.7

The color of the released trityl cation measured spectrophotometrically at 498 nm ($\varepsilon=7.6 \times 10^4$) determines the loading of amino functionality on the starting resin, which serves as a reference. Kinetic study with the support-bound unblocked isoleucine showed complete tritylation of amino groups in 30 seconds. The detritylation with 2% dichloroacetic acid (DCA) was complete within two minutes. 2% DCA/CH_2Cl_2 did not release any BOC protecting group even after three hours of treatment. The solutions of DMTCl/CH_2Cl_2 and $(nBu)_4NClO_4$/CH_2Cl_2 were stable at room temperature for at least 45 days without any detectable loss of activity, as determined by comparative trityl assay with the freshly prepared reagents.

The advantages of this approach include the following:

1. The stable covalent attachment of the monitoring reagent allows extensive washings to remove the excess reagent, which makes the assay consistent and reliable.

2. The monitoring procedure is non-destructive and reversible in nature. Hence, the operations can be performed in the reactor itself, which makes the desired automation simpler.
3. Sensitivity is increased compared to the aliquot monitoring methods, because the total resin is employed for monitoring. Although the results were correlating closely with the Kaiser ninhydrin test, there were a few instances where this assay could detect unreacted amino groups of much less than 1%, which could not be detected by the Kaiser test.

To quantitate the nonbasic or relatively hindered amines, nitrophenylisothiocyanate-*O*-trityl (NPIT, Scheme 5.8) has been applied to measure the amount of free amine group on resin based on the concentration of the released trityl cation concentration in solution (Chu & Reich, 1995).

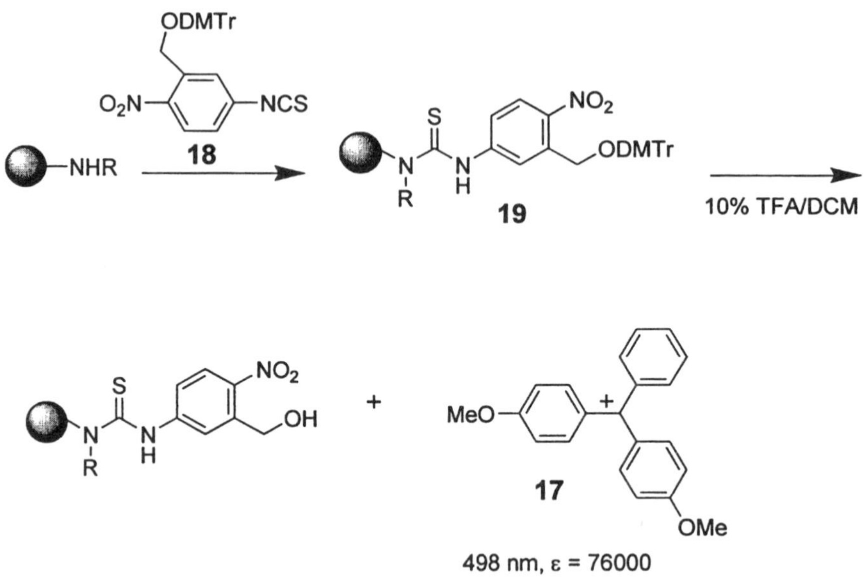

Scheme 5.8

A colorimetric method (Tyllianakis et al., 1994) for the direct determination of total solid-supported sulfhydryl, aldehydo, hydrazido, and N-hydroxysuccinimido carboxylate groups as well as of immobilized cysteine, tyrosine, and thyroxin using only the commercially available bicinchoninic acid/copper protein assay reagent was described (Scheme 5.9). The method is based on the ability of these groups to

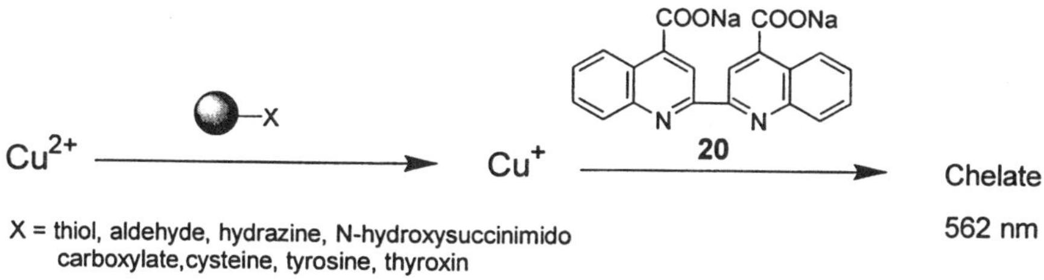

Scheme 5.9

reduce Cu^{2+} to Cu^{+}, which forms a chelate complex with bicinchoninic acid absorbing at 562 nm. Each assay requires only one incubation step of the solids with the reagent for one hour at 60° C. The quantitation of the different groups is finally carried out through standard curves of appropriate substances.

The assays were found to be accurate, reproducible (inter assay CV less than 3%), and very sensitive, allowing the determination of nanomole quantities of functional groups per assay tube.

A simple colorimetric method (Ngo, 1986) has been developed for the quantitative determination of reactive amino groups of a solid support. The method uses a reagent that reacts specifically and facilely with amino groups and concurrently introduces free sulfhydryl groups (Scheme 5.10). The latter can then be titrated colorimetrically by using the well-known Ellman's reagent. Thus, an excess of 2-iminothiolane

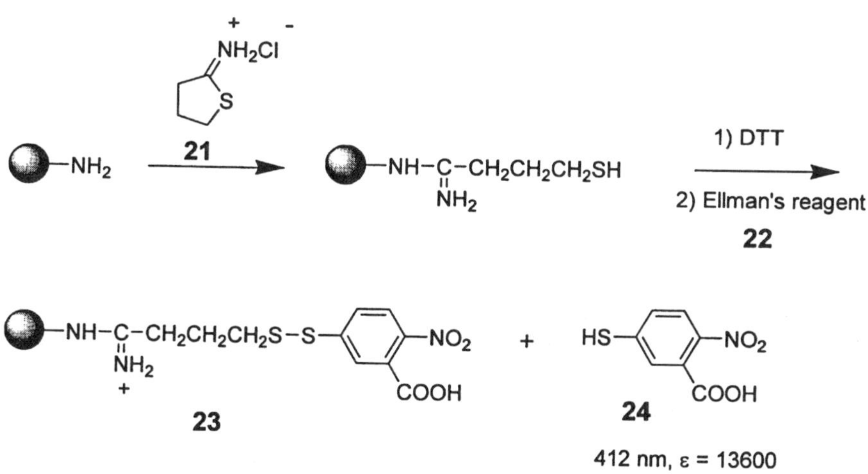

Scheme 5.10

(Traut's reagent) was reacted with amine-carrying solid supports. After removing unreacted 2-iminothiolane and keeping the resultant solid-phase sulfhydryl groups in a reduced state by using dithiothreitol, the solid supports, after being thoroughly washed, were then reacted with 5,5′-dithiobis-(2-nitrobenzoic acid) (DTNB) to quantify the sulfhydryl groups that were generated from reacting solid-phase amino groups with Traut's reagents. The DTNB reacts with sulfhydryl groups at pH 8 to produce 1 mol of 2-nitro-5-thiobenzoic acid per mole of sulfhydryl group. The 2-nitro-5-thiobenzoic acid is quantitated by UV spectroscopic measurement.

Fmoc SPPS also offers a unique opportunity to noninvasively monitor N-α-amino group deprotection (Scheme 5.11). A fulvene-piperidine adduct is formed during piperidine deprotection. This adduct has UV absorption maxima at 267 nm (ε=17500–18950 $M^{-1}cm^{-1}$), 290 nm (ε=5280–5800 M^{-1} cm^{-1}) and 301 nm (ε=6200–7800 M^{-1} cm^{-1}). These absorption properties permit reasonably accurate determination of deprotection efficiency. Quantitative spectrophotometric monitoring of the deprotection of the Fmoc amino acid resins can also be used to determine the substitution level of amino acid per gram of resin (Heimer et al., 1981; Wade et al., 1991).

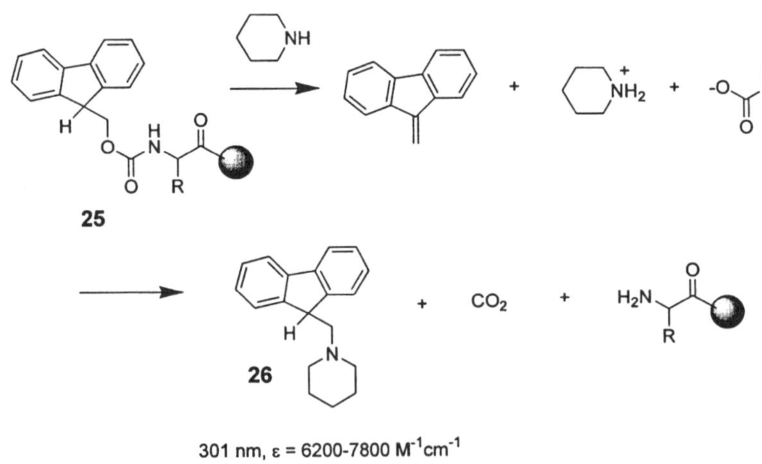

Scheme 5.11

It has also been applied to the continuous-flow, solid-phase peptide synthesis (Meienhofer et al., 1979). The procedures of Fmoc synthesis and quantitative measurement have been analyzed in detail (Fontenot et al., 1991). An excellent review has summarized more specific information on Fmoc chemistry (Fields & Noble, 1990).

The rate of attachment of Fmoc-containing Knorr linker to solid supports was quantitatively determined for various crosslinked polystyrene resins by Fmoc quantitation (Li et al., 1999). Results indicate that amide formation is favored on amino resins of smaller size and those with polyethyleneglycol (PEG), while both reaction rate and percentage of completion drop significantly when the reaction is performed in DMF in place of DCM.

Similar to the Fmoc-peptide chemistry, 4-nitrophenylethyloxycarbonyl (NPEOC) has been used as a protection group of amines in peptidylphosphonates synthesis (Campbell & Bermak, 1994). The NPEOC group can be cleaved to form a stable nitrostyrene that has an intense absorbance at 308 nm with an extinction coefficient of 13,200 $M^{-1}cm^{-1}$.

5.3 QUANTITATIVE SPECTROSCOPIC METHODS FOR ORGANIC FUNCTIONAL GROUPS

Methods previously developed for amine quantitation in SPPS can also be used in SPOS. However, SPOS involves many more organic functional groups. Quantitative analysis of these groups are needed in the reaction optimization stage.

5.3.1 Quantitation of Aldehyde/Ketone Groups

On-resin quantitation of the loading in the starting resin and the yield of synthetic intermediate and product are required for reaction optimization in polymer-supported combinatorial and high-throughput parallel synthesis. A general principle for direct quantitation of the absolute amount of organic functional groups in resin-bound starting materials, intermediates, as well as products was developed (Yan & Li, 1997). The method employs fluorescence or UV-Vis spectroscopy and the availability of a large number of fluorescent dyes, which can be attached to various organic functional groups, to quantify organic reactions directly on resin. As the first application of this principle, a fluorescence method for rapid determination of the absolute amount of aldehyde or/and ketone groups on polystyrene and PS-PEG resins (Scheme 5.12) was reported (Yan & Li, 1997).

To a suspension of formylpolystyrene or PS-PEG-CHO resin in DMF was added an approximately twofold excess of dansylhydrazine. The fluorescence spectra in Figure 5.1a and Figure 5.1b show the decrease

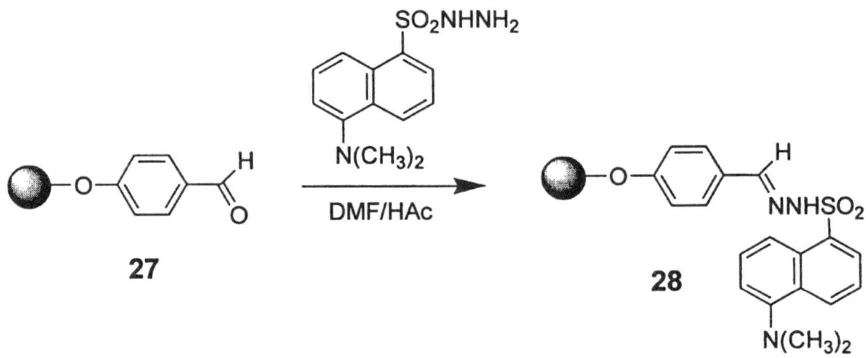

Scheme 5.12

of hydrazine concentration in the supernatant for reactions with formylpolystyrene and PS-PEG-CHO resins. After the reaction is completed, the consumed amount of dansylhydrazine is equal to the absolute amount of dye molecules covalently and non-covalently bound to the resin beads. The amount of hydrazine molecules non-covalently trapped in the resin matrix was corrected by subtracting the amount of dye molecules trapped in the same amount of 1-%-divinylbenzenepolystyrene or PS-PEG-Br resin under identical conditions.

To get the real-time reaction information during the course of this reaction, single-bead IR spectra at various times during the reaction were obtained (Figure 5.2). For formylpolystyrene resin (Figure 5.2a), intensities of the C-H stretch band at 2728 cm^{-1} and the carbonyl band at 1700 cm^{-1} from the aldehyde decrease, and intensities of the $-N(CH_3)_2$ stretch band at 2790 cm^{-1} and the N-H band from the hydrazone at 3218 cm^{-1} increase with time. These data unambiguously confirmed that the decrease in hydrazine concentration in supernatant shown in Figure 5.1a is due to its reaction with the resin-bound aldehyde to form hydrazone on resin. Similarly, reaction on PS-PEG resin was followed by the decrease of the intensity of the aldehyde carbonyl band at 1723 cm^{-1} (Figure 5.2b).

This method is simple to perform and accurately quantitates the absolute amount of aldehyde and aliphatic ketone groups on resin. It can be applied to 3 to 10 mg of resin sample (for loading 1.0 to 0.3 mmol/g) containing low micromolar amounts of aldehyde or ketone groups.

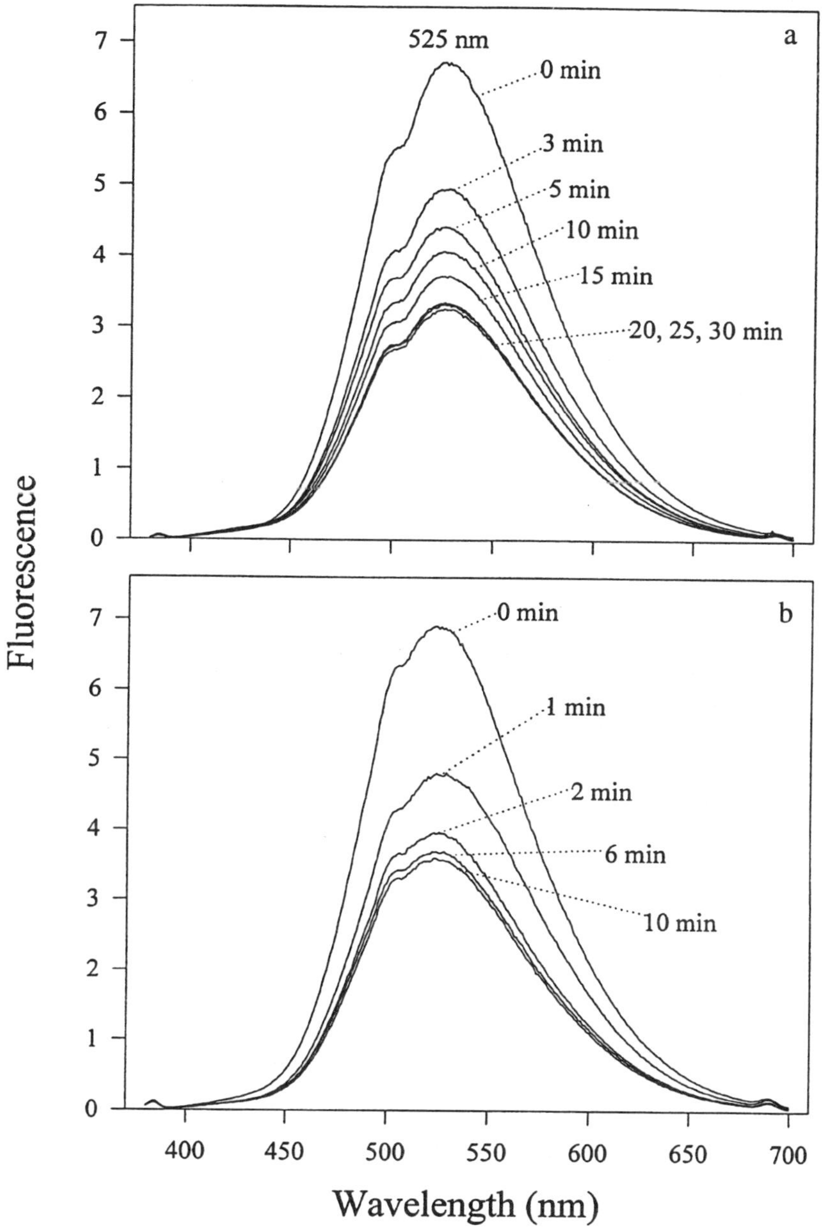

Figure 5.1 *Fluorescence emission spectra of hydrazine in the supernatant of the reaction mixture at various times for (a) formylpolystyrene resin and (b) PS-PEG-CHO resin.*

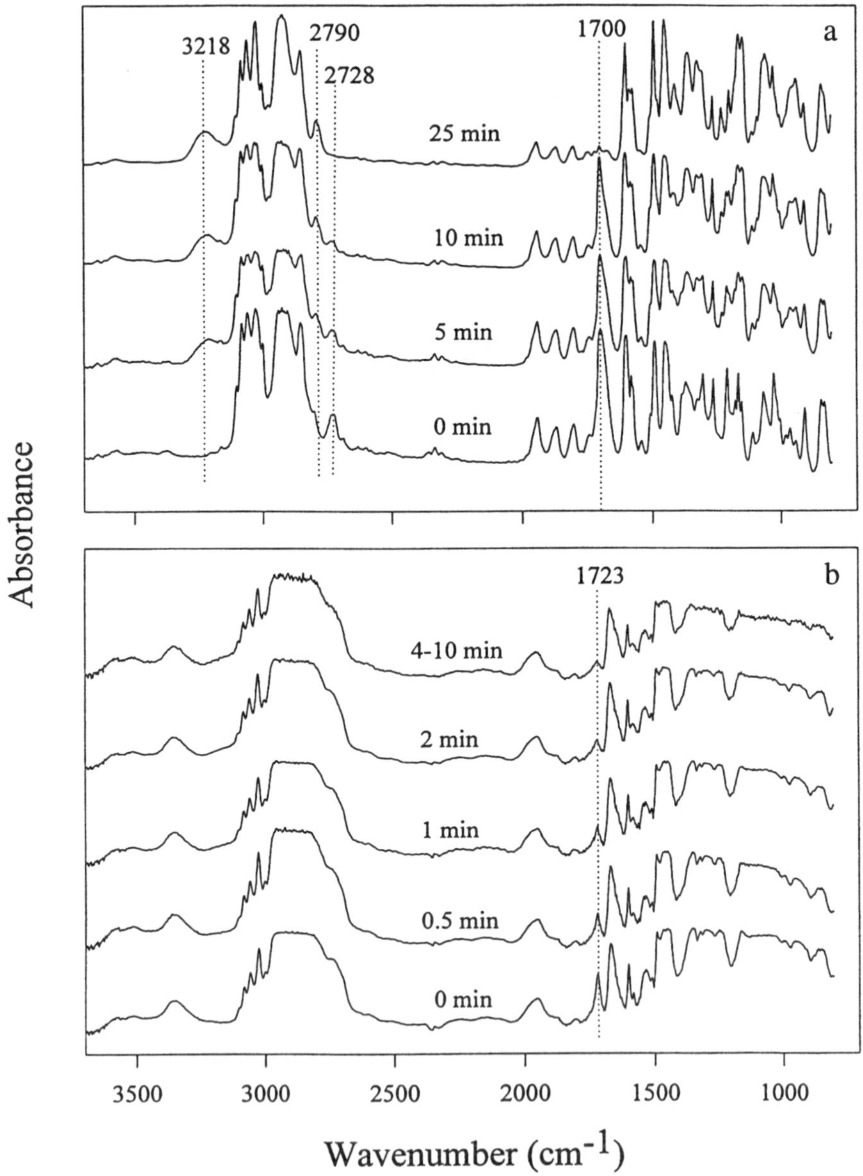

Figure 5.2 *IR spectra of the single bead taken out of the reaction mixture at various times during the reaction for (a) formylpolystyrene resin and (b) PS-PEG-CHO resin.*

5.3.2 Quantitation of Hydroxyl Groups on Resin

Because the UV-Vis spectrometer is more stable, less expensive, and widely available, a UV spectroscopic method for rapid determination of the absolute amount of hydroxyl groups on polystyrene resin using 9-anthroylnitrile (Scheme 5.13) was reported (Yan et al., 1999).

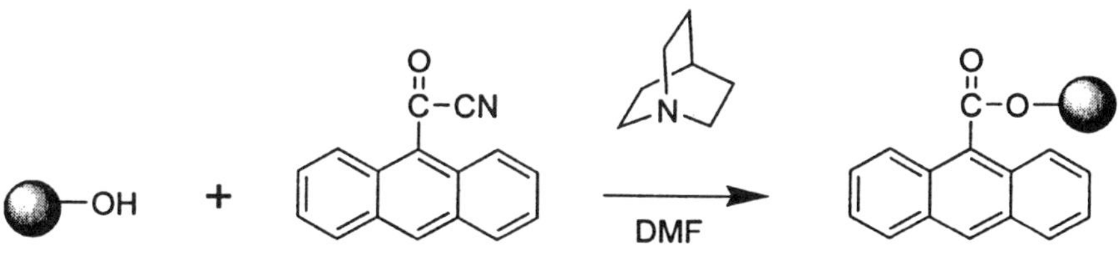

29 **30**

9-anthroylnitrile

Scheme 5.13

An approximately twofold excess of 9-anthroylnitrile in DMF was added to a suspension of Wang resin and the same amount of plain polystyrene resin control (5–9 mg in 0.2 mL DMF with an excess amount of quinuclidine). The UV spectra of the supernatant for polystyrene control and the Wang resin were taken after 20 min. Figure 5.3a shows a decrease in dye concentration in the supernatant. Reaction occurs in the solid phase as shown by the single-bead FTIR spectra taken before and after a 20-min reaction. Figure 5.3b shows the disappearance of hydroxyl bands at 3458 and 3578 cm^{-1} and the formation of a new band at 1722 cm^{-1} after 20 min. The band at 1722 cm^{-1} is attributable to the ester carbonyl in the product **30.** Single-bead IR spectra at various times during the reactions were also measured (Figure 5.4). Intensities of the O-H stretch bands at 3578 cm^{-1} and 3458 cm^{-1} decrease gradually, and the carbonyl band at 1722 cm^{-1} from the product increases with time. The time course again shows that the esterification is complete in 20 min.

The absolute amount of hydroxyl groups in various PS-based samples was then determined by the dye consumption deduced from UV spectroscopic measurements. The amount of dye molecules noncovalently trapped in the resin matrix was simultaneously corrected by subtracting the amount of dye molecules absorbed in the same amount

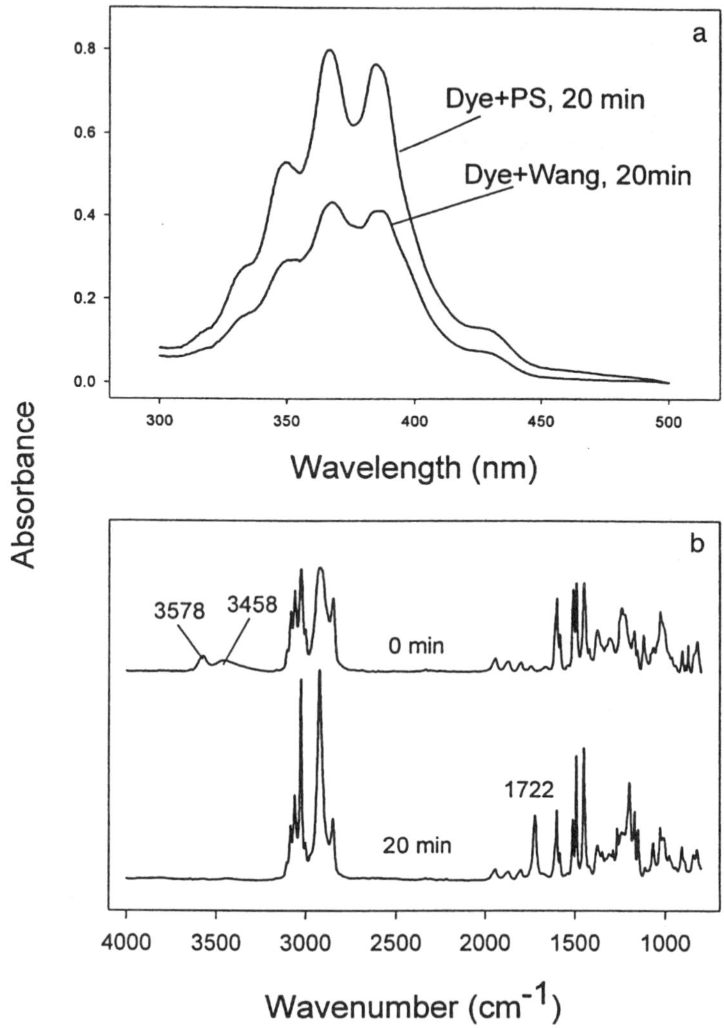

Figure 5.3 *UV-Vis absorption of dye adducts spectra and single-bead FTIR spectra taken from Wang resin. (a) UV-Vis spectra of the supernatant that contains 9-anthroylnitrile-quinuclidine adduct mixed with polystyrene resin or Wang resin for 20 min. A 5 μL of supernatant was diluted into 2 mL DMF for measurement. DMF was used as reference and the pathlength was 1 cm. (b) Single-bead FTIR spectra taken from the starting material Wang resin and the same resin after 20-min reaction with 9-anthroylnitrile. The IR results show the formation of resin-bound ester carbonyl functional group and the disappearance of hydroxyl functional group. All spectra were taken using the transmission mode at room temperature as described in the Experimental Section.*

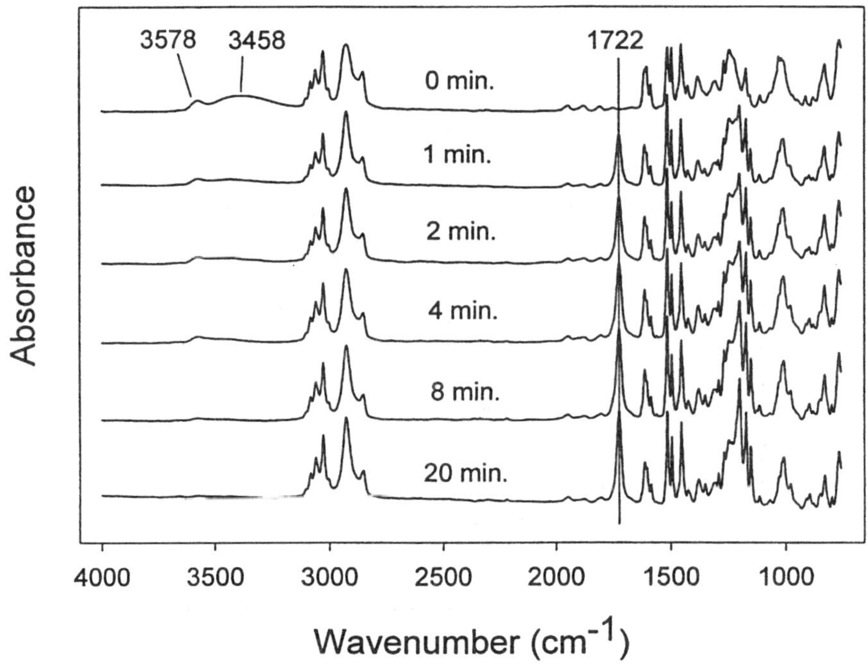

Figure 5.4 *Single-bead FTIR spectra at various times during the reaction shown in Scheme 5.13. Spectra were taken from a single flattened bead at 0, 1, 2, 4, 8, and 20 min after the initiation of the reaction. The hydrogen-bonded and unbounded hydroxyl stretching at 3458 cm^{-1} and 3578 cm^{-1} disappear as the ester carbonyl band at 1722 cm^{-1} increases with time.*

of 1% divinylbenzene polystyrene resin under the identical conditions. The loading of hydroxyl was calculated based on the decreased UV-Vis absorbance. The results agreed with the loading values determined by Fmoc-derivatization method. The advantages of this method are its simplicity, speed, and the requirement of low milligram amount of samples.

5.3.3 Quantitation of Carboxyl Groups on Resin

PDAM reacts with carboxyl groups to form an ester product (Scheme 5.14). In the study on the solid-phase version of this reaction (Yan et al., 1999), the reaction was found to be fast and efficient, allowing the quantitative determination of solid-phase-bound carboxyl groups.

The time course of this reaction was monitored by single-bead FTIR method as shown in Figure 5.5. The IR spectrum of the starting carboxyl

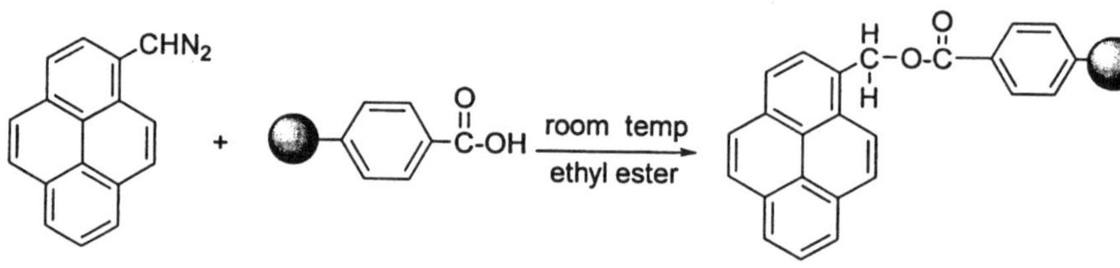

Scheme 5.14

resin (Figure 5.5, spectrum at 0 min) exhibits two carbonyl bands at 1688 and 1722 cm^{-1} corresponding to associated and free carboxyl groups. After reaction for 33 min, the band at 1688 cm^{-1} disappears while an ester carbonyl band in the product appears coincidentally also at 1722 cm^{-1}. Besides these spectral changes, the acid -OH group stretching signals at 2500–3500 cm^{-1} also disappear after reaction for 33 min. Therefore, the reaction was complete in 30–40 min.

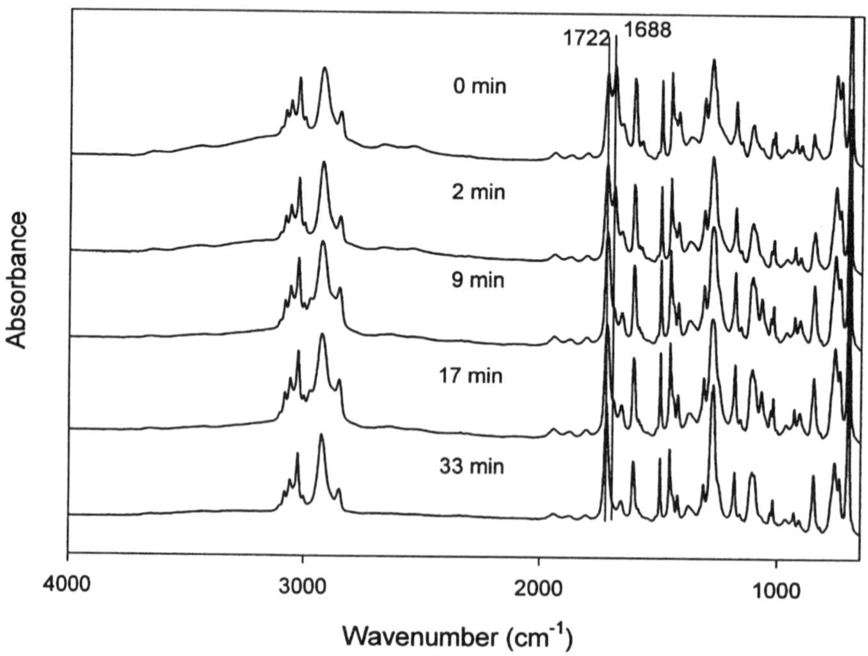

Figure 5.5 *Single-bead IR spectra taken at various times during the reaction in Scheme 5.14.*

In order to ensure the completion of the reaction with carboxyl resins from various sources, all quantitation reactions were carried out for 50 min using an approximately twofold excess of PDAM. To calibrate non-covalently-bound PDAM molecules in the resin, plain polystyrene resin was used as a control. Figure 5.6 shows the UV-Vis absorption spectra of PDAM after mixing with carboxyl resin (lower spectrum) or polystyrene resin (top spectrum) for 50 min. A variety of resins (aminomethyl, prolinol-2-chlorotrityl-, oxime-, Rink acid-, and Wang resins) were tested for their reactivity with PDAM, and no spectral change in the starting resins or formation of ester or amide products were detected by single-bead IR. Phenol resin has low reactivity toward PDAM.

In summary, a method using PDAM and UV-Vis spectroscopy for rapidly quantifying carboxyl groups in both PS- and PS-PEG resins has been developed. Most organic functional groups except phenol do not interfere with the quantitation. The method provides a useful tool for assisting reaction optimization in combinatorial synthesis.

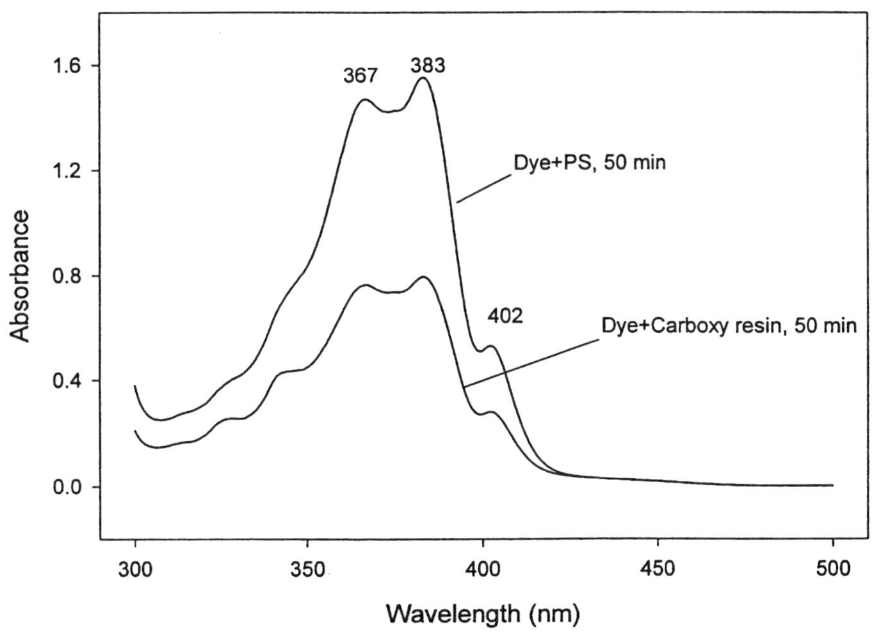

Figure 5.6 *UV-Vis absorption spectra of PDAM in the supernatant obtained before and after the reaction.*

5.3.4 Quantitation of Polymer-Supported Sulfhydryl Groups

A sensitive and simple method is described for the quantitative determination of free sulfhydryl (SH) groups on polymer supports (Sharma et al., 1990). The method includes the reaction of 4,4′-dimethoxytrityloxy-S-(2-thio-5-nitro-pyridyl)-2-mercaptoethane (DTNPME) with polymer-supported sulfhydryl groups. After removal of excess reagent through washing, the polymer support is treated with perchloric acid to release the 4,4′-dimethoxytrityl cation from the polymer support into the solution. The dimethoxytrityl cation (λ_{max} = 498 nm, ε_{498} = 70,000 $M^{-1}cm^{-1}$) is then quantified spectrophotometrically.

A method (Ellman, 1959) has been reported for the qualitative and quantitative determination of sulfhydryl groups in simple (low molecular weight) sulfhydryl (SH) compounds and biological samples. Under certain conditions, this method can be used in the analysis of SPOS resins (e.g., on PS-PEG resins).

5.4 COMBUSTION ELEMENTAL ANALYSIS METHODS

Elemental analysis of resin-bound molecules has often been used in solid-phase synthesis papers (Frechet & Schuerch, 1971; Mitchell et al., 1978; Frechet et al., 1979; N'Guyen & Boileau, 1979; Forman & Sucholeiki, 1995). However, the scope and value of this method is not very clear. A systematic investigation on the role of combustion elemental analysis in the quantitative monitoring of solid-phase organic synthesis was reported (Yan et al., 1998). In this section, quantitative elemental analysis results obtained for eight resin-bound organic compounds are compared with those obtained from well-established spectrophotometric methods (Fmoc method). In addition, a six-step peptoid synthesis and a dansylhydrazone synthesis were quantitatively monitored using elemental analysis. All results were corroborated by well-established spectrophotometric methods.

5.4.1 Analysis of Resin-Bound Organic Compounds

A series of resin-bound compounds were quantitatively analyzed by combustion elemental analysis methods (Diagram 5.2). In a carbon-hydrogen-nitrogen (CHN) analysis, resin compounds were combusted and oxidized completely to form CO_2, H_2O, and N_2, which are then

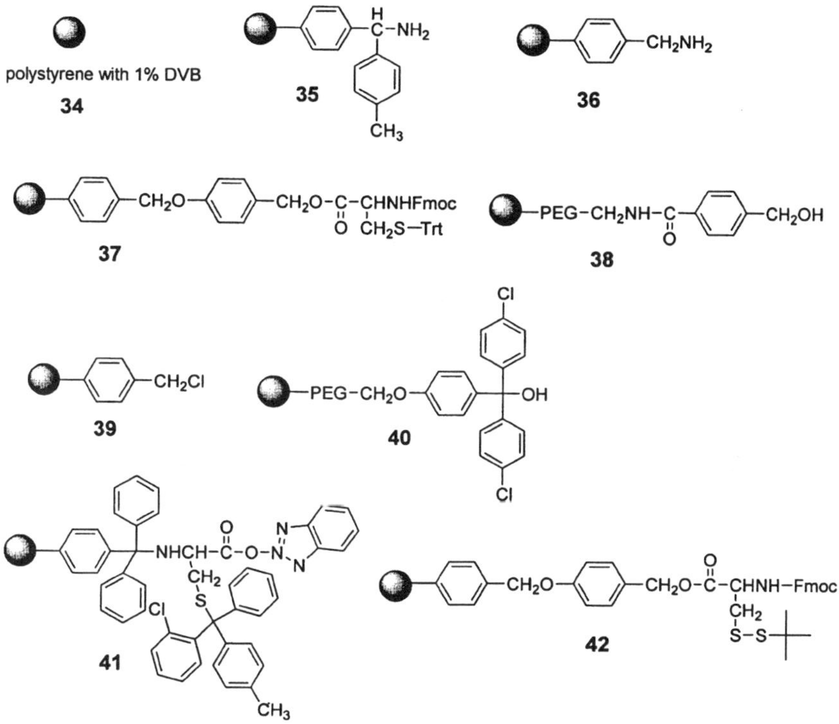

Diagram 5.2

quantitatively determined simultaneously on the basis of their different thermoconductivities relative to the carrier gas helium. All samples for CHN analysis were in reference to the standard acetanilide. 1% Divinylbenzene (DVB)-polystyrene copolymer is the basic matrix for many resins used in SPOS. The theoretical C/H ratio of the plain 1% DVB-polystyrene resin based on the ratio of starting materials used for polymerization is 92.3/7.7. Elemental analysis of this resin in quadruple gave a ratio of 92.0±0.2/7.7.±0.1, showing excellent reproducibility and accuracy. To further investigate the reproducibility of the method, a multiple analysis of compounds **35–38** shown in Diagram 5.2 was carried out for their nitrogen contents (Table 5.1). The reproducibility of the data is very good as shown by the very small variations. It is important to point out that only 2 mg of resin is required for a CHN analysis and each analysis takes less than 20 min in an automated fashion. In routine analysis, a single analysis or a duplicate measurement should be sufficient.

Table 5.1 Elemental analysis of resin-bound compounds: reproducibility and accuracy.

Compd	Weight (mg)	Element	# of expt.	Value (mg/g)	Expected[a] (mg/g)
34	2	C/H	4	920±2/77±1	923/77
35	2	N	4	6.7±0.2	6.44
36	2	N	4	14.4±0.3	14.7
37	2	N	4	9.9±0.2	9.1
38	2	N	4	3.8±0.2	3.9
39a[b]	15–20	Cl	2	36.5±0.01	36.1
39b	15–20	Cl	2	25.7±0.01	26.6
40	15–20	Cl	2	13.6±0.1	13.3
41	15–20	Cl	2	22.9±0.2	22.1
41	20–35	S	2	22.5±0.3	20.2
42	20–35	S	2	34.8±0.4	32.0

Note:[a] Theoretical value for 34 and spectrophotometric measurement after derivatization.
[b] 39a and 39b are from two different batches.

In addition to nitrogen quantitation, a series of resin samples containing sulfur and chlorine (**39–42**) were also analyzed (Table 5.1). All the expected values, including those for nitrogen quantitation mentioned above, came from well-established spectrophotometric analysis, such as analysis of the cleaved fluorenylmethoxy-carbonyl (Fmoc) and N′-hydroxylbenzotriazole (HOBt) groups except that the value for plain polystyrene is based on the ratio of starting materials for polymerization. Standard deviations from the expected value for most samples are below 5%, a level acceptable for quantitative analysis in SPOS.

5.4.2 Quantitatively Monitoring SPOS Reactions

To apply the elemental analysis method in quantitatively monitoring of multistep SPOS, the synthesis of a dansylhydrazone from a resin-bound aldehyde was investigated (Scheme 5.12). The nitrogen content was increased from <0.2 to 22.7 mg/g after the reaction (Table 5.2), showing a 93.4% conversion, which is also confirmed by single-bead IR and a dye-coupling quantitation method.

The successful monitoring of a series of peptoid synthesis reactions (Scheme 5.15) further substantiated the potential of the method for quantitatively monitoring solid-phase organic synthesis. The starting material Rink amide AM resin (**43,** 0.44 mmol/g) was first activated and quantitated by the cleavage of the Fmoc chromophore. The loading of the resulting amine resin is increased to 0.48 mmol/g due to the

Table 5.2 Elemental analysis of resin-bound compounds: reaction monitoring.

Compd	Weight (mg per expt)	t Elemen	Value (vs. theor) (mg/g resin)	Yield (N or Br Analysis)	Yield NPIT
27	2	N	<0.2	—	—
28	2	N	22.7 (24.3)	93.4%[o]	—
43	25–35	Br	—	—	—
44	25–35	Br	0 (0.0)	—	109%[o]
45	25–35	Br	37.1 (36.7)	101%[o]	negative[s]
46	25–35	Br	0.12 (0.0)	97%[s]	100.5%[o]
47	25–35	Br	32.3 (34.9)	92.5%[o]	negative[s]
48	25–35	Br	0.08 (0.0)	complete[s]	78%[o]
49	25–35	Br	24.3 (32.3)	75.2%[o]	negative[s]
50 + 51	25–35	Br	0.02 (0.0)	complete[s]	30%[o]

Note: o—overall yield. s—one-step yield.

altered resin/functional group ratio after the leaving of Fmoc. In all SPOS, it is necessary to re-calculate the loading after each synthetic step due to the changed resin/functional group ratio. The increased weight from the new group was added to the base weight, and the original loading in term of mmol/g was re-calculated after each reaction step. The six-step synthetic reaction was quantitatively monitored by both the bromine quantitative analysis and nitrophenylisothiocyanate-O-trityl (NPIT) quantitation (Chu & Reich, 1995). The disappearance of the primary and the secondary amines was quantitatively correlated with the formation of the brominated compounds as determined by both the NPIT test and the bromine quantitation (Table 5.2).

In the last step, the bromine test suggests a quantitative conversion of **49** to **51**. However, the NPIT test suggests the presence of only partial amine product. In order to investigate the inconsistency, end products were cleaved from resin and analyzed by MS and HPLC. MS confirmed the presence of both **50** and **51**, and HPLC indicated the ratio of **50/51** is 1/1.82. The cause for the formation of the side product **50** is the presence of 6% impurity (phenylethyamine) in the reagent for the last step synthesis. This result indicates that the purity of reagents is crucial when the impurity is more reactive. This type of on-resin analysis effectively helps facilitate the reaction optimization steps required in developing on-resin synthetic processes.

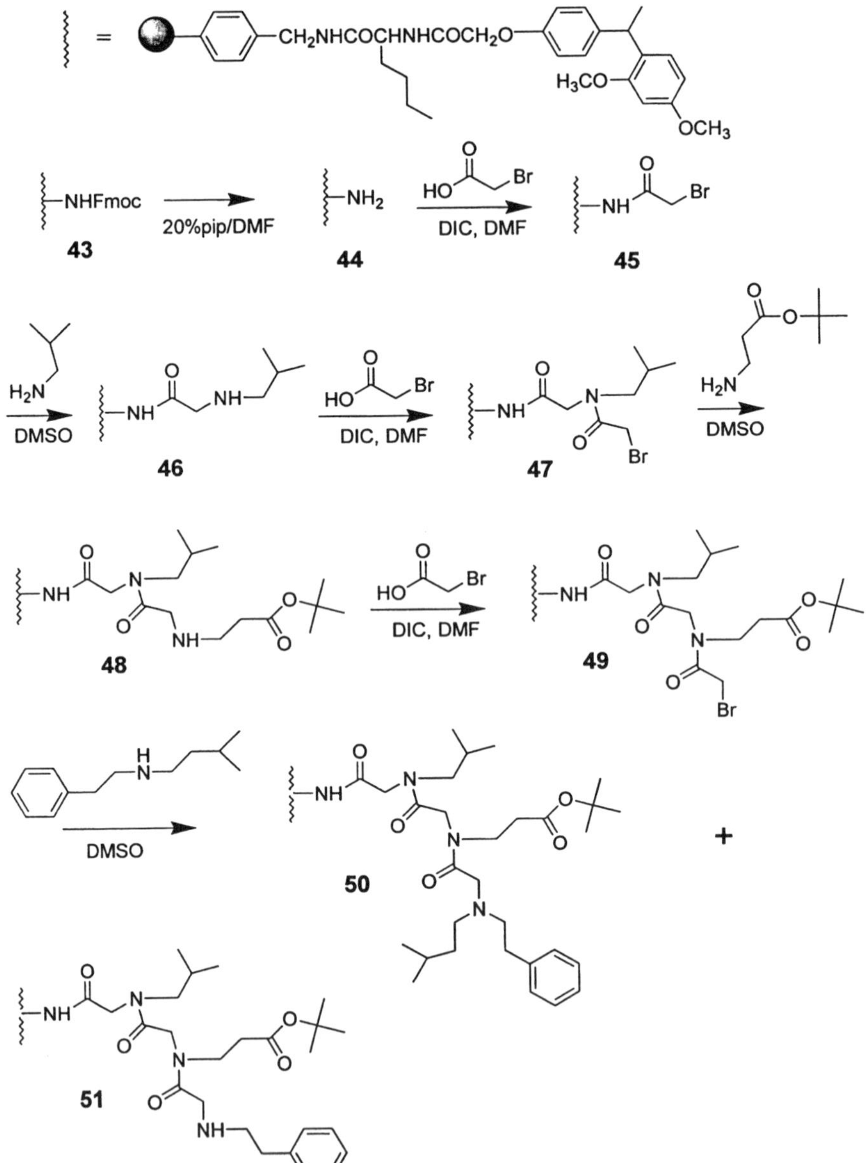

Scheme 5.15

Based on these data, the elemental analysis method is, in general, reproducible and accurate in the quantitative analysis of resin-bound organic compounds. Considering the severe shortage of on-resin quantitation methods at present, it adds a valuable tool for monitoring solid-phase organic synthesis steps that cannot be quantitated by current methods.

It should be emphasized that it is important to wash and dry the sample completely before the elemental analysis, because the resin-trapped reagent or solvent molecules will cause the variable results. In practical considerations, although the combustion elemental analysis method is a destructive method and is not always sample efficient, it will prove certain utility in SPOS.

5.5 ELECTROCHEMICAL METHODS

A noninvasive method for quantitative determination of basic functionality (pK_{HB} + >7) on solid support is described (Patek et al., 1998). This method is based on complete protonation of basic functionality after treatment of the resin with a large excess of 1% $HClO_4$, which provides an easily detectable perchlorate anion. After thorough washing, the bound anion is eluted with a suitable base and is quantified potentiometrically with perchlorate ion-selective electrode (ISE). The amount of basic component is calculated from output potential that is logarithmically proportional to the activity of the selected ion in the sample via the Nernst equation.

Using the mV-meter with resolution of 0.1 mV, one should be able to read changes in concentration of 0.4%. However, for practical purposes, a resolution of 1 mV will suffice. Due to the dependence of electrochemical potential on temperature, samples and standards must be at the same temperature at the instant of the measurement. A 1° C difference in temperature will result in 3% error at the 10^{-4} M level.

It was noticed that there are some limitations in quantitative determination of basic functionality. One obvious problem inherent in this method is a salt hydrolysis effect. Because the procedure for perchlorate determination involves repetitive washings with distilled water, solvolysis of ammonium salt will occur, which is dependent upon the dissociation constant of the protonated base and the autoprotolytic constant (pK_{Ap}) of the solvent used. In general, as can be expected from the corresponding pK_{HB+}'s, most primary, secondary, and tertiary amines, amidines, guanidines, imidazoles, and electron rich dialkylaminobenzenes behave normally in aqueous solution, whereas pyridines and anilines are not suitable basic functionalities for quantitative determination. Notably, indol-type compounds are completely "invisible" with such a technique in an aqueous solution. In addition, attempts to extend the method to aminomethylphenyl and p-methylbenzhydrylamine (MBHA) type of resins was not successful. Unfavorable swelling properties of these resins in aqueous media and a low tolerance of the ISE to organic solvents are the main reasons for the failure of this method.

In conclusion, ISE methodology can be used to quantify a variety of basic functionality and their relative changes during a reaction on solid support. Limitations of this methodology include the use of water-compatible solid supports and basic functionality possessing a pK_{HB+} within a certain range. A significant advantage of the described method is that it is simple and noninvasive when compared to traditional "cleave-and-analyze" techniques. The concept of using ISE to monitor progress of reactions on solid phase offers another attractive tool to solid-phase organic and peptide chemists.

Conventionally, the loading of Merrifield resins is determined by elemental analysis of incorporated chlorine or titration of pyridinium chloride after hot pyridine treatment of Merrifield resin. Because most of the mass of the MicroTube™ reactors or Chiron crowns is polypropylene and the percentage of incorporated chlorine in Merrifield MicroTube™ reactors is extremely small, elemental analysis would not be accurate. Titration is more tedious and requires more Merrifield MicroTube™ reactors for analysis. A potentiometric method that permits a measurement of chloromethylation using chloride ion selective electrode (ISE) was developed (Li et al., 1998). The method involves the formation of pyridinium chloride after hot pyridine treatment of the Merrifield MicroTube™ followed by the quantitation of chloride ion by chloride ion selective electrode. This method is simple, convenient, and very accurate. It gives accurate analytical results in a linear range from 5 to 150 nmol chloride per Merrifield MicroTube™ reactor. Typical Merrifield MicroTube™ reactors have a loading of 30 μmol ± 10% RSD (Relative Standard Deviation).

Another method for monitoring amine in solid-phase peptide synthesis was described (Brunfeldt et al., 1969, 1972). The method is based upon a potentiometric titration of the whole batch of resin-bound peptide suspended in a mixture of methylene chloride and glacial acetic acid 1/1 (v/v). Perchloric acid dissolved in glacial acetic acid was used as titrant. The analysis was done in an automated fashion by connecting to the synthesizer.

Classic organic analysis methods can be used when the resin-bound compounds can be converted into the proper form in solution. One example is the Volhard titration analysis of chlorine content in Merrifield resins (Lu et al., 1981).

5.6 ON-RESIN X-RAY, EPR, AND FLUORESCENCE METHODS

SPOS reactions were monitored by energy-dispersive X-ray spectroscopy (EDS) in the environmental scanning electron microscope (ESEM) (Carlton et al., 1997). EDS in the ESEM has the advantage of minimal sample size, speed, and simplicity because the analyses are performed without special specimen preparation. The progress of a two-step synthetic transformation was followed using EDS-ESEM by the presence of a sulfur and bromine peaks in products of each synthetic step. In another study, the elemental analysis of single-bead-bound library member by EDS was shown to be a reliable and fast method of analysis (Neilly & Hochlowski, 1999). X-ray photoelectron spectroscopy (XPS) has also been used to monitor reactions on solid-phase (Yoo et al., 1999). These x-ray-based methods require the involvement of heteroatom-containing molecules.

Polymer-bound peptide aggregation and its dependence on solvent and resin loading were studied by EPR (Cilli et al., 1997). Model low and highly peptide-loaded resins containing an aggregating sequence were labeled with a paramagnetic amino acid derivative and were studied regarding their solvation behavior in different solvent systems. EPR spectroscopy has allowed the detection of differentiated levels of peptide chain aggregation as a function of solvent and resin loading.

Fluorescent beads were studied using bead suspension in conventional cuvette method (Scott & Balasubramanian, 1997). The range of linearity of on-bead fluorescence is shown dependent on both the extent of loading and the Stokes shift of the fluorophore. Fluorescence quenching was found in this study. However, suspension measurement allows heavy inter-bead quenching. To avoid such complications, fluorescence was recorded from a single bead (Yan et al., 1999). This study unambiguously demonstrated the intra-bead fluorescence self-quenching when the Stokes shift of the fluorophore is small (when the absorption and emission spectra overlap).

5.7 SUMMARY

Lack of quantitation methods is the major weakness of SPOS. The optimization of combinatorial chemistry, however, requires the development of more effective on-support quantitative methods. These methods should have the following features:

1. They target a diverse set of organic functional groups.
2. They have high sensitivity and, therefore, require as small amount of a sample as possible.
3. They are fast and convenient for ordinary laboratories.

Although most organic functional groups can react with a reagent containing an Fmoc group to form an Fmoc derivative, this process takes a long time, requires more samples and two-step reactions, and is not applicable for normal SPOS scale. Dye consumption method has proved to be a useful method for quantitation.

Many classic analytical methods, such as electrochemical methods, EDS, EPR and elemental analysis, can also be used to perform quantitative analysis.

5.8 REFERENCES

Brunfeldt, K., Christensen, T., Villemoes, P. Automatic monitoring of solid phase synthesis of a decapeptide. *FEBS Lett.* 1972, 22, 238.

Brunfeldt, K., Roepstorff, P., Thomsen, J. Process control in the solid phase peptide synthesis by titration of free amino groups. *Acta Chem. Scand.* 1969, 23, 2906.

Campbell, D. A., Bermak, J. C. Solid-phase synthesis of peptidylphosphonates. Rapid fluorescence determination of the absolute amount of aldehyde and ketone groups on resin supports. *J. Am. Chem. Soc.* 1994, 116, 6039.

Carlton, R. A., Orton, E., Lyman, C. E., Roberts, J. E. Qualitative analysis of solid phase synthesis reaction products by X-ray spectrometry. *Microsc. Microanal.* 1997, 3, 520.

Christensen, T. A qualitative test for monitoring coupling completeness in solid phase peptide synthesis using chloranil. *Acta Chemica Scandinavica B,* 1979, 33, 763.

Chu, S. S., Reich, S. H. NPIT: A new reagent for quantitatively monitoring reactions of amines in combinatorial synthesis. *Bioorg & Med Chem. Lett.* 1995, 5, 1053.

Cilli, E. M., Marchetto, R. Schreier, S., Nakaie, C. R. Use of spin label EPR spectra to monitor peptide chain aggregation inside resin beads. *Tetrahedron Lett.* 1997, 38, 517.

Dorman, L. C. A non-destructive method for the determination of coupling reactions in solid phase peptide synthesis. *Tetrahedron Lett.* 1969, 2319.

Ellman, G. L. Tissue sulfhydryl groups. *Arch. Biochem. Biophys.* 1959, 82, 70.

Esko, K., Karlsson, S., Porath, J. A method for determining free amino groups in polymers with particular reference to Merrifield synthesis. *Acta Chem. Scand.* 1968, 22, 3343.

Felix, A. M., Jimenez, M. H. Rapid fluorometric detection for completeness in solid phase coupling reactions. *Anal. Biochem.* 1973, 52, 377.

Fields, G. B., and Noble, R. L. Solid phase peptide synthesis utilizing 9-fluorenylmethoxycarbonyl amino acids. *Int. J. Peptide Protein Res.* 1990, 35, 161.

Fontenot, J. D., Ball, J. M., Miller, M. A., David, C. M., Montelaro, R. C. A survey of potential problems and quality control in peptide synthesis by the fluorenyllmethoxycarbonyl procedure. *Peptide Res.* 1991, 4, 19.

Forman, F. W., Sucholeiki, I. Solid-phase synthesis of biaryls via the stille reaction. *J. Org. Chem.* 1995, 60. 523.

Frechet, J. M. J., de Smet, M. D., Farrall, M. J. Application of phase-transfer catalysis to the chemical modification of cross-linked polystyrene resins. *J. Org. Chem.* 1979, 44, 1774.

Frechet, J. M., Schuerch, C. Solid-phase synthesis of oligosaccharides. I. Preparation of the solid support. Poly[p-(l-propen-3-ol-1-yl)styrene]. *J. Am. Chem. Soc.* 1971, 93, 492.

Garden II, J., Tometsko, A. M. Fluorometric method for quantitative determination of free amine groups in peptide-containing Merrifield resins. *Anal. Biochem.* 1972, 46, 216.

Gisin, B. F. The monitoring of reactions in solid-phase peptide synthesis with picric acid. *Anal. Chim. Acta.* 1972, 58, 248.

Hancock, W. S., Battersby, J. E. A new micro-test for the detection of incomplete coupling reactions in solid-phase peptide synthesis using 2,4,6-trinitrobenzene-sulphonic acid. *Anal. Biochem.* 1976, 71, 260.

Heimer, E. P., Chang, C., Lambros, T., Meienhofer, J. Stable isolated symmetrical anhydrides of N^{α}-9-fluorenylmethyloxycarbonylamino acids in solid-phase peptide synthesis. *Int. J. Peptide Proteins Res.* 1981, 18, 237.

Kaiser, E., Bossinger, C. D., Colescott, R. L., Olsen, D. B. Color test for terminal prolyl residues in the solid-phase synthesis of peptides. *Anal. Chim. Acta.* 1980, 118, 149.

Kaiser, E., Colescott, R. L., Bossinger, C. D., Cook, P. I. Color test for detection of free terminal amino groups in the solid-phase synthesis of peptides. *Anal. Biochem.* 1970, 34, 595.

Krchnak, V., Vagner, J., Safar, P., Lebl, M. Noninvasive continuous monitoring of solid-phase peptide synthesis by acid-base indicator. *Collect. Czech. Chem. Commun.* 1988b, 53, 2542.

Krchnak, V., Vagner, J., Lebl, M. Noninvasive continuous monitoring of solid-phase peptide synthesis by acid-base indicator. *Int. J. Pept. Prot. Res.* 1988a, 32, 415.

Li, R., Xiao, X., Czarink, A. W. Merrifield MicroTube™ reactors for solid phase synthesis. *Tetrahedron Lett.* 1998, 39, 8581.

Li, W., Xiao, X., Czarnik, A. W. Kinetic comparison of amide formation on various crosslinked polystyrene resins. *J. Combi. Chem.* 1999, in press.

Lu, G. S., Mojsov, S., Tam, J. P., Merrifield, R. B. Improved synthesis of 4-alkoxybenzyl alcohol resin. *J. Am. Chem. Soc.* 1981, 46, 3433.

Meienhofer, J., Waki, M., Heimer, E. P., Lambros, T. J., Makofske, R. C., Chang, C. Solid phase synthesis without repetitive acidolysis, preparation of Leucyl-Alanyl-Glycyl-Valine using 9-fluorenylmethyloxycarbonylamino acids. *Int. J. Peptide Protein Res.* 1979, 13, 35.

Mitchell, A. R., Kent, S. B. H., Engelhard, M., Merrifield, R. B. A new synthetic route to *tert*-butyloxycarbonylaminoacyl-4-(oxymethyl)phenylacetamidomethyl-resin, an improved support for solid-phase peptide synthesis. *J. Org. Chem.* 1978, 43, 2845.

Neilly, J. P., Hochlowski, J. E. Elemental analysis of individual combinatorial chemistry library members by energy-dispersive x-ray spectroscopy. *Appl. Spectr.* 1999, 53, 74.

Ngo, T. T. Colorimetric determination of reactive amino groups of a solid support using Traut's and Ellman's reagents. *Appl. Biochem. Biotech.* 1986, 13, 213.

N'Guyen, T. D., Boileau, S. Phase transfer catalysis in the chemical modification of polymers. Part II. *Tetrahedron Lett.* 1979, 28, 2651.

Patek, M., Bildstein, S., Flegelova, Z. Monitoring solid phase reactions with ion-selective electrode. *Tetrahedron Lett.* 1998, 39, 753.

Reddy, M. P., Voelker, P. J. Novel method for monitoring the coupling in solid phase peptide synthesis. *Int. J. Peptide Protein Res.* 1988, 31, 345.

Sarin, V. K., Kent, S. B. H., Tam, J. P., Merrifield, R. B. Quantitative monitoring of solid-phase peptide synthesis by the ninhydrin reaction. *Anal. Biochem.* 1981, 117, 147.

Scott, R. H., Balasubramanian, S. Properties of fluorophores on solid phase resins; Implications for screening, encoding and reaction monitoring. *Bioorg. Med. Chem. Lett.* 1997, 7, 1567.

Shah, A., Pahman, S. S., de Biasi, V., Camiller, P. Development of colorimetric method for the detection of amines bound to solid support. *Anal. Commun.* 1997, 34, 325.

Sharma, P., Sathyanarayana, S., Kumar, P., Gupta, K. C. A spectrophotometric method for the estimation of polymer-supported sulfhydryl groups. *Anal. Biochem.* 1990, 189, 173.

Troll, W., Canan, R. K. A modified photometric ninhydrin method for the analysis of amino and imino acids. *J. Biol. Chem.* 1953, 200, 803.

Tyllianakis, P. E., Kakabakos, S. E., Evangelatos, G. P., Ithakissios, D. S. Direct colorimetric determination of solid-supported functional groups and ligands using bicinchoninic acid. *Anal. Biochem.* 1994, 219, 335.

Vojkovsky, T. Detection of secondary amines on solid phase. *Peptide Research,* 1995, 8, 236.

Wade, J. D., Bedford, J., Sheppard, R. C., Tregear, G. W. DBU as an N^{α}-deprotecting reagent for the fluorenyl-methoxycarbonyl group in continuous flow solid-phase peptide synthesis. *Peptide Res.* 1991, 4, 194.

Yan, B., Jewell Jr., C. F., Myers, S. W. Quantitatively monitoring of solid-phase organic synthesis by combustion elemental analysis. *Tetrahedron.* 1998, 54, 11755.

Yan, B., Li, W. Rapid fluorescence determination of the absolute amount of aldehyde and ketone groups on resin supports. *J. Org. Chem.* 1997, 62, 9354.

Yan, B., Liu, L., Astor, C. A., Tang, Q. Rapid determination of the absolute amount of resin-bound hydroxyl or carboxyl groups for the optimization of solid-phase combinatorial and parallel organic synthesis. 1999, *Anal. Chem.* in press.

Yan, B., Martin, P., Lee, J. Single bead fluorescence spectroscopy: The detection of self quenching. *J. Combin. Chem.* 1999, 1, 78.

Yoo, S.-E., Gong, Y.-D., Seo, J.-S., Sung, M.-M., Lee, S. S., Kim, Y. X-ray photoelectron spectroscopy analysis of solid-phase reactions using 3-Brominated Wang resin. *J. Combin. Chem.* 1999, 1, 177.

6

Quality Control of Combinatorial Libraries

THE CHARACTERIZATION OF all chemical library members may seem unnecessary, because it may be argued that only an examination of the active compounds is necessary. Although it may be economical to do so, the lack of hits may mean the tested compounds are not active or the synthesis is failed in the first place. Such operations could mislead researchers into overestimating the diversity of the molecules they explore. Therefore, it is important to confirm the desired diversity present in a library. The application of MS in such a confirmation study has recently been reviewed (Loo et al., 1996, Swali et al., 1999, Süßmuth & Jung, 1999).

Although biological assays can be carried out on beads directly, in-solution assays are more practical. If the in-solution assay is the approach, the characterization of library members should be performed after cleavage from solid supports. The compound libraries to be analyzed are either discrete compound libraries produced via parallel synthesis or one-bead-one-compound libraries made through combinatorial synthesis, such as split-and-pool synthesis (Furka et al., 1988, 1991; Lam et al., 1991; Houghten et al., 1991). In split-and-pool synthesis, discrete compounds are synthesized at the single-bead level. If the pooled libraries are examined as a whole, the analysis of mixtures is encountered. Various approaches have been developed to analyze discrete compounds and mixtures of compounds.

6.1 ANALYSIS OF DISCRETE COMPOUND LIBRARIES

6.1.1 MS Analysis

The validation of libraries from high-throughput, parallel, solid-phase synthesis is mostly achieved after cleaving compounds from solid support. Therefore, the samples for analysis are in a format similar to libraries made by high-throughput, parallel, solution-phase synthesis. Discrete (parallel synthesis) compound libraries are well suited for high-throughput MS analysis, in contrast to split libraries. First, multiple parallel synthesis should generate a single end product with a known molecular weight (MW) in a specific rack well. Second, these end products are delivered for analysis in solution and in standardized cassettes suitable for automated laboratory handling, e.g., 96-well plates. Third, the sensitivity is rarely a problem for MS from the amount of compounds generated by combinatorial synthesis.

High-throughput electrospray ionization mass spectrometry (ESI-MS) analysis for molecular weight determination of synthetic peptides including characterization of byproducts arising from synthesis has been described (Smart et al., 1996). The data from electrospray mass spectrometry are processed with an algorithm that calculates the contribution of the target peptide and each of the identifiable contaminants to the total ionizable material in a sample of synthetic peptide. All essential data are obtained in less than three minutes per sample. In this work, the advantage of this technique was the rapid analysis of the many hundreds of peptides/peptidomimetics required in systematic structure-activity studies.

The ESI LC/MS interfaces developed in the early 1990s proved to be timely technologies for rapidly providing molecular weight information for diverse samples. The only challenge is that the daunting high-throughput needed (e.g., 50 to 100 K analyses per year) and the resulting data interpretation problem.

An automated open-access flow-injection analysis (FIA) MS system has been developed for the QC of combinatorial libraries (Hegy et al., 1996). Using this system, an analysis of compounds from a 96-well plate can be completed in three hours as of 1996. Three tasks are automatically accomplished in each run: the confirmation of the presence of the expected compound, the detection of any inter-sample carryover, and the automated reporting of results to the chemist's bench.

Enhancements to the FIA-MS system (Gorlach et al., 1998) were specifically designed to help the combinatorial chemist improve the end-product purity by pinpointing the synthetic inefficiencies. A 3-D spectral map of the entire combinatorial chemistry rack assists in reaction optimization by color highlighting contamination ions that are detrimental to overall purity.

The size of some large libraries originally deterred full characterization in a reasonable time period. In this situation, analysis of a statistically relevant subset of the combinatorial library by FIA-MS was done prior to biological testing. However, the recent development of sub-minute duty cycles for FIA-MS can now support complete or near complete analytical survey of even large libraries.

6.1.2 MS-Guided Purification

Although the parallel synthesis (one compound per well) strategy eliminates the deconvolution steps, which is necessary when pooled libraries are screened, parallel synthesis products are usually screened as crude mixtures because purification slows the process of lead discovery. However, screening crude products can produce misleading results due to active side products, unreacted material, or synergistic interaction of multiple components. Data from "active" wells can only be evaluated following resynthesis and purification, and data from "inactive" wells may be unreliable. Although a total purification approach may not be necessary for lead discovery, screening pure compounds is the only sure route to immediate, reliable structure-activity relationships. Therefore, there is a clear advantage to purify individual products obtained from parallel synthesis. Such an effort has recently been reported (Kiplinger et al., 1998). In the approach using serial purification of samples made from parallel synthesis, the speed is a key issue. In this work, a 10 mm × 5 mm C-18 HPLC column was used. Due to the moderate backpressures generated, operation at elevated flows is feasible. The separations were performed at a flow rate of 5 ml/min. While use of shorter HPLC column length sacrifices theoretical plates, the vast majority of separations involving crude reaction mixtures do not require high efficiency, and the reduction in column length with an increased flow allows for sharply increased throughput. This purification procedure produces material of acceptable purity with a good recovery rate, and product compounds are readily quantified by weighing.

Rapid reverse-phase analytical and preparative HPLC methods have been developed for application to parallel-synthesis libraries (Weller et al., 1997). Gradient methods, short columns, and high flow rates allow analysis of more than 300 compounds per day on a single analytical system or purification of up to 200 compounds per day on a single preparative system.

Among other multiple-column HPLC systems, an automated parallel analytical/preparative HPLC/MS workstation (Analyt/PrepLCMS) has been developed to increase the analysis speed for characterizing and purifying combinatorial libraries (Zeng & Kassel, 1998). The system incorporates two columns operated in parallel for both LC/MS analysis and preparative LC/MS purification. This system has a multiple sprayer IonSpray interface to support flows from multiple columns and a valving configuration, which is under complete software control for delivering the crude samples to the two HPLC columns from a single autosampler. By incorporating fast chromatography and mass-based fraction collection on the new parallel Analyt/PrepLCMS system, more than 200 compounds can be characterized per instrument per day, and more than 200 compounds can be purified per instrument per night.

An even higher throughput analysis of compounds synthesized by parallel synthesis has been reported recently (Wang et al., 1998). Compounds can be processed eight at a time using this eight-probe autosampler coupled with electrospray ionization. The system makes it possible to analyze a microtiter plate in approximately 10 minutes and significantly improves the efficiency of analysis as compared to single-probe autosamplers. Eight samples are injected at the same time and are analyzed by a mass spectrometer sequentially in time. Both serial and parallel configurations for the eight-probe autosampler were evaluated. There was no obvious advantage of one configuration over the other.

6.1.3 High-Throughput NMR

There are two reasons to focus on the analysis of solution-phase samples: first, as stated in a previous section, most drug discovery programs perform the critical bioassay studies on solution state (cleaved) compounds, and second, the automated manipulation of liquid samples is much easier (and more reliable) than the automated manipulation of heterogeneous SPOS resin slurries.

Although HT MS has been applied to characterize compounds from combinatorial libraries, NMR has not been used in a similar fashion. There are four reasons for this:

1. The samples are normally dissolved in protonated solvents that generate large background signals that complicate the NMR spectra.
2. The volumes of solute are often too small for conventional NMR spectroscopy (< 500 μL). The extreme case is a "one bead one compound" library.
3. The samples are often in 96-well microtiter plate format, which cannot be used directly by regular NMR probe.
4. Conventional NMR analysis is too slow.

HT NMR will be possible only when these problems are solved. The solvent interference has been overcome by a solvent suppression method called "WET." WET uses a series of four frequency-selective (shaped rf) pulses, each having simultaneous ^{13}C decoupling and each followed by a pulsed field gradient (PFG); the resulting four-element pattern has been optimized to create a highly effective solvent suppression pulse sequence element that has been incorporated into a number of 1D- and 2D-NMR sequences (Smallcombe et al., 1995).

The problem of the small sample volume has been addressed by the development of highly sensitive on-flow LC-NMR probes designed for small volume samples, such as HPLC fractions. The sensitivity of such probes allows them to detect nanomole amount of samples. These developments have elevated LC-NMR into a high-performance technique. This technique is shown as technique 1 in Figure 6.1 as LC-NMR setup. Although it is referred to as "on-flow" NMR, the solvent flow may always be temporarily paused to trap the sample in the detector coil during acquisition to allow averaging to be used to improve sensitivity.

The problem of 96-well plate format and low NMR throughput have been solved by incorporating an HPLC autosampler at the front end of the LC-NMR. This creates an "autosampled LC-NMR" (technique 2 in Figure 6.1). The critical aspect in these two formats is the delivering of a sample to the NMR probe by using a flowing liquid.

Removing the HPLC column simplifies the operation. The autosampler injects an aliquot of sample from a 96-well plate. The solvent will sweep it through the tubing and into the NMR probe (technique 3 in Figure 6.1). Signal can be averaged in the NMR coil. This process, when combined with certain scan technique and optional locking and shimming, is quite fast. A ^{1}H NMR spectrum requires one minute of signal averaging (32 scans), and the entire "deliver/acquire/flat-out" process for each sample may take three minutes.

The system in technique 3 can be further simplified by removing the HPLC pump and the mobile phase (technique 4 in Figure 6.1). In this direct-injection system, the autosampler takes a sample aliquot from a 96-well plate (100–300 μL) and, by using the injector, loads it directly into the detector cell of the NMR probe—without the use of any mobile phase or dilution. After completion, the process is reversed to return the intact sample directly back into the microtiter plate. This technique was reviewed by Keifer (1997, 1998). Examples of spectra recorded using techniques 3 and 4 are shown in Figure 6.2.

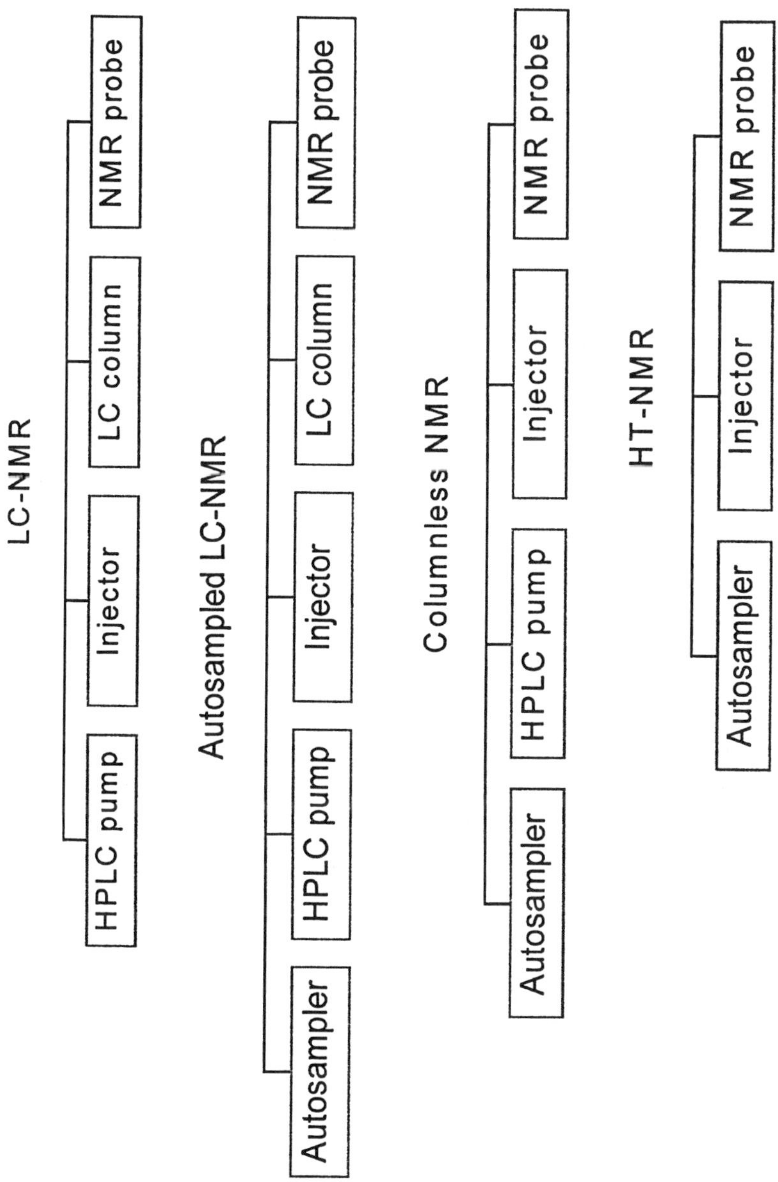

Figure 6.1 *Scheme showing the evolution of HT-NMR spectroscopy.*

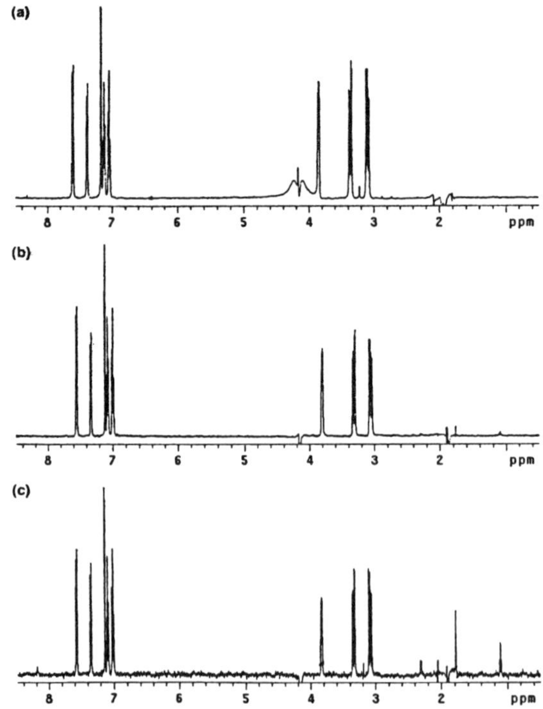

Figure 6.2 *Direct-injection and "on-flow" ${}^{1}H$ spectra acquired on tryptophan solution (1 mg/mL). WET solvent suppression was used for both CH_3CN resonance (at 1.95 ppm) and for the $D_2O/HOD/H_2O$ resonance (at 4.18 ppm); WET also selectively suppressed the ${}^{13}C$ satellites of acetonitrile. The top spectrum (a) was acquired with HT NMR, with the tryptophan dissolved in a fully protonated solvent mixture (1 mg/mL tryptophan dissolved in 50:50 $CH_3CN:H_2O$). This is possible because WET suppression does not require the NMR to maintain a ${}^{2}H$ "lock" for stability of the magnetic field (so a deuterated solvent is not required) and WET is fully capable of high-quality H_2O-signal suppression. The middle spectrum (b) was also acquired by HT NMR to inject a tryptophan solution directly into the probe (and return it directly to the sample container when finished), but a different tryptophan solution (1 mg/mL in 50:50 $CH_3CN:D_2O$) was used. The bottom spectrum (c) was acquired using "columnless" LC-NMR, with 100 μL of tryptophan solution (1 mg/mL in 50:50 $CH_3CN:D_2O$). In "columnless" LC-NMR, the mobile-phase sweeps the sample from the injector loop into the NMR probe for detection, then rinses out the NMR probe at the end of signal averaging (analogous to a "stopped-flow" version of a "carrier gas" in GC). All spectra were acquired at 500 MHz, using 1 min acquisitions (32 scans each), without any form of chromatographic separation. (Reproduced with permission from* Drug Discovery Today. *1997, 2, 468–478. Copyright© 1997. Elsevier Science Ltd.)*

Techniques 1 and 2 have the advantage of being able to provide chromatographic separation for mixtures, but only at a disadvantage in increase in analysis time. Techniques 3 and 4 use the opposite approach—no chromatographic separation—so the analysis is fast. Technique 4 produces higher sensitivity than "columnless NMR" because there is no dilution. Because a sample can be analyzed in less than three minutes, this technique can be viewed as the first generation of HT NMR spectroscopy. This HT NMR spectra acquisition will make the spectra interpretation the next bottleneck. However, recently developed NMR techniques, such as proton chemical shift spectra from filtered J-spectra, reference deconvolution, spectral calculations, and automated spectral analysis, offer the potential that a fully automated real-time microtiter plate NMR acquisition and data analysis system for combinatorial chemistry libraries might some day be developed.

6.2 ANALYSIS OF POOLED LIBRARIES

Combinatorial synthesis using split and pool protocol (Furka et al., 1988, 1991; Lam et al., 1991; Houghten et al., 1991) produces one-bead-one-compound libraries. Although discrete compounds are actually made at the single-bead level, the analysis of a mixture is encountered when the library is to be analyzed as a pool.

Characterization of a pooled library is not trivial. Any destructive analytical methods will not be useful because, theoretically, only one bead carries one unique compound. The huge number of compounds in the final library made by this method precludes the possibility of a full characterization. The focus has to be on the synthetic route optimization and the characterization of small-size trial libraries.

6.2.1 Theoretical Calculation of Mass Distribution

The feasibility of probing combinatorial library diversity by MS is addressed theoretically using computer-generated mass distributions of combinatorial synthesized peptide libraries containing between two and seven amino acids (Demirev & Zubarev, 1997). The behavior of several "global" (integral) parameters of such mass distributions-mass centroid, dispersion, skewness, and kurtosis was studied. The centroid and dispersion are shown to carry information that may characterize the completeness of the synthetic effort.

Based on determining global (integral) as well as local characteristics of the overall mass distributions of all individual library components, the implementation of the computer simulation seems feasible for libraries containing from several hundred up to several thousand individual components. Such an approach can also be useful for MS monitoring the diversity of reaction products containing around 10^3 individual members at each subsequent step along the combinatorial synthetic pathway. Appropriate statistical criteria still have to be formulated in order to prescribe the adequate sampling rate, resulting in a predetermined tolerance limit within a given confidence interval.

Molecules have different ESI-MS sensitivities depending on the functional groups present. It is beneficial to examine both a positive ion and negative ion spectrum to obtain a more realistic picture of the components. In summary, for larger libraries where mass resolution of each individual component is prohibitive, one can compare a calculated theoretical mass distribution with the measured spectrum to gain insight. The shape and position of the unresolved distribution provide useful information regarding the success of the synthesis.

6.2.2 MS Analysis

In combinatorial synthesis, the formation of byproducts during the synthesis cannot be completely avoided. Major reasons for the occurrence of byproducts, for example, are incomplete coupling or deprotection steps and undesirable oxidation by the oxygen in the air. Because the reaction is applied not to a single compound but to an entire population of different substrate molecules, there is always the potential of encountering some selectivity toward the generation of certain molecules at the expense of others. Impurities or missing components in library mixtures used in biological assays can lead to misinterpretation of the obtained results. Therefore, quality control of pooled libraries must accomplish two tasks:

1. To confirm whether all of the expected compounds were produced during the combinatorial synthesis
2. To identify byproducts.

The analysis of compound mixture library is not so straightforward. Techniques such as NMR, IR, UV, and fluorescence spectroscopy cannot provide meaningful structural information due to severe peak overlaps. However, ESI-MS and matrix-assisted laser desorption/time of flight mass spectrometry (MALDI-TOF MS) produce essentially fragmentation-free spectra with a pseudomolecular ion peak for every component. They have been widely used for mixture analysis.

A general limitation in the use of common ionization techniques for the analysis of mixtures is that ionization of some components in a mixture may be partially or completely suppressed, even though the same compounds when analyzed alone give intense molecular ions.

ESI-MS was used to establish that a sizable fraction of the expected compounds are produced in the synthesis of compound libraries (Dunayevskiy et al., 1995). Experiments were designed to determine if the combination of a xanthene acid chloride core with a set of amines results in the formation of a mixture in which all of the expected compounds are present above a certain concentration threshold. Direct observation of 65,341 compounds in the final library was out of the question, and even a library of 136 compounds (only four building blocks) appeared daunting in size. Therefore, smaller yet representative trial libraries were constructed. In a library of 55 expected compounds, 43 (78%) were detected. Missing were predominantly those compounds that contained the arginine-p-benzylamide building block (Dunayevskiy et al., 1995).

Even though the analysis of mixtures with a diversity of 10^4 to 10^5 is not yet possible, the preparation of specifically designed model libraries and their investigation by ESI mass spectrometry have proven to be a useful technique for the analysis of mixtures of structurally related compounds. With the mass-spectrometric data in hand, one is confident that the synthetic methodology produced highly diverse libraries of well-defined composition, and one is able to put faith in the results of subsequent screening assays.

Because MS identifies the synthetic compounds based on the molecular mass, the presence of a large number of isobaric isomers in a mixture is the major problem. In a 48-component octapeptide library with three variable amino acids, only 16 different mass values were detected (Metzger et al., 1994). No matter that it is a small pool or a large pool of compounds, no structural information can be extracted if isobaric problems exist. Coupling MS to other separation techniques or the use of high-resolution MS or tandem MS are required to resolve isobaric mixtures in small-library analysis.

For a compound mixture, mass peaks within the m/z range for the quasi-molecular ions of the lightest and heaviest peptides of the mixture are expected. The presence of peaks at lower or higher m/z values immediately indicates byproducts (for example, deletion peptides or peptides containing uncleaved side-chain protecting groups). In this way, it is possible to rapidly reveal byproducts even in libraries containing billions of peptides.

Compound mixtures can be conveniently introduced into the ESI source via an autosampler (60 samples per hour). Experimental data are then compared with the characteristic mass and intensity distribution of the library or sublibrary with the interactive help of a computer program that predicts mass distribution of a compound library. This kind of program, as discussed in Section 6.2.1, is based on a linear relationship between the number of isobaric peptides and intensity. Byproducts can be revealed very quickly. Structural information on byproducts can be obtained by using tandem mass spectrometry with daughter ion, parent ion, or constant neutral loss scans. For smaller libraries, on-line HPLC-MS helps to distinguish isobaric peptides or to identify individual components in the mixture.

ESI-MS, tandem MS, and on-line RP-HPLC-ESI-MS were used to evaluate the composition and purity of three different aryl ether mixtures consisting of 10 and 45 aryl ethers synthesized on solid support by Williamson etherification (Haap et al., 1997). The library features two potential pharmacophores connected with three different spacers and serves as a model for a detailed component analysis. When libraries were made by two different synthetic methods, the split-and-pool method and the premix method, different mass distributions in the ESI-MS spectra were observed. Components that could not be detected by ESI-MS method were found by MS/MS experiments. Parent ion and constant neutral loss scans allowed the identification of components with common structural features.

The incorporation of the high-throughput LC-MS method (Daley et al., 1997) further increased the efficiency of the analysis of pooled compounds. For example, if 10 96-well plates are logged in, and each well contains 10 components, then 9600 compounds can be confirmed in 16 hours.

A quality-control analysis of an encoded combinatorial library produced by split synthesis was reported (Lewis et al., 1998). Because the library was made to have at least three copies of each component, the randomly selected individual beads were first photo-cleaved for analysis with capillary HPLC/ESI-MS analysis. Then, the dialkylamine tags used to record the synthesis steps were chemically cleaved from the bead and were decoded by HPLC with fluorescence detection. In this way, they could compare the expected products from the decoding results with the detected products by LC-MS. These results confirmed the success of the library synthesis and allowed chemists to carry this library confidently into high-throughput screening for biological activity.

6.2.3 High-Resolution MS

Many times, however, two or more components in a mixture cannot be unambiguously identified because they have nearly identical masses that cannot be distinguished by conventional mass spectrometers. For example, the substitution of lysine for glutamine in a peptide yields a molecular weight difference of 0.036 mass units.

ESI Fourier transform mass spectrometry (FTMS) can be used to address such common challenges in characterizing complex mixtures (Winger & Campana, 1996). For example, a phenylalanine-substituted octapeptide ion (F) at M/Z 964 had a measured mass (963.5346) that was in error by 17 ppm.

The utility of Fourier transform ion cyclotron resonance (FTICR) mass spectrometry for analysis of selected small-peptide combinatorial libraries containing 10^2 to 10^4 individual components was demonstrated (Nawrocki et al., 1996). The FTICR technique offers many advantages over conventional mass spectrometric methods. These include ultra-high mass resolution, high mass accuracy, multi-stage tandem mass spectrometry (MS^n), and the ability to trap ions for extended periods of time. As libraries become larger and more diverse, it will be increasingly difficult to separate individual peaks with the same nominal mass-to-charge ratio. However, with the high mass resolution afforded by the FTICR technique, it is possible to mass resolve individual compounds in isobaric mixtures. The high mass-resolution spectra obtained by FTICR MS can reveal details about libraries that would otherwise be obscured in data from lower resolution mass spectrometric techniques. A study of pooled libraries with 36, 78, and 120 components showed that 70–80% of the library components can be identified without separation (Fang et al., 1998). Furthermore, the FTICR technique can be combined with collisionally activated dissociation (CAD) or infrared multiphoton dissociation (IRMPD) of isolated species to yield a range of information about the structures of individual components. The mass resolution can further be enhanced because the mass resolution increases linearly with the field strength of the magnet.

6.2.4 Tandem MS

Small libraries are often constructed as a model study for final synthesis of a large library. The quality of such libraries can be assessed by ESI-MS and tandem MS. A study of xanthene derivatives clearly

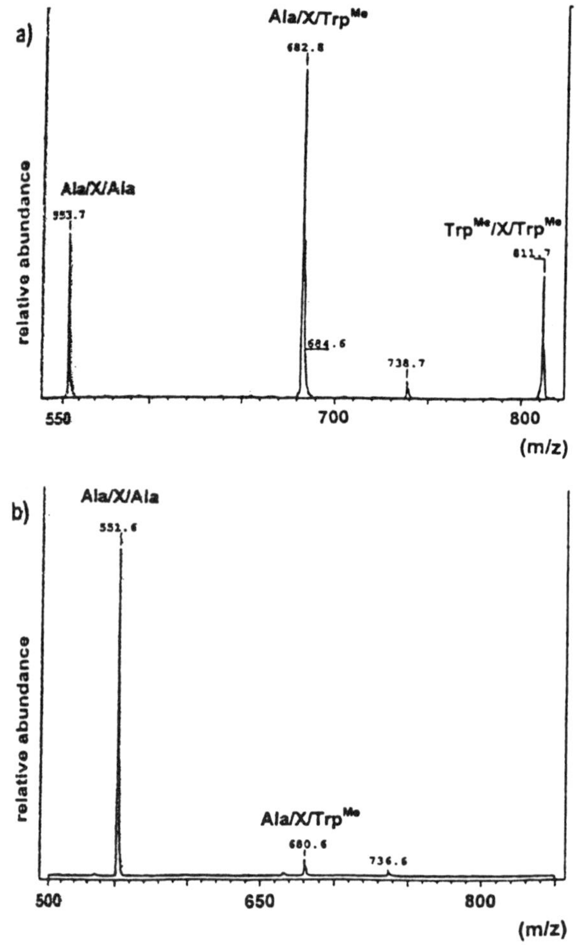

Figure 6.3 *Mass spectra of the mixture of Trp^{Ma}/X/Trp^{Me}, Ala/X/Trp^{Me}, and Ala/X/Ala. (a) Positive ion and (b) negative ion ESI. (Reproduced with permission from* Anal. Chem. *1995, 67, 2909. Copyright© 1995, American Chemical Society.)*

shows the complementary roles of positive and negative ion ESI analysis (Dunayevskiy et al., 1995). Figure 6.3 compares the mass spectrum of the same set of compounds differing in acidic groups on the molecule in positive and negative ion mode. The compound lacking acidic residues, such as TrpMe/X/TrpMe, disappears completely in the negative ion spectrum. In general, the presence of those compounds, which were obscured in positive (negative) ion mass spectra, could be verified by negative (positive) ion detection.

A tandem mass spectrometer consists of an ion source, two analyzers separated by a fragmentation region, and an ion detector. The ionization of the mixture produces ions characteristic of the individual components. The parent ion undergoes collisionally activated dissociation (CAD) through collisions with neutral gas molecules in the fragmentation region to yield various daughter ions. Subsequent mass analysis of the daughter ions by the second mass analyzer permits identification of the separated components.

Although the specific operating mode chosen for a particular analysis will depend on the information desired, the three most common MS/MS operating modes of tandem mass spectrometer are daughter ion scan, parent ion scan, and neutral loss scan. A daughter ion scan consists of selecting a parent ion characteristic of the analyte by the first mass analyzer, fragmenting it by CAD in the fragmentation region, and scanning the second mass analyzer to obtain a daughter mass spectrum. Analogous to a normal mass spectrum, the daughter mass spectrum can be used for identification of an analyte by standard mass spectral interpretation. In the parent ion scan mode, a specific daughter ion is selected with the second mass analyzer, and the first mass analyzer is scanned over a specific mass range, selecting parent ions of different m/z that fragment to yield the specific daughter ion. This mode can be used to screen for a class of compounds that fragments to yield a common substructure. In a neutral loss scan, both mass analyzers are scanned with constant difference in mass. The resulting neutral loss spectrum contains the daughter ions that arise from the loss of a specific neutral fragment from the parent ion. This mode is useful for screening for a class of compounds characterized by a common fragmentation pathway.

The accurate mass and MS/MS data can be rapidly and easily obtained by an automated MALDI FT MS procedure for multiple samples. Multiple ionization techniques can also be used simultaneously in order to introduce sample and mass reference ions into the FTICR cell. In this way, 20 samples from a combinatorial libraries were analyzed automatically in one hour, with errors of less than 5 ppm for the masses of both parent and fragment (Tutko et al., 1998).

A "basket in a basket" method based on a multistage accurate mass spectrometric (MAMS) technique was developed and demonstrated (Wu, 1998) by obtaining a unique elemental composition of a compound (with a molecular weight of 517) from combinatorial synthesis.

The accurate masses for the parent and the fragment ions were obtained with up to five stages of MAMS using FTICR-MS.

If the higher mass ions are dissociated into smaller fragments, it is possible to use the uniquely defined compositions of these fragment ions to "reconstruct" the elemental compositions of the parent ions. Not only the composition of the smaller fragments served to determine those for the larger fragments, but also the composition of these large fragments (larger rings) set the upper limit for the next generation of ions. This is similar to putting a smaller basket into a bigger one—a mutual confinement would occur so that only limited shapes of the baskets would fit into each other. Clearly, such a "basket in a basket" approach can be best implemented by combining multistage fragmentation techniques (MS^n) with accurate mass measurements, i.e., using multistage accurate mass spectrometry (MAMS). The "basket in a basket" approach also greatly reduces mass accuracy requirements because the original larger ions were dissociated into smaller fragments, which could be determined using only a moderate mass accuracy.

The coupling of a reliable nanoelectrospray ionization technique with MAMS is important for its practical usage. In general, the amount of sample available for analytical work is limited, at least in the lead generation stage of combinatorial chemistry. There have been many cases where biological activities came from unexpected products of the combinatorial synthesis. In such cases, structural determination becomes extremely important, while the amount of sample available for analysis is usually small. In the experiments using from MS^2 to MS^5 (Wu, 1998), a total of 3 μL of the solution was loaded into the flow injection source, i.e., the amount of the analyte compound injected was ~30 pmol. Such experiments would be impossible without the nanoelectrospray technique.

6.2.5 Liquid Chromatography (LC)/MS and Capillary Electrophoresis (CE)/MS

6.2.5.1 LC-MS

The combination of HPLC with MS divides a complicated mixture library into many small mixture groups, and identification of compounds becomes easier (Metzger et al., 1993). For an octapeptide library of 48 components, individual components cannot be identified from the MS spectrum. An LC-MS measurement was performed, and the data are shown in Figure 6.4. Panel a) displays the HPLC trace of all molecular ions, and Panel b) shows only components with the mass value of 1040.6. Due to the isobaric isomers, at least eight components have the mass value of 1040.6. In each fraction at various retention

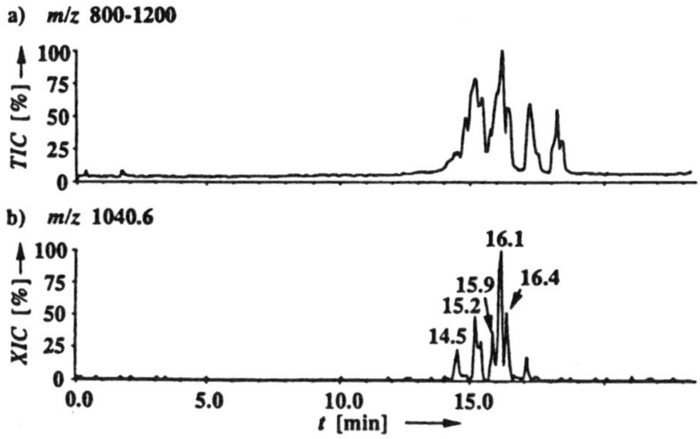

Figure 6.4 *HPLC-MS of a peptide library. (a) Reconstructed total ion chromatogram (m/z 800–1200) and (b) ion chromatogram of m/z 1040.6 corresponding to the monoisotopic (M + H)*$^+$ *ions of eight isobaric peptides of the mixture (TIC: total ion current; XIC: extracted ion current). (Reproduced with permission from Angew.* Chem. Int. Ed. Engl. *1993, 32, 894–896. Copyright© 1993, WILEY-VCH Verlag Gmbh.)*

times, multiple compounds were detected. This example demonstrates how the complexity of the mixture library can be resolved/reduced using LC-MS techniques.

LC/ESI MS with selected-ion monitoring makes possible the visualization and positive identification of a family of isobaric ions from libraries comprising hundreds of peptides (Lambert et al., 1997). For larger libraries (up to 10^5 peptides), a given peptide, known to be unique at a given m/z, was detected, but the full characterization of others at other m/z values from the library is mainly limited by LC separation efficiency. However, this approach offers new possibilities to evaluate the completeness of combinatorial libraries.

6.2.5.2 CE-MS

Capillary electrophoresis (CE) generates rapid, high-resolution separations based on differences in the electrophoretic mobility of charged species in an electric field in small-diameter fused-silica capillaries. CE is mostly suitable for compounds containing groups with diverse acid/base properties. Application of CE before mass spectrometric detection provides fast and efficient separation of library components for their subsequent characterization by ESI-MS. When voltage is applied across the capillary, electroosmotic flow (EOF) moves the sample from the positive end (anode) to the negative end (cathode). EOF becomes significant at pH values >5 and drives the analytes toward the cathode, regardless of their charge. The mobility of the library compounds and, therefore, their migration times are determined by the difference between the mobility of the EOF and the electrophoretic mobility of the species. Buffer composition, pH, and ionic strength can be varied to optimize electrophoretic separation. The best electrophoretic condition could be chosen to separate a complex mixture into several groups of analytes, each migrating according to their charge. Within each such group, one can expect significantly fewer isobaric compounds compared with the unresolved library mass spectrum. A small model library of xanthene derivatives (Diagram 6.1) was analyzed by CE-ESI MS to test the possibility of making a large library based on xanthene core (Dunayevskiy et al., 1996). Figure 6.5 shows the electropherogram of the library and extracted ion electropherograms with selected ion current. In this characterization, 160 of 171 compounds in the library were identified.

With reversed-phase, high-performance liquid chromatography (HPLC), separation of physicochemical diverse mixtures presents a significant challenge. The separation in CE is based on differences in charge and size and provides significantly higher efficiencies than HPLC. The applicability of CE as a separation tool for N-(substituted)-glycine-peptoid combinatorial mixtures was demonstrated (Robinson et al., 1998). A multiple-background electrolyte additive strategy provided enhanced selectivity over the additives used individually. An ion-pairing agent, 1-heptane sulphonic acid (HSA), presumably reduced the propensity to form strong intramolecular interactions between the peptoids and disrupted electrostatic interactions with the capillary wall. Methyl-β-cyclodextran was used as a partitioning component to the separation to aid in the resolution of the more hydrophobic peptoids. The addition of organic solvents to the background electrolyte, alone and in conjunction with HSA, has been shown to enhance the separation of peptoids.

CE offers potential advantages over traditional HPLC for the analysis of complex peptoid mixtures. Only nanoliters of sample are injected for each analysis, and as little as 5 μL of sample is needed for reproducible injection. This can be a tremendous advantage when extremely small amounts of sample are available, as is typical with combinatorial chemistry. Separations are inherently based on charge-to-size ratios, and partitioning can be incorporated, providing greater flexibility in developing a separation.

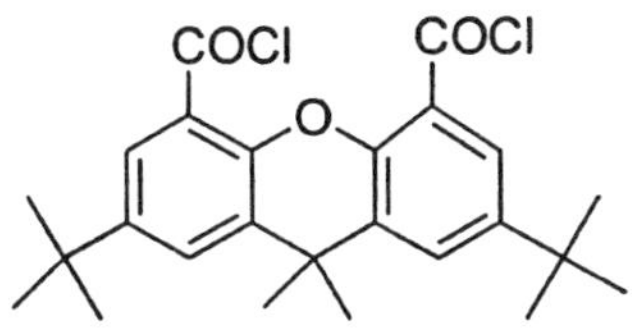

Amino Acid	Protected Reagent Used
1. (Gly)-OMe	Glycine-methyl ester
2. (Ala)	L-alanine-tert-butyl ester
3. (Ser)	O-tert-butyl-L-serine-tert-butyl ester
4. (Pro)	L-proline-tert-butyl ester
5. (Val)	L-valine-tert-butyl ester
6. (Leu)	L-leucine-tert-butyl ester
7. (Ile)	L-isoleucine-tert-butyl ester
8. (Asn)	L-asparagine-tert-butyl ester
9. (Asp)	L-aspartic acid-ß-tert-butyl-a-tert-butyl ester
10. (Thr)-OMe	O-tert-butyl-L-threonine-methyl ester
11. (Glu)	L-glutamic acid-γ-tert-butyl-α-tert-butyl ester
12. (His)	N^{im}-trity-L-histidine
13. (Lys)-OMe	N^{E}-Boc-L-lysine-methyl ester
14. (Met)-OMe	L-methionine-methyl ester
15. (Phe)	L-phenylalanine-tert-butyl ester
16. (Arg)	N^{g}-4-methoxy-2,3,6-trimethylbenzene-sulfonyl-L-arginine
17. (Trp)-OMe	L-tryptophane-methyl ester
18. (Tyr)-OMe	O-tert-butyl-L-tyrosine-methyl ester

Diagram 6.1

6.2.6 NMR Methods for Mixture Analysis

In order to achieve purification and structural identification simultaneously for compound mixtures from either the natural sources or from split synthesis, HPLC-NMR has been studied as a useful analytical tool.

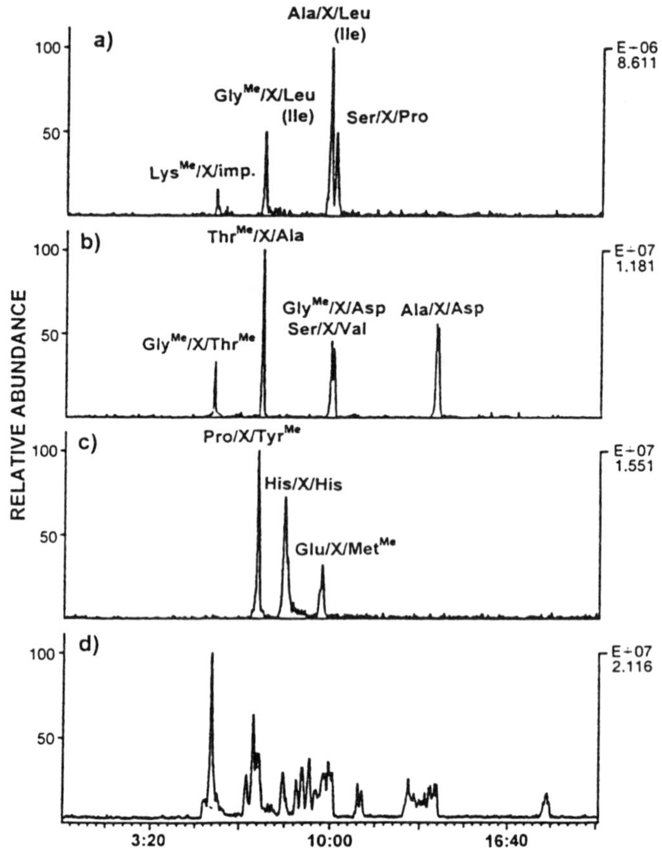

Figure 6.5 *CE-MS electropherogram of a library with extracted ion electropherograms. (a) m/z = 595.5; (b) m/z = 597.5; (c) m/z = 685.5; (d) Total ion current electropherogram. (Reproduced with permission from* Proc. Natl. Acad. Sci. USA. *1996, 93, 6155. Copyright© 1996, National Academy of Sciences, USA.)*

In HPLC-NMR operations, it is possible to achieve ^{1}H NMR spectroscopic detection of HPLC elution in three modes of operation:

1. Direct on-flow detection of NMR spectra in real time is possible in favorable circumstances.
2. Stopped-flow analysis allows the full range of NMR experiments to be performed on a defined chromatographic fraction.
3. It is possible to collect fractions of the HPLC eluate into capillary loops for later off-line full NMR spectroscopic evaluation.

A model system based on a mixture of 27 tripeptide combinations of alanine, methionine, and tyrosine as the C-terminal amides has been studied (Lindon et al., 1995). The ^{1}H NMR spectra were obtained in real time using on-flow HPLC-NMR spectroscopy at 600 MHz, and 21 of the 27 peptides were identified on the basis of chemical shifts and coupling constants, including the use of diagnostic values derived from ^{1}H NMR spectra and ^{1}H-^{13}C and ^{1}H-^{15}N HMQC studies on authentic AAA-OH and YYY-OH.

A model four-component mixture of aromatic ring positional isomers was studied (Chin et al., 1998). These isomers cannot be identified by MS without time-consuming MS/MS studies. The mixture (100 μg) was first analyzed by stopped-flow HPLC-NMR. For these data, 128 scans were more than sufficient to obtain suitable S/N. The structure determination for each compound is obtained by consideration of the chemical shifts and the coupling patterns. In an on-flow HPLC/NMR analysis, approximately 200 μg of four-component mixture was used.

An NMR strategy for mixture analysis without physical separation of the components was reported (Lin & Shapiro, 1997). The method is based on the combination of HMBC with TOCSY. As in the case with NMR assignments for peptides and small proteins, a structure verification rather than a *de novo* structure determination is usually involved in combinatorial chemistry problems. The strategy used for structure verification is similar to that employed to perform structure assignment in peptides and small proteins. A TOCSY spectrum was initially obtained to establish the individual spin system connectivity in an analogous manner to the peptide example. However, in general, the total spin connectivity cannot be accomplished due to non-reporting centers. For most organic compounds, the spin systems are broken into several fragments, which must be connected to regain molecular identity. Further connectivity can be made by HMBC. The HMBC experiment connects protons to carbons via two or three bond couplings. This allows the connection of protonated carbons to non-protonated carbon centers, thereby allowing the spin systems derived from the TOCSY experiment to be linked. Applying this principle, the TOCSY spectrum of the mixture of six esters was collected. Following chemical shift information, the spin fragmentation of organic groups can be obtained. The following HMBC data allow the connection of separate molecular groupings to identify molecules. If the ^{13}C chemical shifts for the common connecting groups are insufficiently resolved to allow assignment, the diffusion-encoded NMR methods can be used to aid the structural assignment (Lin & Shapiro, 1996).

6.3 HPLC FOR ANALYZING COMPOUNDS FROM DISCRETE AND MIXTURE LIBRARIES

An important early step in the characterization of the compound is to obtain an accurate concentration of the compound in solution in order to generate an accurate pharmacological activity measurement. Traditionally, the concentration of a chemical product is obtained by purifying milligram quantities to a purity greater than 95%, drying the purified material, and weighing it as accurately as possible. A solution of known concentration is then prepared by adding a defined volume solvent to the known mass of the compound. This method is too restrictive and slow to keep pace with the perpetual acceleration of the drug discovery process and is inaccurate in the presence of impurities, counter-ions, unevaporated solvents, and unidentified particulate, particularly when compounds are available only in minute amounts as in combinatorial chemistry. An alternative quantitation method that is sensitive, specific, and suitable for high-throughput analysis is needed.

Application of a nonselective detector to the analysis of compound libraries, in theory, would improve the speed and accuracy with which these analyses are performed, by allowing the quantitation of the entire library of compounds with a single external standard. While no truly universal detection method for HPLC exists, there are several detection methods that have been demonstrated to provide similar response to compounds within a given class. Two of these methods are discussed below.

6.3.1 HPLC with a Chemiluminescence Nitrogen Detector (CLND)

Reversed-phase, high-performance liquid chromatography (RP-HPLC) with UV detection provides a rapid, reproducible, and sensitive means of assessing the presence and sometimes the purity of compounds. The use of MS detection with RP-HPLC adds compound identity and structure verification to the analysis. However, it is difficult and cumbersome to use either of these detectors for on-line compound quantitation because of the requirement for a proper calibration standard, which is either the authentic compound or a closely related analogue.

HPLC with chemiluminescent nitrogen detection (HPLC-CLND) was used for the quantitative analysis of peptides and organic compounds synthesized by solid-phase peptide synthesis (Bizanek et al., 1996; Fitch et al., 1997; Taylor et al., 1998). CLND provided quantitative information regarding the nitrogen distribution of peptide samples following HPLC separation. This technique permits the universal quantitation of the nitrogen-containing compounds in an on-line mode without pre- or post-column derivatization and free of interference from non-nitrogen-containing UV chromophores. Approximately 90% of the >65,000 developmental and marketed drugs in the commercial database MDL Drug Data Report (MDDR) contain nitrogen, indicating that the CLND is widely applicable as a detector for analyzing pharmaceuticals.

The nitrogen detection mechanism, which is based upon a chemiluminescence reaction, is elucidated in Scheme 6.1. When each analyte reaches the nitrogen detector, it is mixed with oxygen and undergoes oxidation at high temperature (1000°C–1100°C) in the furnace. Nitrogen-containing components, with the exception of diatomic nitrogen (N_2), are converted to nitric oxide (NO). The NO is dried and reacted with ozone (O_3) in the reaction chamber to form nitrogen dioxide in the excited state ($^*NO_2$), which rapidly decays to the stable ground state (NO_2) and releases a photon (hυ). The emitted photons were detected by a photomultiplier tube (PMT), and the amplified signal was recorded using an integrator.

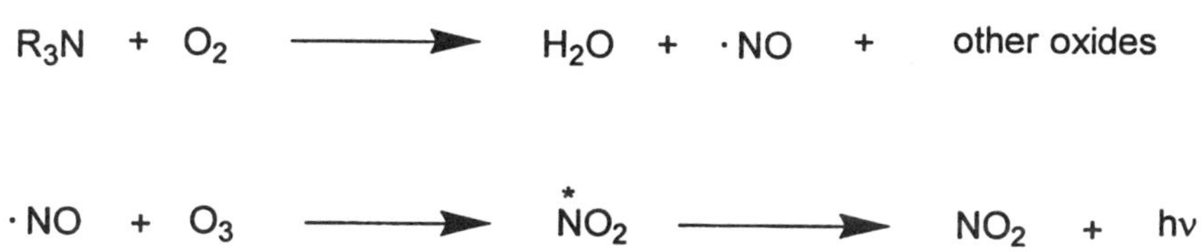

Scheme 6.1

By flow injection analysis (FLA) and in conjunction with reversed-phase (RP) chromatography using gradient elution, the CLND produced a linear response from 25 to 6400 pmol of nitrogen, and that was equivalent for a set of chemically and structurally diverse compounds. An average error of approximately ±10% was determined among the compounds. In addition, the response was independent of mobile-phase composition.

CLND detection for HPLC is an important new high-throughput technique for measuring yields and purities in organic chemistry and combinatorial chemistry. This detector can be calibrated with any nitrogen-containing compound and, thus, allows quantitative analysis when authentic reference standards are not available. The application to solid-phase synthesis is illustrated with a diketopiperazine synthesis (Fitch et al., 1997), a mixture of 16 crude synthetic peptides produced by SPPS (Fujinari et al., 1996), and peptides that were isolated from casin hydrolysate (Fujinari & Manes, 1994).

The measurement of yield is the premier application of HPLC/CLND quantitative analysis, but other applications are clear. In the early stages of a lead development project, weighable quantities of authentic pure samples of a compound are not available, and yet quantitative measurements such as IC_{50}, solubility or plasma stability need to be made. HPLC/CLND can be used to calibrate solutions made from sub-milligram synthetic samples. In the split/pool method of combinatorial synthesis, mixtures of compounds are made that are difficult to characterize. The HPLC/CLND of a nominally equimolar pool (based on nitrogen) should yield equal-sized peaks. The CLND will be especially useful in conjunction with splitting to a mass spectrometer.

The precision of the CLND response when used with gradient elution RP-HPLC was measured by quadruplicate injections of a mixture containing the nine calibration compounds at equal nitrogen concentration (Taylor et al., 1998). Under the chromatographic conditions used, the compounds were all baseline-resolved and exhibited acceptable peak shapes in the water/MeOH/IPA mobile phase (Figure 6.6). Peak widths in the CLND trace were nearly identical to those in the UV trace. The baseline remained flat over the entire gradient, making

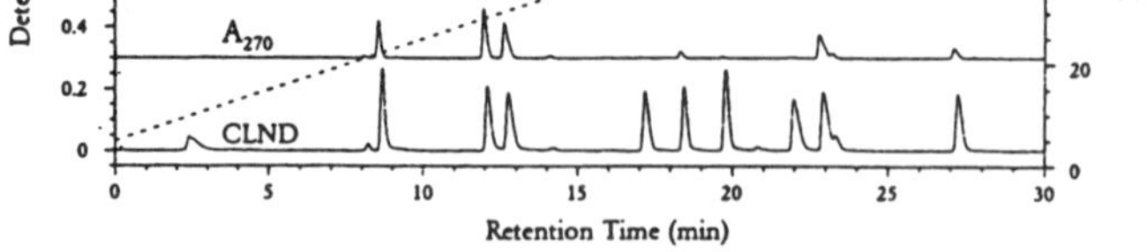

Figure 6.6 Chromatogram of a mixture of the calibration compounds separated by RP-HPLC with gradient elution and detected by UV absorption and chemiluminescent nitrogen. The injected sample contained 2 nmol each of caffeine (1), triprolidine (2), chloropheniramine (3), diphenylalanine (4), 6-nor-6-allyllysergic acid diethylamine (5), angiotensin II (6), diphenhydramine (7), doxepin (8), and dibucaine (9). (Reproduced with permission from Anal. Chem. *1998, 70, 3344. Copyright© 1998, American Chemical Society.)*

quantitation relatively simple. For each compound, peak areas were measured and averaged. All compounds exhibited RSDs of ≤2% for the quadruplicate injections. In addition, peak areas in the CLND trace showed less variability than those in the UV trace. In the CLND trace, peak areas exhibited an average deviation of ~9% around the mean peak area, whereas, after adjusting for mass of each compound injected, UV peak areas deviated ~15% at 214 nm and ~120% at 270 nm. These results demonstrate good reproducibility of the CLND response for replicate injections and between different compounds.

6.3.2 Evaporative Light Scattering Detector

Ford and Kennard (1996) were the first to demonstrate equivalent mass response for a series of low-molecular-weight polymers using an evaporative light-scattering detector (ELSD). Briefly, the column effluent enters the detector through a narrow bore tube where it is mixed with a high-velocity stream of nitrogen gas and is nebulized. The nebulized sample stream then enters a heated drift tube. The drift tube is maintained at a sufficient temperature to cause rapid evaporation of the mobile-phase solvent, leaving a narrow mist of the nonvolatile solute to be swept toward the detection region by a stream of nitrogen carrier gas. The detection system consists of a 670-nm laser and a silicon photodiode oriented at an angle of 90° and perpendicular to the central axis of the cylindrical drift tube. Passage of the narrow band of solute particulate through the laser beam results in scattering of the source. The intensity of the scattered light registered by a photodiode is proportional to the total amount of sample present in the eluting solute band.

The advantages of ELSD over UV detection for the quantitation of organic combinatorial libraries composed of small organic compounds by HPLC are described (Kibbey, 1995, Hsu et al., 1999). The detector's response is independent of the sample chromophore, which makes it well suited to chromatographic analyses of mixtures of dissimilar solutes. Thus, HPLC with ELSD offers the potential for reducing false-positive or false-negative results in screening assays, because of its ability to detect the presence of impurities that absorb poorly in the UV (e.g., those impurities originating from the polymeric support). Furthermore, the ELSD exhibits a nearly equivalent response to compounds of similar structural class. Hence, rapid quantitation for compound libraries may be carried out with the use of a single external standard. The quantitation errors for amino acids by normal-phase HPLC with ELSD is on average approximately ±10%.

Successful application of the ELSD as a nonselective detector in HPLC has been used for the analysis of a wide variety of compounds, including alkyl- and alkylaryl surfactants (Bear, 1988), triglycerides (Stolyhwo et al., 1985), phospholipids (Sotirhos et al., 1985), sugars (Macrae & Dick, 1981), pharmaceutical salts (Peterson & Risley, 1995), and excipients (Lafosse et al., 1992). In each of these applications, the advantages of the ELSD have been its near-equivalent response to a variety of compounds within a given structural class, its compatibility with gradient elution, and its overall reliability and ease of operation.

6.4 SUMMARY

In most cases, library members should be characterized after cleavage from solid support. It is evident from above discussions that MS is the major technique in the quality control of soluble combinatorial libraries, while HT NMR is still in its infancy. Speed and sensitivity are main advantages of the MS techniques, and the quantitation is a drawback.

Both qualitative and quantitative analysis, at least for a statistically representative set, of combinatorial library members are crucial for confirming the chemical diversity desired and giving confidence on screening results. Efficient quantitation methods in high-throughput format are currently not available, but quantitative detectors, such as CLND and ELSD, give some possibilities for quantitation of combinatorial library members both in FIA and on-line format.

Mixture analysis is at the very early stage of study. Comparison of mass distribution obtained from MS measurements and that from computer simulation gives very crude approximation because the ionization efficiency of a diverse set of compounds is unknown. This opens a field for future research and development.

In general, the application of serial analytical methods to deal with products made from parallel synthesis is definitely a mismatch. A breakthrough in parallel qualitative and quantitative analysis is badly needed.

6.5 REFERENCES

Bear, G. R. Universal detection and quantitation of surfactants by high-performance liquid chromatography by means of the evaporative light-scattering detector. *J. Chromatogr.* 1988, 459, 91.

Bizanek, R., Manes, J. D., Fujinari, E. M. Chemiluminescent nitrogen detection as a new technique for purity assessment of synthetic peptides separated by reversed-phase HPLC. *Peptide Research.* 1996, 9, 40.

Chin, J., Fell, J. B., Jarosinski, M., Shapiro, M. J., Wareing, J. R. HPLC/NMR in combinatorial chemistry. *J. Org. Chem.* 1998, 63, 386.

Daley, D. J., Scammell, R. D., James, D., Monks, I., Raso, R., Ashcroft, A. E., Hudson, A. J. High-throughput LC-MS verification of drug candidates from 96-well plates. *Am. Biotech. Lab.* 1997, 15, 24.

Demirev, P. A., Zubarev, R. A. Probing combinatorial library diversity by mass spectrometry. *Anal. Chem.* 1997, 69, 2893.

Dunayevskiy, Y., Vouros, P., Carell, T., Wintner, E. A., Rebek Jr., J. Characterization of the complexity of small-molecule libraries by electrospray ionization mass spectrometry. *Anal. Chem.* 1995, 67, 2906.

Dunayevskiy, Y. M., Vouros, P., Wintner, E. A., Shipps, G. W., Carell, T., Rebek Jr., J. Application of capillary electrophoresis-electrospray ionization mass spectrometry in the determination of molecular diversity. *Proc. Natl. Acad. Sci. USA.* 1996, 93, 6152.

Fang, A. S., Vouros, P., Stacey, C. C., Kruppa, G. H., Laukien, F. H., Wintner, E. A., Carell, T., Rebek Jr., J. Rapid characterization of combinatorial libraries using electrospray ionization Fourier transform ion cyclotron resonance mass spectrometry. *Combi. Chem. & High Throughput Screening.* 1998, 1, 23.

Fitch, W. L. Szardenings, A. K., Fujinari, E. M. Chemiluminescent nitrogen detection for HPLC: An important new tool in organic analytical chemistry. *Tetrahedron Lett.* 1997, 38, 1689.

Ford, D. L., Kennard, W. Vaporization analyzer, *J. Oil Color Chem. Assoc.* 1996, 49, 299.

Fujinari, E. M., Manes, J. D. Nitrogen-specific detection of peptides in liquid chromatography with a chemiluminescent nitrogen detector. *J. Chromatog. A.* 1994, 676, 113.

Fujinari, E. M., Manes, J. D., Bizanek, R. Peptide content determination of crude synthetic peptides by reversed-phase liquid chromatography and nitrogen-specific detection with a chemiluminescent nitrogen detector. *J. Chromatog. A.* 1996, 743, 85.

Furka, A., Sebestyen, F., Asgedom, M., Dibo, G. In *Highlights of Modern Biochemistry: Proceedings of the 14th International Congress of Biochemistry* (VSP, Utrecht, The Netherlands), 1988, 5, 47.

Furka, A., Sebestyen, F., Asgedom, M., Dibo, G. General method for rapid synthesis of multicomponent peptide mixtures. *Int. J. Pept. Protein Res.* 1991, 37, 487.

Gorlach, E., Richmond, R., Lewis, I. 1. High-throughput flow injection analysis mass spectroscopy with networked delivery of color-rendered results. 2. Three-dimensional spectral mapping of 96-well combinatorial chemistry racks. *Anal. Chem.* 1998, 70, 3227.

Haap, W. J., Metzger, J. W., Kempter, C., Jung, G. Composition and purity of combinatorial aryl ether collections analyzed by electrospray mass spectrometry. *Mol. Diversity.* 1997, 3, 29.

Hegy, G., Gorlach, E., Richmond, R., Bitsch, F. High throughput electrospray mass spectrometry of combinatorial chemistry racks with automated contamination surveillance and results reporting. *Rapid Commun. Mass Spectr.* 1996, 10, 1894.

Houghten, R. A., Pinilla, C., Blondelle, S. E., Appel, J. R., Dooley, C. T., Cuervo, J. H. Generation and use of synthetic peptide combinatorial libraries for basic research and drug discovery. *Nature.* 1991, 354, 84.

Hsu, B. H., Orton, E., Tang, S.-Y., Carlton, R. A. Application of evaporative light scattering detection to the characterization of combinatorial and parallel synthesis libraries for pharmaceutical drug discovery. *J. Chromatogr. B.* 1999, 725, 103.

Keifer, P. A. High-resolution NMR techniques for solid-phase synthesis and combinatorial chemistry. *Drug Discovery Today.* 1997, 2, 468.

Keifer, P. A. New methods for obtaining high-resolution NMR spectra of solid-phase synthesis resins, natural products, and solution-state combinatorial chemistry libraries. *Drugs of the Future.* 1998, 23, 301.

Kibbey, C. E. Quantitation of combinatorial libraries of small organic molecules by normal-phase HPLC with evaporative light-scattering detection. *Molecular Diversity.* 1995, 1, 247.

Kiplinger, J. P., Cole, R. O., Robinson, S., Roskamp, E. J., Ware, R. S., O'Connell, H. J., Brailsford, A., Batt, J. Structure-controlled automated purification of paralleled synthesis products in drug discovery. *Rapid Commun. Mass Spectr.* 1998, 12, 658.

Lafosse, M., Elfakir, C., Morin-Allory, L., Dreux, M. The advantages of evaporative light-scattering detection in pharmaceutical analysis by high-performance liquid chromatography and supercritical fluid chromatography. *J. High-Res. Chromatogr.* 1992, 15, 312.

Lam, K. S., Salmon, S. E., Hersh, E. M., Hruby, V. J., Kazmierski, W. M., Knapp, R. J. A new type of synthetic peptide library for identifying ligand-binding activity. *Nature.* 1991, 354, 82.

Lambert, P. H., Boutin, J. A., Bertin, S., Fauchere, J.-L., Volland, J.-P. Evaluation of high performance liquid chromatography/Electrospay mass spectrometry with selected ion monitoring for the analysis of large synthetic combinatorial peptide libraries. *Rapid Commun. Mass Spectr.* 1997, 11, 1971.

Lewis, K. C., Fitch, W. L., Maclean, D. Characterization of a split-pool combinatorial library. *LC-GC.* 1998, 16, 644.

Lin, M., Shapiro, M. J. Mixture analysis in combinatorial chemistry. Application of diffusion-resolved NMR spectroscopy. *J. Org. Chem.* 1996, 61, 7617.

Lin, M., Shapiro, M. J. Mixture analysis by NMR spectroscopy. *Anal. Chem.* 1997, 69, 4731.

Lindon, J. C., Farrant, R. D., Sanderson, P. N. Separation and characterization of compounds of peptide libraries using on-flow coupled HPLC-NMR spectroscopy. *Magne. Reson. In Chem.* 1995, 33, 857.

Loo, J. A., DeJohn, D. E., Loo, R. R. O., Andrews, P. C. Application of mass spectrometry for characterizing and identifying ligands from combinatorial libraries. *Annual Reports in Medicinal Chemistry.* 1996, 31, 319.

Macrae, R., Dick, J. Analysis of carbohydrates using the mass detector. *J. Chromatogr.* 1981, 210, 138.

Metzger, J. W., Kempter, C., Wiesmuller, K.-H., Jung, G. Electrospray mass spectrometry and tandem mass spectrometry of synthetic multicomponent peptide mixtures: Determination of composition and purity. *Anal. Biochem.* 1994, 219, 261.

Metzger, J. W., Wiesmuller, K.-H., Gnau, V., Brunjes, J., Jung, G. Ion-spry mass spectrometry and high-performance liquid chromatography-mass spectrometry of synthetic peptide libraries. *Angew. Chem. Chem. Int. Ed. Engl.* 1993, 32, 894.

Nawrocki, J. P., Wigger, M., Watson, C. H., Hayes, T. W., Senko, M. W., Benner, S. A., Eyler, J. R. Analysis of combinatorial libraries using electrospray Fourier transform ion cyclotron resonance mass spectrometry. *Rapid Commun. Mass Spectr.* 1996, 10, 1860.

Peterson, J. A., Risley, D. S. Validation of an HPLC method for the determination of sodium in LY293111 sodium, a novel LTB_4 receptor antagonist, using evaporative light-scattering detection. *J. Liq. Chromatogr.* 1995, 18, 331.

Robinson, G. M., Manica, D. P., Taylor, E. W., Smyth, M. R., Lunte, C. E. Development of a capillary electrophoretic separation of an N-(substituted)-glycine-peptoid combinatorial mixture. *J. Chromatog. B.* 1998, 707, 247.

Smallcombe, S. H., Patt, S. L., Keifer, P. A. Wet solvent suppression and its applications to LC NMR and high-resolution NMR spectroscopy. *J. Magn. Reson. A.* 1995, 117, 295.

Smart, S. S., Mason, T. J., Bennell, P. S., Maeij, N. J., Geysen, H. M. High-throughput purity estimation and characterization of synthetic peptides by electrospray mass spectrometry. *Int. J. Peptide Res.* 1996. 47, 47.

Sotirhos, N., Thorngren, C., Herslof, B. Reversed-phase high-performance liquid chromatographic separation and mass detection of individual phospholipid classes. *J. Chromatogr.* 1985, 331, 313.

Stolyhwo, A., Colin, H., Guiochon, G. Analysis of triglycerides in oils and fats by liquid chromatography with the laser light-scattering detector. *Anal. Chem.* 1985, 57, 1342.

Süßmuth, R. D., Jung, G. Impact of mass spectrometry on combinatorial chemistry. *J. Chromatgr. B.* 1999, 725, 49.

Swali, V., Langley, G. J., Bradley, M. Mass spectrometric analysis in Comb. Chem. Curr. Opinion in *Chem. Biol.* 1999, 3, 337.

Taylor, E. W., Qian, M. G., Dollinger, G. D. Simultaneous on-line characterization of small organic molecules derived from combinatorial libraries for identity, quantity, and purity by reversed-phase HPLC with chemiluminescent nitrogen, UV, and mass spectrometric detection. *Anal. Chem.* 1998, 70, 3339.

Tutko, D. C., Henry, K. D., Winger, B. E., Stout, H. Hemling, M., Sequential mass spectrometry and MS^n analyses of combinatorial libraries by using automated matrix-assisted laser desorption/ionization Fourier transform mass spectrometry. *Rapid Commun. Mass Spectr.* 1998, 12, 33.

Wang, T., Zeng, L., Strader, T., Burton, L., Kassel, D. B. A New ultra-high throughput method for characterizing combinatorial libraries incorporating a multiple probe autosampler coupled with flow injection mass spectrometry analysis. *Rapid Commun. Mass Spectr.* 1998, 12, 1123.

Weller, H. N., Young, M. G., Michalczyk, S. J., Reitnauer, G. H., Cooley, R. S., Rahn, P. C., Loyd, D. J., Fiore, D., Fischman, S. J. High throughput analysis and purification in support of automated parallel synthesis. *Mol. Diversity* 1997, 3, 61.

Winger, B. E., Campana, J. E. Characterization of combinatorial peptide libraries by electrospray ionization Fourier transform mass spectrometry. *Rapid Commun. Mass Spectr.* 1996, 10, 1811.

Wu, Q. Multistage accurate mass spectrometry: a "basket in a basket" approach for structure elucidation and its application to a compound from combinatorial synthesis. *Anal. Chem.* 1998, 70, 865.

Zeng, L., Kassel, D. B. Developments of a fully automated parallel HPLC/mass spectrometry system for the analytical characterization and preparative purification of combinatorial libraries. *Anal. Chem.* 1998, 70, 4380.

7

Selection of Active Compounds from Combinatorial Libraries

COMBINATORIAL LIBRARIES ARE commonly tested for their biological activity in solution. Discrete compounds from solution-phase synthesis can be assayed directly or after purification. Discrete compounds made on solid supports have to be cleaved and purified before screening. The throughput of such assays has undergone major improvement, e.g., from traditional 96-well to 1536-well formats to deal with the huge number of compounds synthesized by combinatorial chemistry methods. An alternative way is to select active compounds from mixtures by assays in solution or directly on bead. Mixture assay is particularly useful for screening combinatorial libraries made by "split-and-pool" synthesis. When active compounds or beads are discovered, the structure elucidation of active compounds can only be achieved by various single-bead decoding schemes. The direct structure study at the single-bead level has also been attempted.

7.1 IN-SOLUTION SCREENING OF COMBINATORIAL LIBRARIES

Routine HT screening has been used to screening discrete compound libraries and even split libraries in small pools of 10–20 compounds. It is estimated that fluorescence assay is several-fold faster than the radiolabeled assay. With few exceptions, the final cost per well for the fluorescence screen is less than its radioactive counterpart. Several assay methods are discussed in this section.

7.1.1 Fluorescence Polarization Assay

Fluorescence polarization (FP) allows the discrimination of bound and free states of a fluorophore based on differences in rotation rates between the two states. The rotational correlation time of a molecule is directly proportional to its molecular volume at constant temperature and viscosity. A fluorescent ligand, if free in solution, tumbles with a rotational relaxation time of less than one nanosecond. When the ligand is excited with plane-polarized light of the correct wavelength, the resulting induced dipole will fluoresce in the same plane after its characteristic emission lifetime, typically about 100 nanoseconds. During this time, the dipole will have tumbled randomly with respect to the original plane of excitation. Thus, if one measures the emission through two polarizing filters, first parallel, then perpendicular to the excitation plane of polarization, nearly equal intensities will be observed. After the fluorescent ligand binds to a macromolecule, it rotates at the rotational relaxation time of the complex, typically 100 nanoseconds or more. Therefore, during the nanosecond lifetime of the fluorescence event of the original polarization of the excitation light will be preserved, and one will detect higher fluorescent intensity when the polarizers are set parallel than when they are perpendicular.

7.1.2 Time-Resolved Fluorescence Assay

Homogeneous time-resolved fluorescence (HTRF) has been successfully applied to immunoassays, protein-protein binding, receptor binding, enzyme assays, and DNA hybridization. It has been proven fast and robust when compared to radioisotopic methods.

7.1.2.1 EUROPIUM AS A LONG-LIVED FLUORESCENT LABEL

Eu^{+3} and other rare earth lanthanide are known for the slow decay of their fluorescent signal. Because the interfering background fluorescence has a very short decay time, the long-lived Eu^{+3} fluorescence can be measured after the short-lived background has decayed.

However, Eu^{+3} is not a good fluorescent label by itself due to the quenching effect. In order to be measured, the Eu^{+3} must be complexed in a structure that allows Eu^{+3} fluorescence. This structure must have two important properties. First, the complex must be capable of acting as an antenna to collect the excitation energy and transfer it to the

Eu^{+3}. Second, the complex must protect the Eu^{+3} from fluorescence quenching.

7.1.2.2 EUROPIUM CRYPTATE AS A STABLE FLUORESCENT COMPLEX

The problem of poor Eu^{+3} fluorescence has been solved by the use of macropolycyclic compounds, the cryptates. The cryptates trap Eu^{+3} during chemical synthesis in an essentially irreversible complex. The cryptate has all the necessary properties for enhancing Eu^{+3} fluorescence. It acts as an energy antenna for collecting the excitation energy and transferring it to the Eu^{+3}, protects the Eu^{+3} from quenching, does not interfere with the fluorescent emission, and can be used to conjugate the Eu^{+3} to biological macromolecules.

7.1.2.3 HOMOGENEOUS FORMAT BASED ON FORSTER ENERGY TRANSFER THEORY

Forster's theory predicts that a 50% energy transfer occurs at distance of 9 nm for (Eu) allophycocyanin (XL665) complex. This is the largest distance reported for a donor:acceptor pair. At shorter distances, the energy transfers efficiency increases (75% at 7.5 nm). This transfer distance is compatible with most molecular interactions involved in biological processes and will allow the use of HTRF in many applications.

7.1.2.4 SELECTIVITY IN MEASURING BOUND LABEL IN THE PRESENCE OF FREE LABEL

All the necessary elements are now available to construct an HTRF assay: a long-lived Eu^{+3} fluorescence to eliminate background, a (Eu) cryptate cage to conjugate, protect, and excite the Eu^{+3}, and an energy transfer pair that allows signal amplification and homogeneous measurements without interference from free label.

Of prime importance to screening is that HTRF is a true homogeneous, solution assay. A donor (europium cryptate) and an acceptor, XL665, are used as two labels. When they are in proximity, the cryptate transfers energy to the XL665 that fluoresces with a long decay and at a wavelength that allows it to be selectively measured in the presence of free cryptate and uncoupled XL665.

By measuring the specific XL665 signal and the cryptate signal as internal reference, a ratio can be used that compensates for the presence of colored compounds in the assay. This makes HTRF far more robust to the screening of synthetic combinatorial libraries.

7.1.3 Fluorescence Correlation Spectroscopy (FCS)

In conventional fluorescence spectroscopy, in contrast to FCS, a relatively large volume is illuminated. Therefore, the average fluorescence intensity is recorded against a high background noise of scattered light and other sources of luminescence. These noises limit the resolution and sensitivity in conventional fluorescence spectroscopy.

In FCS (Rigler, 1995), a sharply focused laser beam illuminates a volume element of 10^{-15} liter, the size of an *E. coli* cell. This volume is so small that, at a given point in time, it hosts only one particle out of the many under analysis. The single molecules diffusing through the illuminated volume over time give rise to fluorescence light bursts. Each individual burst, resulting from a single molecule, can be registered. Hereby, confocal imaging eliminates any interference from background signals. The photons are recorded in a time-resolved manner by a highly sensitive single-photon detection device. The measured transition time is related to the size and shape of the particle.

Diffusion events of more than one particle are registered as fluctuations of the emitted fluorescence intensity. The number of molecules in the illuminated volume element, as well as their characteristic translational diffusion times, can be computed from an autocorrelation curve. Therefore, autocorrelation of the time-dependent fluorescence signal allows faster and slower diffusing particles to be differentiated.

Properties such as binding or catalytic activity can be calculated directly from the diffusion times and ratio of faster (smaller) and slower (larger) molecules. Interactions of biological importance that can be analyzed are hybridization between nucleic acid primers and DNA or RNA targets, binding between peptide ligands and isolated as well as cell-bound receptors, or between antigen and antibodies. Because the interaction can be analyzed rapidly in small volumes without the need for separating unbound from bound ligands, an important application of FCS is envisaged in large-scale screening of combinatorial libraries. The sensitivity has been advanced to the point that detection of a single dye molecule is possible in the submillisecond range.

7.1.4 Microphysiometry

Microphysiometry is based on a simple concept that requires a sophisticated technology to implement. The concept is that it is possible to detect receptor activation and other physiological changes in living cells by monitoring the activity of energy metabolism. Extracellular acidification rate is unusually sensitive to ligand-receptor interactions. It is fundamentally governed by catabolism via the excretion of acidic end-products, such as carbon dioxide and lactic acid. In addition, it is substantially influenced in a complementary fashion by the regulation of intracellular pH, a process sensitive to physiological changes such as receptor activation. Such an assay is relatively general: instead of a host of mechanism-specific assays, a single-assay format is used to analyze a variety of pharmacologically distinct systems. The rate of extracellular acidification was measured for both biological and instrumental reasons.

A reason for choosing extracellular acidification rate is the availability of a pH sensor that can make sensitive measurements (~500 (μpH units RMS noise) with good time resolution (one pH measurement per second) in micro-volume (~100 nL) geometry that is suitable for integration into microfluidic systems.

This cellular assay system fulfills the requirement for a high-throughput screen because of the following:

- It detects receptor activation generically, relatively independent of the signal transduction pathway.
- Several targets, for example a series of subtypes of a receptor, can be assayed simultaneously to a single sample.
- Sample requirements as low as ~50 μL.
- Cell consumption is only <1000 cells per data point.

7.1.5 Surface Plasmon Resonance (BIAcore)

BIAcore detection is based on surface plasmon resonance (SPR), which is a phenomenon that occurs between incoming photons and electrons in the surface of the thin metal (e.g., gold) coated onto a glass support. At a specific wavelength and angle of incident, light energy is transferred to electrons in the metal surface causing reflected light to disappear. This angle of non-reflectance, the resonance angle, changes with the refraction index near the metal surface. BIAcore measures the shift in the reflectance minimum caused by modifications of the refraction index, which varies with the mass concentration near the metal surface. The shift in the resonance angle is plotted against time and is displayed in a sensorgram (the resonance angle or resonance unit, RU versus time in seconds), which provides a measurement in real time.

The BIAcore detection system is quantitative, and the signal is proportional to the mass adsorbed on the sensorchip, for instance,

1000 RU corresponds to 1 ng/mm^2. This signal corresponds to 6.7 × 10^{-15} moles/mm^2 of IgG (150,000 Dalton). If the molecular weight is very low, it may not be possible to detect the reaction. In this case, it would be better to couple the low-molecular-weight compound on the sensorchip and to inject the higher molecular-weight ligand in solution.

7.2 AFFINITY SELECTION OF LEAD COMPOUNDS BY MASS SPECTROMETRY

Although in-solution assays are the major methods for the selection of lead compounds from discrete libraries, the success of these assays relies on the screening of well-defined pure components because ambiguity may result from the screening of mixtures. The screening of mixtures is a major challenge.

7.2.1 Fourier Transform Ion Cyclotron Resonance (FTICR) MS

Bruce et al. (1995) described a method called bio-affinity characterization MS (BACMS). The method is based upon the combination of capabilities for trapping intact complexes (formed from ESI of the appropriate solution) in the gas phase and mass analyses using FTICR MS. The BACMS approach eliminates the need for distinct separation and purification steps involving the original mixture solution. The key to this approach is the ability to directly transfer fragile noncovalent complexes into the FTICR mass spectrometer. In the BACMS method, the complex can be formed free in solution at physiologically relevant pH. The advantages of the method are the ability to first recognize the complex in solution, to separate it in the gas phase from other ions (including other complexes), and finally to provide structural information on the complex and selected ligand.

Figure 7.1 shows a conceptual representation of the simplest version of the BACMS method. The complexes to be investigated are electrosprayed directly from the solution containing both the affinity target and the ligand library and, using appropriate ESI source conditions, are transferred to the gas phase and trapped in the FTICR along with all other complexes and non-complexed species. The complexes of interest are first identified in the mass spectrum and are isolated by selected-ion accumulation (SIA), which also simultaneously optimizes the population of selected ions. The process is followed by low-energy dissociation to retain the high binding affinity ligand(s)

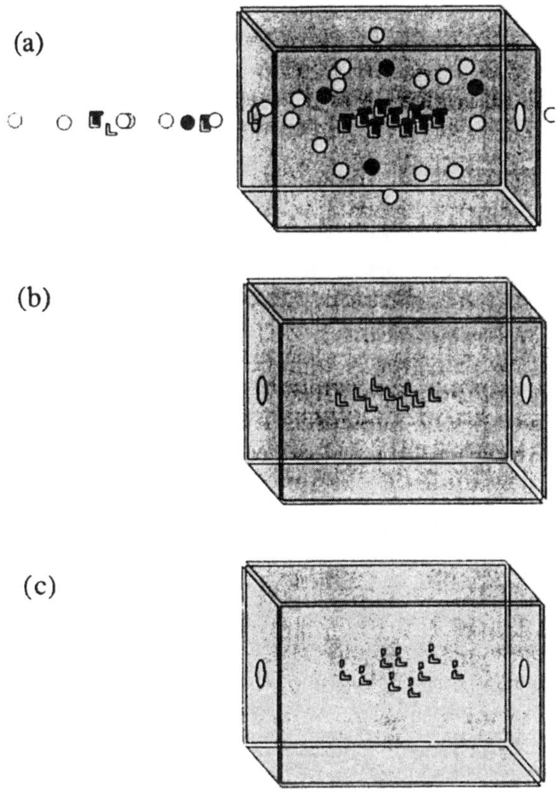

Figure 7.1 *A conceptual representation of the BACMS technique. (a) The mixture solution is ionized by electrospray ionization, and the complex of interest is selectively accumulated in the FTICR trap. (b) The noncovalent complex is then heated to liberate the affinity ligand species. (c) The ligand ions are retained for further CAD studies to present structural information. "T" represents the target biomolecules, and "L" represents ligands, which form high binding affinity-specific noncovalent complexes with the target molecules. Circles (both solid and empty) represent ions from other molecules present in the mixture. (Reproduced with permission from Bruce et al., Rapid Commun. In* Mass Spectr., *9, 644–650. Copyright© 1995, John Wiley & Sons Limited.)*

that contribute to the selected complexes. After selection and dissociation, the high-affinity ligand species isolated in the same experiment can then be studied using collisionally activated dissociation (CAD) to establish structural information. With this method, the time-consuming techniques usually employed for these kinds of studies, such as

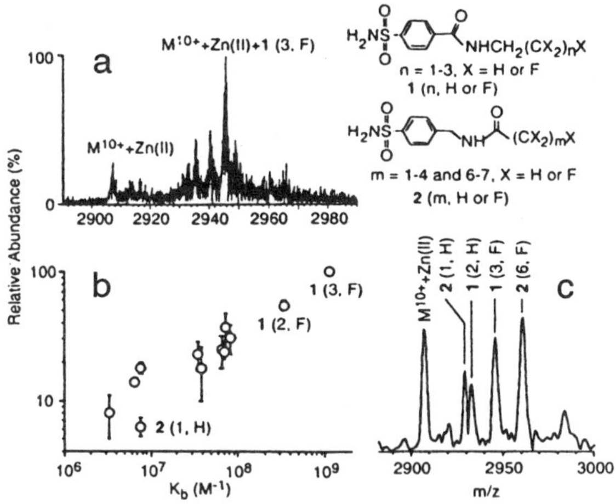

Figure 7.2 *(a) Mass spectra of the 10+ charge state of BCAH (1.0 μM) and 18 inhibitors (0.05 μM each) in 10 mL NH_4OAc (pH 7.0). (b) Correlation of the relative ion abundance with the values of K_b in solution. Six of the K_b values represent the sum of those from two inhibitors with the same mass. (c) Mass spectra of BCAII (1.0 μM) and four inhibitors (2 (1, H) = 60 μM, 1 (2, H) = 3.0 μM, 2 (6, F) = 1.5 μM, 1 (3, F) = 0.1 μM) in 10 mM NH_4OAc (pH 7.0). (Reproduced with permission from* J. Am. Chem. Soc., *1995, 117, 8859–8860. Copyright© 1995, American Chemical Society.)*

affinity chromatography, are eliminated. The enhancement of the dynamic range of the protein detection is achieved with SIA. The amount of protein detected by the method is in 10^{-16} moles level.

The ESI-FTICR MS method was used to study competitive binding of inhibitors (binding constant ranging from 1.7×10^6 to 1.1×10^9 1/M) to bovine carbonic anhydrase II (BCA II, see Cheng et al., 1995). The binding of 18 inhibitors (0.05 μM each) with BAC II (1.0 μM) was detected (Figure 7.2a). The relative ion abundance correlates well with values of K_b in solution (Figure 7.2b). A four-component mixture containing inhibitors with the highest (1.1×10^9) and the lowest affinity (1.7×10^6), with their concentrations inversely proportional to their K_b values, was prepared and studied. The four inhibitors formed 1:1 complexes to BCA II with approximately equal ion abundance (Figure 7.2c).

A mixture of carbonic anhydrase II (CA II, 2.5 μM) and a peptide library (0.5 μM for each inhibitor, 289 compounds) in a 10 mM NH_4OAc solution (pH 7.0) was analyzed using ESI-FTICR (Gao et al., 1996). Major peaks in the mass spectrum correspond to intact complexes in the -7 to -9 charge states (Figure 7.3a). CAD of the isolated -9 charge

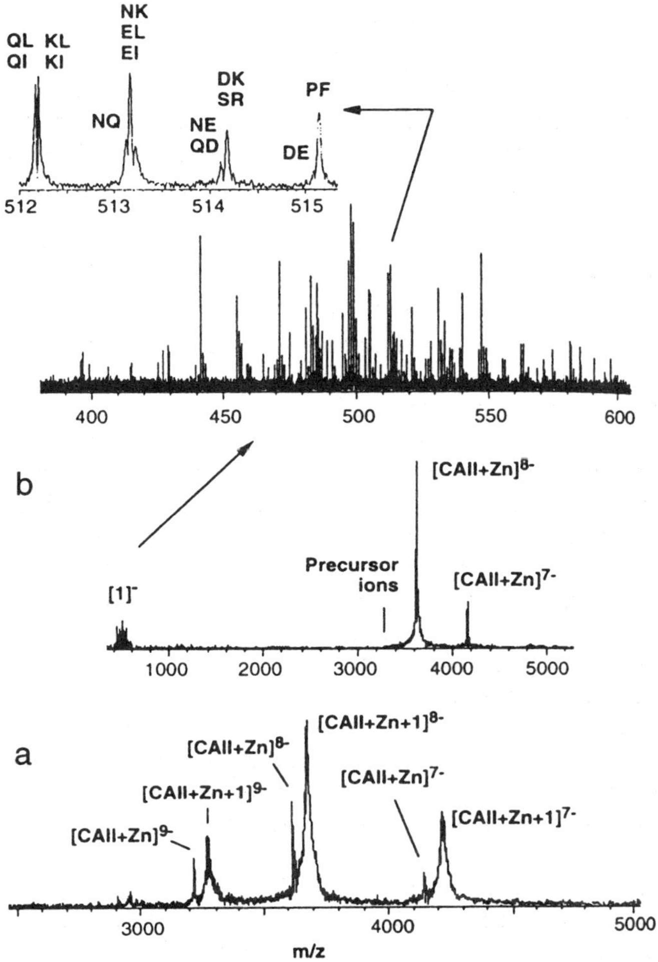

Figure 7.3 *(a) ESI-FTICR mass spectrum from a mixture of the 289-component* L*-library (0.5* μ*M each) and CAII (2.5* μ*M) in 10 mM NH_4OAc (pH 7.0). (b) Collision-induced dissociation (CID) of the isolated complex ions of [CAII + Zn + 1]$^{8-}$. The CID condition was set such that the complex ions were completely dissociated. The insets show expanded views of the single-charged inhibitors, [1]$^{-1}$, as the CID product ion and the separation of inhibitor ions with the same nominal masses. The letters in the figure inset denote the amino acid composition of the inhibitor ions. (Reproduced with permission from* J. of Med. Chem., *1996, 39, No. 10., 1949–1955. Copyright© 1996, American Chemical Society.)*

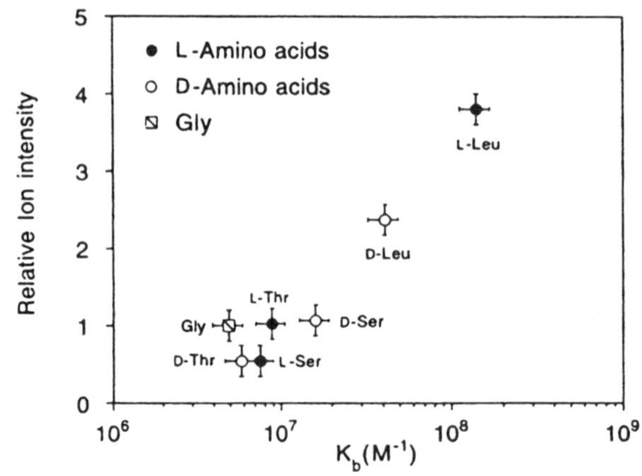

***Figure 7.4** Correlation of relative ion intensities* vs K_b *in solution for seven peptide inhibitors from L- and D-libraries. We selected inhibitors having identical* AA_1 *and* AA_2*, because their ion intensities are free of ambiguities that are caused by sequence isomers. The relative ion intensity for each ligand in both libraries is normalized to that of Gly-Gly, which exists in both libraries. The three-letter codes for amino acids are used to indicate the identity of* AA_1 *and* AA_2*. The values of* K_b *were measured in 20 mM phosphate buffer (pH 7.5) at 37 °C. The uncertainties for the measurement are indicated by the errors bars. (Reproduced with permission from* J. of Med. Chem., *1996, 39, No. 10., 1949–1955. Copyright© 1996, American Chemical Society.)*

state complexes primarily produced the -8 charge state of the protein ($[CA\ II + Zn]^{-8}$) and singly charged negative ions for nearly all the inhibitors (Figure 7.3b). The resulting mass spectra allow the dissociated ligands to be identified based upon their different molecular weights, and their ion intensities reflect the relative binding affinities.

Several inhibitors identified from the two libraries were resynthesized, and their binding affinities to CA II in solution were studied using a fluorescence-binding assay. The binding constants of the seven inhibitors in solution correlated well with the relative intensities of the ions dissociated from the CA II-inhibitor complexes (Figure 7.4). This result provides further support that binding constants in vapor-phase and solution-phase correlate and that the MS data can be used to infer the order of binding affinities in solution. The agreement between vapor and solution data is quantitatively good, but it will be necessary to examine other systems to establish the generality of this type of correlation.

This approach should be most effective for the identification of the tight binding ligands, because non-specific association induced during, or by, the ESI process may limit the quality of information for weaker binding ligands, particularly in larger libraries. In principle, the ESI-FTICR methods should be most useful for determining the relative affinities of a library of compounds to a protein when the binding study can be carried out using volatile buffers such as NH_4OAc, Tris acetate, and ammonium bicarbonate; the protein sample is moderately pure; all the library members will be charged after dissociation of complexes in the gas phase; and the library members have different masses.

A model experiment for the "on-line" screening of substrate libraries by enzymes using combinatorial libraries in combination with ESI-FTICR mass spectrometry has been performed (Wigger et al., 1997). The reaction between the electrophilic substrate 1-chloro-2, 4-dinitrobenzene and components of an H-y-Glu-Cys-Xxx-OH library, catalyzed by glutathione-S-transferase, has been monitored. It shows the feasibility of two-dimensional screening of substrate libraries by ESI-FTICR mass spectrometry.

This technology permits combinatorial substrate libraries to be screened in a simple experiment without tedious workup. The high-resolution analysis by FTICR enables researchers to distinguish between peaks of the same nominal mass and different elemental composition.

The approach is easily applied to a two-dimensional combinatorial experiment, where two functionalized libraries targeting two substrate-binding pockets are presented to a protein. After selection of the best binding substrates of each library, a bond is formed, and the higher mass product is analyzed by ESI-FTICR mass spectrometry.

7.2.2 Size Exclusion Chromatography-MS

Although the ESI-FTICR method does not need the separation of bound from free ligands, most other mixture screening methods need this separation. Size-exclusion chromatography can be used to separate complexes formed between the target protein and the selected ligand molecules, followed by MS identification of the selected compounds (Kaur et al., 1997, 1994, 1995). In this approach, the receptor of interest is incubated in solution with the compound library under receptor-limiting conditions, allowing only compounds with the highest affinities to bind. The receptor/ligand complex is separated rapidly from the unbound library compounds by size-exclusion chromatography. The receptor with bound ligand is excluded from the column, and the excess unbound ligands are eluted in the included volume. An aliquot of the receptor/ligand complex is introduced into the ion source of an electrospray triple-quadrupole mass spectrometer, and the sample is desalted and concentrated on-line in the injector loop, using a peptide reversed-phase cartridge. Under the conditions used for elution off the reversed-phase cartridge, the receptor/ligand complexes dissociate readily. The m/z range of the compound library is scanned to determine the molecular weight of the ligands, which have bound to the receptor in the affinity selection experiment. As a control, the selection experiment is repeated in the presence of a natural receptor ligand competitor with higher binding affinity. The compounds that bind specifically at the binding site are hence identified readily, because they are observed in the selection experiment in the absence of the competitor, but are not observed in the control experiment. Any ligands that are observed both in the absence and presence of the competitor are assumed to bind through general hydrophobic interactions at sites distinct from the receptor-binding site.

A library of 600 compounds was screened by this method (Kaur et al., 1995). The mass spectrum of the ligands that bound to the receptor in the absence of the competitor was obtained, and the molecular ions with m/z 708.4 and 676.4 were identified. The mass spectrum for the control experiments showed that only multiply charged ions from the competitor were observed. The absence of ions with m/z 708.4 and 676.4 indicates that these compounds do not bind in the presence of the competitor. To ensure that the absence of m/z 708.4 in the mass spectrum of the control experiment was not simply due to ionization suppression in the presence of the competitor, once the structure of the compound was established, it was synthesized as a single compound and analyzed in the presence of the competitor. Because compounds with m/z 708.4 and 676.4 compete with the natural ligand, these compounds presumably bind specifically at the receptor-binding site.

Vouros and coworkers (Dunayevskiy et al., 1997) studied the binding between the human serum albumin (HSA) and small molecules with different affinities to the protein. The experimental protocol is comprised of three major steps: isolation and separation of the receptor/ligand complex from the unbound combinatorial library members; dissociation of the complex; and mass spectrometric analysis of the released ligands. The physical separation of the complex is also achieved on a size-exclusion chromatography column. The dilute fractions collected must undergo a 50- to 100-fold preconcentration before further analysis. As an alternative, they chose gel-filtration spin columns, which are commonly used for buffer exchange in protein solutions. This separation method eliminates carryover problems and the need for reconditioning the chromatography stationary phase between runs. The complexes isolated can be dissociated by direct LC/MS operation or by pH denaturation of the protein in a direct CE/MS analysis. Diagram 7.1 shows the experimental procedure. By using this procedure, high-binding-affinity ligands (K_d < 1 μM) can be identified.

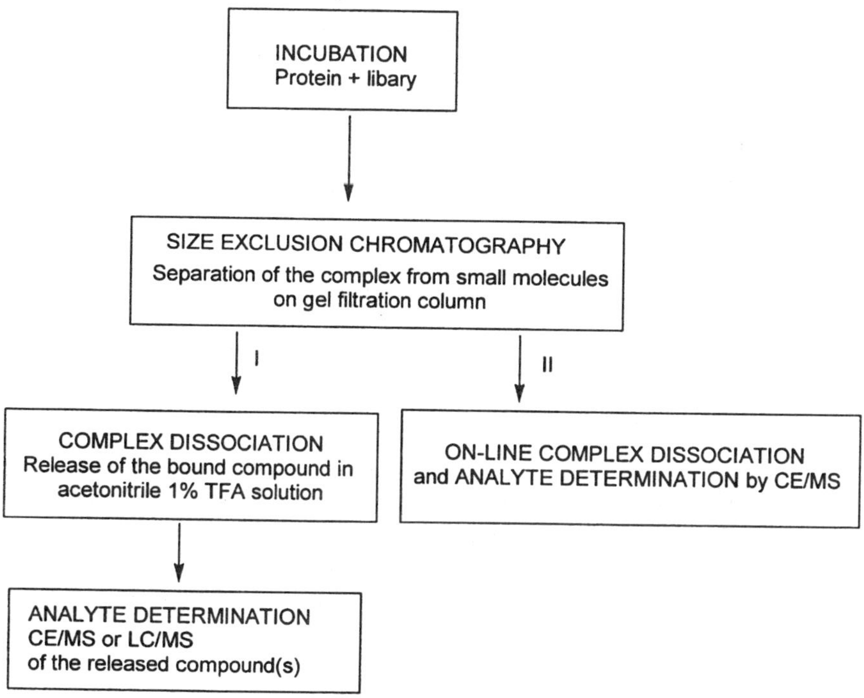

Diagram 7.1

LC/MS and CE/MS methods in structural identification were compared. Both methods provided the required data: the signal corresponding to the active ligand was clearly detected. One possible advantage of LC is its tolerance to the matrix components (salts) present in the sample. The high amount of salt would cause the reduction of the separation efficiency and peak broadening in CE. In LC, the buffer components mostly elute with the solvent in the void volume and are well separated from the ligands. On the other hand, injection volumes of nanoliters in CE allow the performance of multiple analyses from the limited amount of sample. This feature is very important in identification of the active ligands, when several injections have to be made in order to obtain tandem mass spectrometric (MS/MS) spectra for all selected compounds. In general, LC/MS and CE/MS can complement each other for studying diverse compounds from combinatorial libraries.

7.2.3 Immunoaffinity Extraction-HPLC or Ultracentrifugation-MS

An on-line immunoaffinity extraction coupled column HPLC ion-spray MS/MS system has been developed (Nedved et al., 1996). An automated column switching system incorporating an immunoaffinity extraction (IAE) column and two reversed-phase HPLC columns was coupled to a triple-quadrupole mass spectrometer. A Protein G column and antibodies to benzodiazepines were used to screen library components. A pH change in the mobile phase eluted the benzodiazepine-antibody complexes onto a C-18 restricted access media (RAM) column, thereby separating the selected benzodiazepines from the antibody. In a final step, backflushing the RAM column eluted the benzodiazepines onto a C-8 analytical reversed phase column for separation before detection and preliminary structural characterization using ion spray mass spectrometry (MS) and tandem mass spectrometry (MS/MS).

Full-scan IAE/LC/LC/MS and IAE/LC/LC/MS/MS chromatograms were obtained using a known 19-component library and an unknown 20-component benzodiazepine library. Results demonstrated the feasibility of this particular combination of chromatographic techniques to characterize combinatorial libraries.

Individual benzodiazepines are also selected from a multicomponent library mixture in solution by formation of noncovalent immunoaffinity complexes with antibodies raised to therapeutically proven drugs, such as nitrazepam, temazepam, or oxazepam (Wieboldt et al., 1997). Captured compounds are separated from nonspecifically bound library components by centrifugal ultrafiltration. The specifically selected molecules retained on the filter are subsequently liberated from the antibodies by acidification and are analyzed by HPLC coupled with ESI MS. Varying the stoichiometry of the complexation mixture controls competition by the benzodiazepines for limited antibody-binding sites. This procedure selects library components with the greatest affinity for a particular antibody. Affinity ultrafiltration and electrospray mass spectrometry complement each other to enhance screening and identification of pooled drug candidates and potentially can be extended to other small-molecule combinatorial libraries and macromolecular receptor preparations.

Immunoaffinity selection with ultrafiltration provides a method for elucidation of molecular recognition properties for mixtures of related compounds and antibodies. This method could be extended to include other types of receptors and the products of combinatorial library synthesis. Ultrafiltration membranes are currently available with molecular weight cutoffs ranging from 500 to 500,000 Dalton, so it is possible to choose experimental configurations appropriate for many macromolecules and biological receptors.

7.2.4 Pulsed Ultrafiltration-MS

A family of homologues is synthesized in a one-pot reaction, without isolation or purification, and the reaction mixture is screened using a competitive binding assay based on pulsed ultrafiltration/electrospray mass spectrometry (PUF/ESMS) to identify compounds having the highest affinity for a target receptor (Zhao et al., 1997). As a model system to test this approach, a synthetic scheme designed to prepare a series of analogues of the adenosine deaminase inhibitor erythro-9-(hydroxy-3-nonyl) adenine (EHNA), as diastereomeric mixtures, was carried out. Pulsed ultrafiltration screening of the crude reaction mixture against controls without protein detected protonated molecules corresponding to EHNA-type derivatives and three of its linear, alkyl homologues, but did not show protonated molecules for an isobutyl or benzylic EHNA derivative, suggesting the latter was inactive.

An essential aspect of screening large combinatorial libraries is the ability to identify the active components in these complex mixtures, which is usually based on the strength of binding to a selected target macromolecule. During pulsed ultrafiltration, a macromolecular target is trapped in a follow-through cell by an ultrafiltration membrane; then, an aliquot of a ligand is pumped through the ultrafiltration cell. As ligands bind and dissociate from the solution-phase receptors (macromolecules or complexes thereof), their elution profile from the ultrafiltration cell is altered relative to control.

7.2.5 On-Bead Binding Assay/MS Method

Two peptide libraries, Ac-MXXXXXBBRM and Ac-VXXXXXBBRM (B = β-alanine, X = all L-amino acids but C, M, and I), were constructed on TentaGel solid support to search for ligands that bind tightly with the H9724 Lyme antibody (Tu et al., 1998). By using an on-bead ELISA, approximately 120 ligands were selected as candidates for further study. Matrix-assisted laser desorption ionization mass spectrometry analysis of the candidate ligands indicated a high rate of occurrence of certain amino acids at the randomized positions. Based on the initial screening results, a small library was designed and iteratively synthesized. Subsequent library screenings led to the identification of four peptides, Ac-PQEEGX-NH_2 (X=R,K,A,D), that showed specific affinity to the antibody. This combination of solid-phase screening and iterative synthesis is an effective strategy for rapid identification of ligands that bind tightly with disease-specific antibodies and should be applicable to other ligand-receptor systems.

A synthetic chemical library comprised of alkylated and acylated amino acids was synthesized and screened to determine structures that bind to a model target, streptavidin (Stankova et al., 1994). The library was prepared using "split synthesis" and was screened in a solid-phase binding assay. The structure of positively reacting compounds was determined using mass spectroscopy. Positive compounds, together with various structural analogs, were synthesized, and their binding was confirmed.

One limitation of the MS method is the inability to distinguish between isomeric and isobaric species. This problem would be exacerbated as the complexity of the mixture increases. Therefore, if this method is to be used for optimizing enzyme substrates from complex mixtures, the design and synthesis of defined peptide substrate mixtures should be made in the context of the limitations of the particular screening method.

A rule-based procedure for substituent selection allows the design of libraries in which all components have unique nominal molecular weights (Hughes, 1998). Further rules allow the use of isotope patterns to extend the final library size beyond the limits imposed by the desired molecular-weight range. Because each component can be uniquely identified by its molecular weight and isotope pattern, such self-coded combinatorial libraries can, in principle, be applied to any screening approach where sufficient material is available for MS analysis. This strategy imposes no restriction on the synthetic methods used to prepare the library, which is synthesized once only, requires no additional steps for introduction of tag, and therefore is an attractive alternative approach to combinatorial library design.

This approach can only be applied to small libraries. In circumstances where very large combinatorial libraries are studied, a tagging strategy would be more appropriate.

7.2.6 ACE-MS

On-line affinity capillary electrophoresis-electrospray ionization-mass spectrometry (ACE-MS) was used for the simultaneous measurement of multiple constants of an all-D-tetrapeptide library to the model receptor, vancomycin (Dunayevskiy et al., 1998). Determination of K_d values for the 19 peptides in the form of Fmoc-DXYA is demonstrated. The data are compared with the results obtained for individual compounds using ACE-UV, and good correlation between the two detection methods was shown. Simultaneous determination of multiple K_d values by ACE-MS is achieved in one set of experiments, whereas only one K_d value can be obtained by ACE-UV during the same time. ACE-MS measures multiple binding constants in solution in a fast and reliable manner using femtomole amounts of samples.

7.2.7 Other MS Methods

ESI-MS has been used to measure solution-binding constants for the complexes between vancomycin group antibiotics and several different peptide ligands. Furthermore, the solution-phase stability of the ristocetin-diacetyl-L-lysyl-D-alanyl-D-alanine complex as a function of the pH of the solution is well reflected in the ESI mass spectra. The solution-phase stereospecific binding for the studied noncovalent complexes was investigated by ESI-MS using competition experiments with an antibiotic and two stereoisomeric peptide ligands: diacetyl-L-lysyl-D-alanyl-D-alanine and diacetyl-L-lysyl-L-alanyl-L-alanine-d_6. Using appropriate experimental conditions, only the specific binding of the diacetyl-L-lysyl-D-alanyl-D-alanine peptide was observed, as expected, because this stereoisomer is known to bind specifically to the antibiotics in solution. These results suggest that ESI-MS spectra can be, upon careful usage, genuine reflections of structurally specific interactions in solution. The binding of peptide ligands to vancomycin antibiotics involves several intermolecular hydrogen bonds between the amides of the peptide backbone of the antibiotic and the amide groups of the ligand. This type of noncovalent binding is favorable in the gas phase (Wu et al., 1997) and may be the reason for the good correlation observed between complex formation in solution and relative ion abundance in the ESI-MS spectra.

Libraries of ACE inhibitors were prepared by solid-phase methods involving reductive alkylation of resin-bound dipeptides. Affinity selection of the library components identified hydrophilic N-alkylated dipeptides as potential inhibitors (Blackburn et al., 1997).

ESI-MS analysis of adducts formed between limiting amounts of enzyme and a mixture of irreversible inhibitors made the identification of activated dipeptide inactivators of HAV 3C proteins possible (N-iodoacetyl-Ser-Phe-NH_2, Mckendrick et al., 1998).

Quantitative MALDI MS of libraries of porphyrins and peptide-DNA hybrids consisting of up to 35 compounds is described (Berlin et al., 1997). Porphyrin libraries were subjected to *in vitro* selections for liposome incorporation and binding to a protein pocket. It was shown that mesohydroxyphenyl-substituted porphyrins, known as high-activity photosensitizers of tumors, are preferentially incorporated into liposome membranes. A mixture of peptide-DNA hybrids was assayed for the nuclease stability of its components.

7.3 AFFINITY CAPILLARY ELECTROPHORESIS

Chu and Cheng (1998) recently reviewed ACE methods. In ACE, the electrophoretic mobility, μ, of a ligand is altered upon complexation with a receptor in solution. The principle of the method is illustrated schematically in Figure 7.5. The electrophoretic mobility of a protein is related to its mass (M) and net charge (Z) by a relationship of the form $\mu \sim Z/M^{2/3}$. If the protein binds to a charged ligand of relatively small mass, the change in μ due to the change in mass [from $M^{2/3}$ to $(M + m)^{2/3}$] is small relative to the change in μ due to the change in charge (from Z to Z ± z). Thus, the protein-ligand complex will migrate at a different rate than the uncomplexed protein (see Chu et al., 1992).

The running buffer containing the receptor is introduced into the CE capillary followed by the injection of the mixture of ligands. In the present system, kinetic off-rates are fast relative to analyte migration

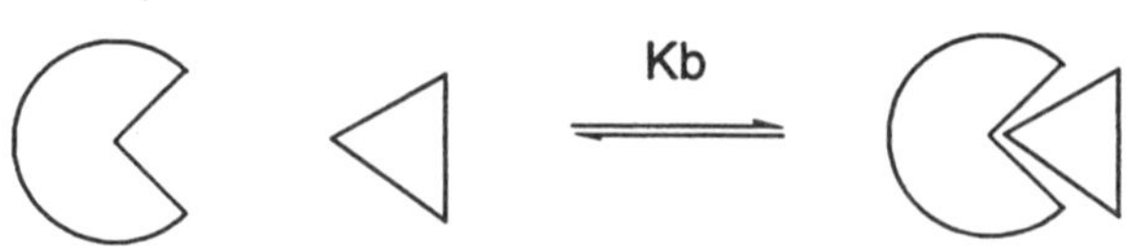

Mass	M	m	M + m
Net Charge	Z	±z	Z ± z
Electrophoretic Mobility μ	$\sim Z/M^{2/3}$		$\sim(Z \pm z) / (M + m)^{2/3}$

Figure 7.5 *The principle of the affinity capillary electrophoresis.*

time. The receptor-ligand complex will undergo multiple events of association-dissociation during the electrophoretic run. Consequently, complexation with the receptor shifts the mobility of the ligand toward the electrophoretic mobility of the complex. This mobility shift corresponds to the ligand's binding affinity and can be used to calculate the dissociation constant of complex (K_d) according to

$$K_d = [R]\{(\mu_t - \mu_c)/(\mu_o - \mu_t)\}$$

where subscript o represents the free ligand in the absence of the receptor, t the free ligand with receptor present in the running buffer at a concentration of [R], and c the ligand complexed with receptor.

ACE experiments allow the qualitative ranking of library members according to their binding affinity to the receptor. When the best ligands are selected, they can be individually resynthesized, and their binding constants can be accurately measured in separate experiments by ACE. A background electrolyte containing a fixed concentration of vancomycin is introduced into the capillary as a plug of varying length by changing the time of injection. Then, a small plug of the library (sample) is introduced. This approach, referred to as the "plug-plug" approach, is illustrated in Figure 7.6 for a library of 100 Fmoc-DDXX peptides. With increasing length of the vancomycin-containing plug, ligands increase the numbers of interaction events during migration through the capillary. With longer receptor plugs, the most tightly bound ligands were retained for greater time compared to migration with no receptor. With the longest plugs, retention of lower affinity peptides could also be observed (Chu et al., 1996). When the receptor and ligands were of the same charge and, thus, migrated in the same direction, the electrophoretic conditions could be chosen so that the receptor and ligands have significantly different mobilities.

Scatchard analysis of the change in the electrophoretic mobility of the protein as a function of the concentration of the ligand in the electrophoresis buffer allows the determination of the binding constant (K_b) for the interaction. Figure 7.7b shows the electropherograms predicted by the model for the migration of CA in an electrophoresis buffer containing 50 mM of charged ligand **1** (Diagram 7.2) and increasing amounts of neutral ligand **2.** The binding constants $K_b^{\pm}$ and K_b^{o} used as input parameters in the simulation were 1.5×10^5 M^{-1} and 1.1×10^6 M^{-1}, respectively. Scatchard analysis of the simulation-generated electropherograms (Figure 7.7b) using equation $R_f/[L^o] = (K_b^o) - (K_b^o)\, R_f$ did not return to the original value of the binding constant used in the simulation but instead gave a value of 1.3×10^5 M^{-1} ($r > 0.99$); this value was equal to the apparent K_b^o measured experi-

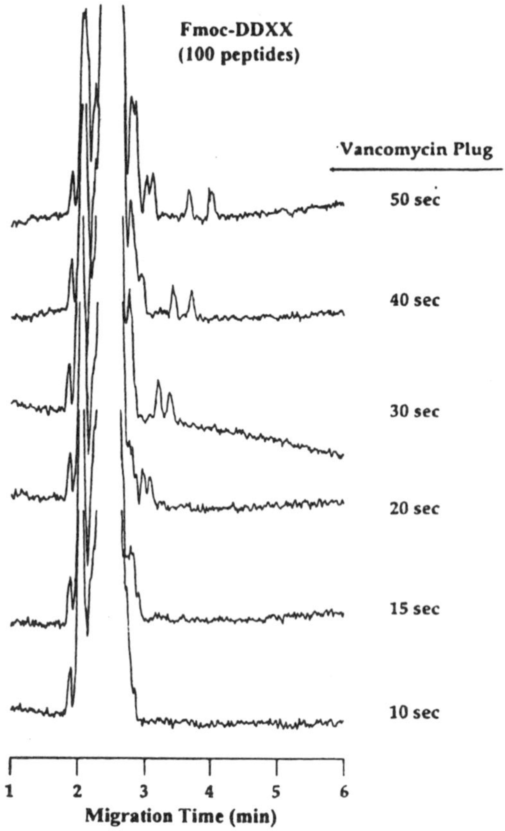

Figure 7.6 *ACE-UV analysis of the Fmoc-DDXX library (100 peptides) with different length plugs of background electrolyte containing 70 μM vancomycin. 50 mM Tris-acetate, pH 8.1, 13 kV, 16 μA; PAV coated capillary, 50 μm i. D., 20 cm effective length, 27 cm total length. (Reproduced with permission from* J. Am. Chem. Soc., *1996, 118, 7835. Copyright© 1996, American Chemical Society.)*

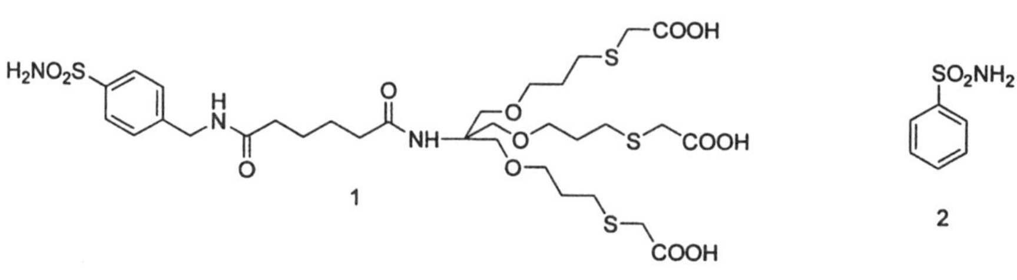

1

2

Diagram 7.2

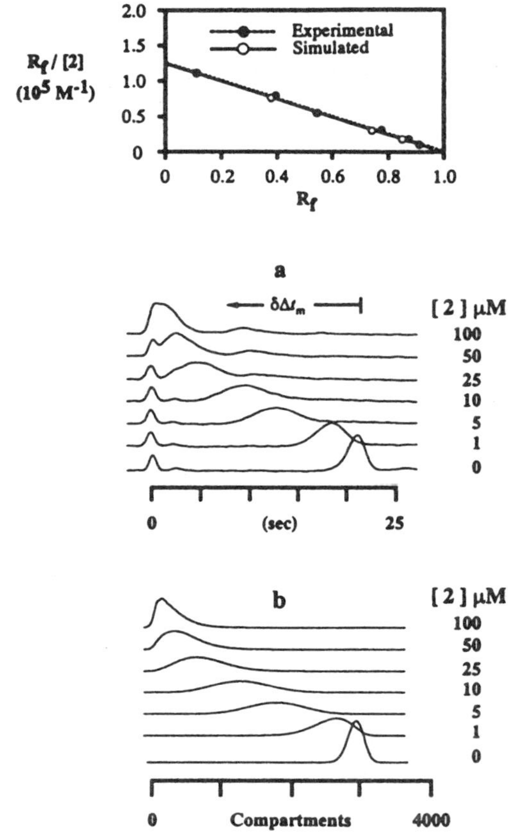

Figure 7.7 *The migration time of CA changes with increasing concentration of neutral 2 in the presence of 50 μM of charged ligand 1 in the electrophoresis buffer. The nonmobile peak is due to horse heart myoglobin. Stacked electropherograms (a) were obtained experimentally ($[CA]_0$ = 10 μM; 192 mM glycine—25 mM Tris buffer, pH 8.3) and (b) were generated by simulation using the parameters $k_{off} = 0.1\ s^{-1}$ and $k_{on} = 1.1 \times 10^5\ M^{-1}\ s^{-1}$ for the neutral ligand. CAL_0 is not mobile within the reference frame. The graph is a Scatchard plot of the data. (Reproduced with permission from* J. Med. Chem., *1993, 36, 129. Copyright© 1993, American Chemical Society.)*

mentally (Figure 7.7a). The original value of the binding constant (K_b^o) used in the simulation was obtained using equation $K_b^o = K_{b,}^o$, $(1 + K_b \pm [L_\pm])$ (Avila et al., 1993).

Interaction of a protein and a ligand that induces a change in the migration time of the protein may also broaden the peak due to the protein in the region of concentrations corresponding to migration times intermediate between that of free protein and that of its saturated complex with ligand. This broadening is most pronounced if the dissociation time ($1/k_{off}$) is of the same magnitude as the migration time of the protein. Analysis of this broadening makes it possible to estimate values for k_{off} and k_{on} in certain cases.

ACE is a useful method to study protein-ligand interactions. The attractive features of ACE include its ability to provide assessment of protein-ligand interactions using very small amounts of samples and ligands in relatively short time. The concentration of the protein in the sample does not appear in the equations used for analysis. The protein need not be pure, and simultaneous determination of individual binding constants using a sample of a mixture of isozymes or different proteins is possible. For example, ACE allowed simultaneous measurement of the binding constant of ligands to CA through changes in migration time and of the off rate for the binding interaction from the peak width.

A major problem in using ACE is the tendency of proteins to adsorb on the wall of uncoated capillaries. This problem becomes more pronounced when the pH of the electrophoresis buffer is close to or lower than those of the protein. A number of approaches are available to control this interaction between protein and capillary wall including the use of additives to the buffer and chemical modification of the silanol groups of the capillary wall. Furthermore, ACE requires that at least one of the ligands have a sufficient charge to cause a measurable change in the migration time of protein-ligand complex.

7.4 AFFINITY NMR

In early 1980s, Stilbs developed a technique called "size-resolved NMR," which was based on diffusion-related echo attenuation. He and coworkers have used this technique to probe the association of different alcohols to α-cyclodextrin (α-CD) in aqueous solution (Rymden et al., 1983).

In the early 1990s, Johnson and coworkers reported a noninvasive analytical method named diffusion-ordered 2D-NMR spectroscopy (DOSY). This method is based on a 2D experiment that displays an NMR chemical shift spectrum in one direction and a "spectrum" of diffusion coefficients or particle sizes in the other (Morris et al., 1992). NMR spectra are obtained with the FT pulsed field gradient method, and data sets are collected as the gradient pulse areas are incremented through approximately 50 values. Then, powerful data inver-

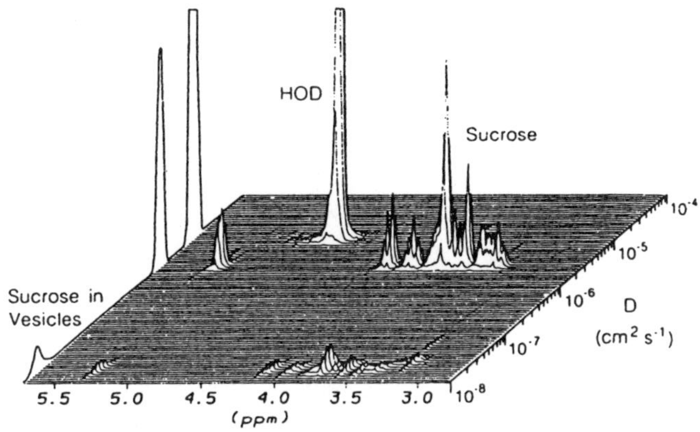

Figure 7.8 *The diffusion-ordered 2D-NMR spectrum of a sample containing sucrose and phospholipid vesicles in D_2O. (Reproduced with permission from* J. Am. Chem. Soc., *1992, 114, 3140. Copyright© 1992, American Chemical Society.)*

sion programs are used to generate diffusion spectra at each chemical shift. The diffusion coefficients spanning five orders of magnitude with a lower limit of less than 10^{-10} cm^2s^{-1} can be measured. Figure 7.8 provides an illustration of DOSY for sucrose in the presence of phospholipid vesicles. Thirty-nine NMR spectra were collected. The diffusion peaks for free sucrose and sucrose entrapped in the vesicles are easily resolved with diffusion coefficients of 3.00×10^{-6} and 1.86×10^{-8} cm^2s^{-1}, respectively.

A related technique called diffusion encoded spectroscopy (DECODE), which involves pulse field gradient (PFG) NMR combined with total correlation spectroscopy (TOCSY), was developed (Lin et al., 1997a). This method relies on PFG NMR, which spatially encodes molecules in solution and enables structure determination of the individual components due to differences in diffusion coefficient.

PFG conditions can be established by which all the resonance of small molecules in the molecular weight (MW) of 200–400 will disappear from the NMR spectrum while the spectra of molecules associated with larger partner molecules (smaller diffusion coefficient) will reappear.

In Figure 7.9, spectra of a nine-component mixture are shown without PFG (a). Applying PFG condition (b) eliminates the normal 1D ^1H NMR spectrum for nine components. When hydroquinine-9-phenanthryl ether was added, the spectra of two compounds (1 and 2) that bind to hydroquinine-9-phenanthryl ether reappear (c).

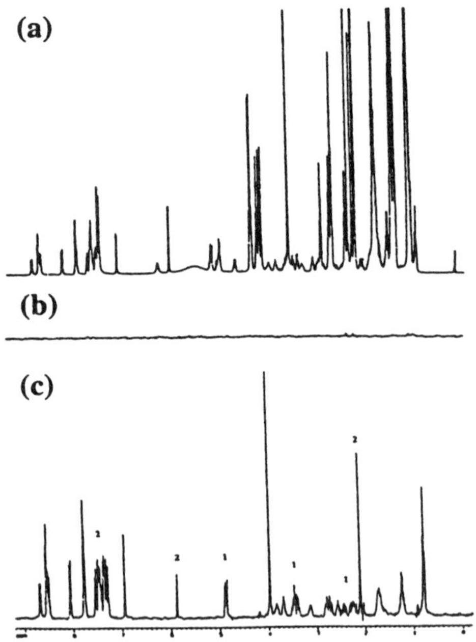

Figure 7.9 *(a) 1D 400 MHz ^1H NMR spectrum of the nine-component mixture in $CDCl_3$. The concentration of each component is 10 mM. Components: (1) DL-isocitric lactone, (2) (S)- (+)-O-acetylmandelic acid, (3) DL-N-acetylhomocysteine thiolactone, (4) (±)-sec-butyryl acetate, (8) buty levulinate, (9) hydroquinine 9-phenanthryl ether. (b) 1D PFG ^1H NMR spectrum of the mixture without hydroquinine 9-phenanthryl ether using LED sequence. (c) 1D PFG ^1H NMR spectrum of the nine-component mixture. Chemical shifts arising from compounds 1 and 2 are labeled. All other shifts are from compound 9. The PFG conditions are the same as in the middle spectrum. (Reproduced with permission from* J. Am. Chem. Soc., *1997, 119, 5249–5250. Copyright© 1995, American Chemical Society.)*

Using this method, a small mixture of compounds can be selectively edited to find and identify components involved in molecular interactions. The structure of the interacting ligand can be deduced directly, without separation of the mixture. However, the application of this method in pharmaceutical research will likely be limited by the total compound concentration tolerable to the biological target, because the relatively high concentration of each component required by NMR quickly adds up to a high total concentration of compounds in the mixture.

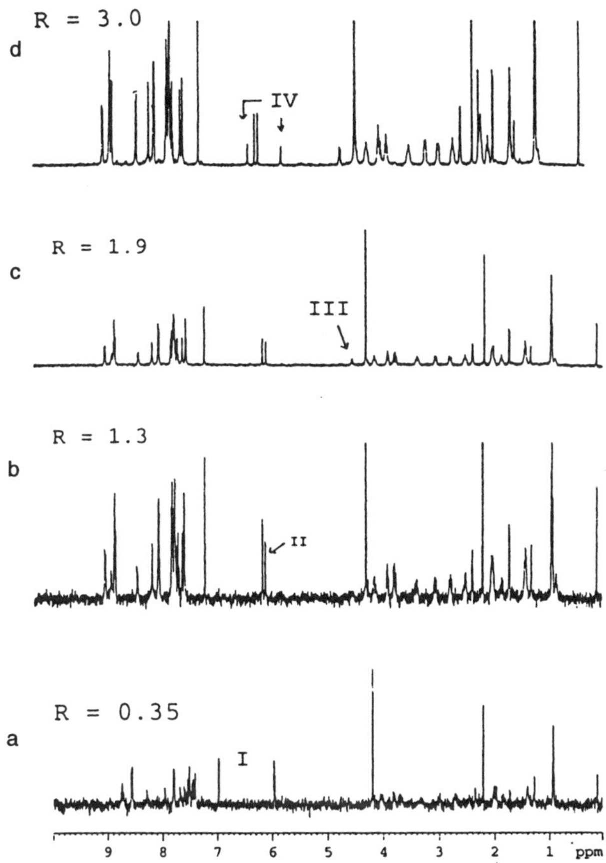

Figure 7.10 *Titration of the 4-carboxylic acid mixture with hydoquinine 9-phenanthryl ether. R is the ratio of hydroquinine to the total equimolar mixture of acids. Important resonance arising from compounds I–IV are shown: I, dichloroacetic acid; II, s-(+)-O-acetylmandelic acid; III, 2-chloropropionic acid; and IV, methacrylic acid. (Reproduced with permission from* J. Org. Chem., *1997, 62, 8930–8931. Copyright© 1997, American Chemical Society.)*

Because the detection of binding by the PFG diffusion NMR method is based on the observation of the ligand NMR spectrum rather than the changes in the receptor NMR spectrum, this should allow the experiment to be "tuned" to the binding affinity of the ligand by changing the relative receptor concentration. This idea was tested by making a mixture of four carboxylic acids and using the hydroquinine-9-phenanthryl ether as the receptor (Figure 7.10) (Lin et al., 1997b). The

NMR spectral tuning is obtained by changing the concentration of the hydroquinine-9-phenanthryl ether relative to the mixture of carboxylic acids. As shown in Figure 7.10, when the concentration of hydroquinine-9-phenanthryl ether is only 0.35 equivalent, only the strongest binding dichloroacetic acid is observed in the NMR spectrum. As the concentration of hydroquinine-9-phenanthryl ether is increased, the other carboxylic acids, II–IV, sequentially begin to appear in their order of binding affinity. All four acids are observed in Figure 7.10d.

This experiment proves that the system can be tuned to a desired sensitivity level and that mixtures of compounds containing only a few components can be screened to select components that bind at a desired level of affinity. Alternatively, mixture compounds could be directly rank ordered by binding affinity.

The complexation reactions between 12-crown-4 (12C4), cyclene (12N4), and 1,4,7,10-tetrathiacyclododecane (12S4) with γ-cyclodextrin (γ-CD) were studied by NMR diffusion measurements (Gafni & Cohen, 1997). It was suggested that, when the host and the guest are both organic species, then NMR diffusion measurement is an attractive method for studying complexation. The larger the difference in size between the guest and the host, the more accurate are the values obtained from NMR diffusion measurements because the diffusion coefficient is a parameter that is intuitively related to the formation of complexes. NMR is not subject to misinterpretations that can be caused by minor impurities and provides better structural information on the investigated species. However, NMR diffusion measurements are relatively time-consuming as compared to other methods. Because this method is based on spin echo, it is difficult to apply on systems that are characterized by short spin-spin relaxation time (T_2). The considerable differences in observed translational diffusion coefficients between binding and nonbinding ligands are required to distinguish them. The free and bound ligands are in fast exchange on the diffusion time scale; thus, the observed diffusion coefficients are the weighted average of the free and bound species. Ligands in the free form should diffuse at a much faster rate than those in the bound state, so that even a small amount of the ligand in the free state causes the apparent diffusion rate of binding ligand to increase considerably.

A new method, diffusion-assisted nuclear Overhauser effect (NOE) pumping method, was reported (Chen & Shapiro, 1998). The concept is to distinguish from nonbinding ligands, by NOE, those molecules capable of binding to the macromolecule. Unlike the traditional affinity NMR experiment, which uses diffusion editing to eliminate signals from the unbound ligands, the NOE-pumping method relies on NOE to

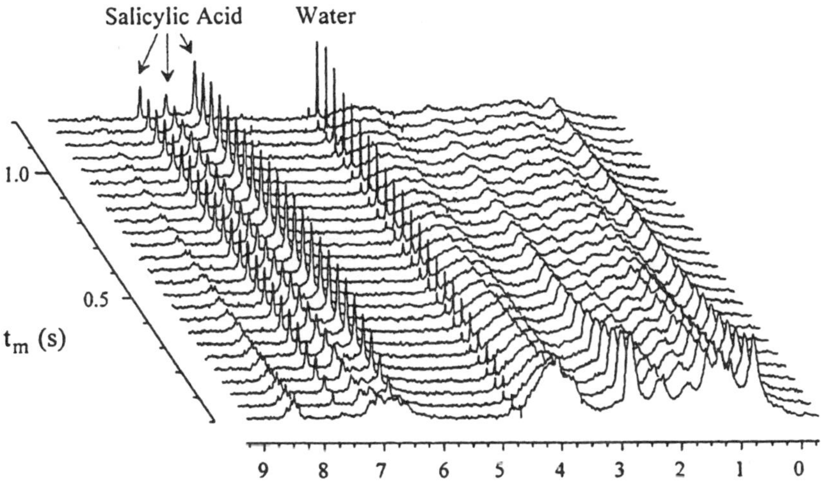

Figure 7.11 *Stack plot of 1D ^1H spectra of human serum albumin with three ligands (salicylic acid, 10 mM; glucose, 10 mM; L-ascorbic acid, 10 mM) as the mixing time t_m increases from 5 ms to 1.2 s in intervals of 0.05 s at a temperature of 300 K, where g = 34.8 g/cm, δ = 3 ms, T = 50 ms, τ = 3.5 ms, 512 scans were collected, and the experiment time for each spectrum is 30 to 40 min. Signal from bound ligand salicylic acid and water are indicated. (Reproduced with permission from* J. Am. Chem. Soc., *1998, 120, 10258–10259. Copyright© 1998, American Chemical Society.)*

transfer the signal from the receptor to the bound ligands. When three ligands (salicylic acid, glucose, and L-ascorbic acid) are added to human serum albumin in D_2O, the binding-ligand signals disappear under the condition where a large enough gradient is applied to suppress the nonbinding ligand signals. However, the bound ligand signals, including the macromolecule-associated water signal, grow substantially as the mixing time t_m increases (Figure 7.11). Signal loss with an increase of the mixing time is much more severe for the protein than for the binding ligand. This results not only from the longer T_1 of the small ligand favoring conservation of the signal but also from constant signal pumping from the protein to the ligands. Consequently, a longer mixing time can be used to suppress the protein signal, giving a cleaner 1D NMR spectrum consisting of mainly the binding ligand signals.

The 1D NOE-pumping experiment offers several advantages. Because the detected signals of "binding" ligand are transferred from the receptor, false positives can be greatly eliminated. Binding activities are observed by directly detecting binding-ligand signals; therefore, any chemical shift changes in the spectrum due to any chemical environment variation will not affect the result. The NOE-pumping experiment can be applied to very large biomolecules without well-resolved macromolecule signals and without the need for isotope labeling. Compared to the simple diffusion-edited affinity NMR experiment, this method does not require that the apparent diffusion coefficients of binding ligands be substantially different from the non-binding ligands.

7.5 OTHER AFFINITY SELECTION METHODS

Affinity selection methods described in the previous section can only screen small libraries. For large compound mixtures, other affinity selection methods have to be developed. On-resin assay can identify active compounds from very large libraries. In some cases, active beads can be selected simply by color changes (Ohlmeyer et al., 1993; Yu et al., 1997). Some on-bead assay methods are shown below.

7.5.1 Fluorescence-Activated Cell (Bead) Sorting (FACS or FABS)

The binding of an antibody, mAb D32.39, to the bead-bound compound libraries and the control beads was analyzed by fluorescence-activated cell (bead) sorting (FACS or FABS) (Needels et al., 1993). Figure 7.12a showed that beads carrying the positive control sequence RQFKVVT and a 69-mer oligonucleotide tag are strongly stained by the antibody, where blank beads are unstained. By contrast, only a small fraction of the encoded library bound mAb D32.39 (Figure 7.12b). Analysis of 105 events indicated that ~2% of the library stained above background levels. Significantly, this binding to mAb D32.39 was specific for the binding site, as it could be completely blocked by pre-incubating the mAb with soluble RQFKVVT peptide (Figure 7.12c). Individual beads from the library having fluorescence intensities comparable with the positive control beads were sorted into microcentrifuge tubes for tag amplification by PCR.

This method was also applied to identify high-affinity ligands to SH2 domains (Muller et al., 1996; Zaegel et al., 1996). Recombinant SH2 domains, expressed as glutathione S-transferase (GST) fusion proteins, were incubated with a phosphotyrosine (Y)-containing peptide library.

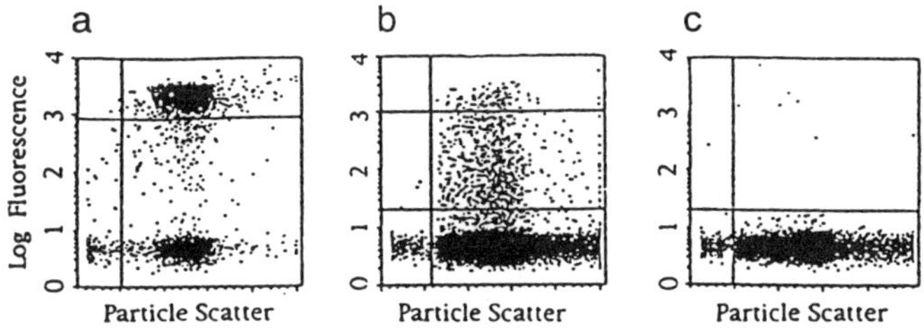

Figure 7.12 *Flow cytometric analysis of binding of mAb D32.39 to 10-µm beads bearing peptide and oligonucleotide. Approximately 10^5 events are recorded in each experiment. Fluorescence intensity is shown on vertical axis. (a) A 1:1 mixture of underivatized (lower population) and RQFKVVT (upper population) beads as negative and positive controls. (b) Binding of mAb to library. (c) Specific binding to library blocked by pre-incubation of mAb with 10 µM RQFKVVT. (Reproduced with permission from Needels et. al.,* Proc. Natl. Acad. Sci. USA, *1993, 90, 10703. Copyright© 1993, National Academy of Sciences, USA.)*

Using the "split and pool synthesis" approach by which each bead carries a unique peptide sequence, 6.4×10^5 individual peptides of nine amino acids in length (EPX_6 YX_{19} X_7 X_{19} X_7 X_6) were synthesized on beads (10 µm diameter). Binding of peptides to a fluorescence-labeled SH2 domain generated fluorescent beads. Flow cytometric sorting isolated high-fluorescence beads. Subsequent pool sequencing of the selected beads revealed a distinct pattern of phosphotyrosine-containing motifs for each individual SH2 domain: the SH2 domain of the adapter protein Grb2 predominantly selected beads with the sequence YENDP, whereas the C-terminal SH2 domain of the tyrosine kinase Syk selected YEELD. The pool sequencing on the isolated beads was used to obtain consensus information. High sample throughput is the major advantage of the flow cytometry method. A sorting rate of more than 1000 beads per second is easily achieved with conventional instruments. Short analytical runs of the library beads incubated with the binding partner at different concentrations (Figure 7.13) immediately yield information on the range of affinity interactions present on the beads.

7.5.2 Radio Activity Assay

A ^{14}C-labeled macrobicyclic receptor was assayed against an encoded combinatorial tripeptide library to establish the efficacy of radioactivity assay (Nestler et al., 1996). The described microradiographic

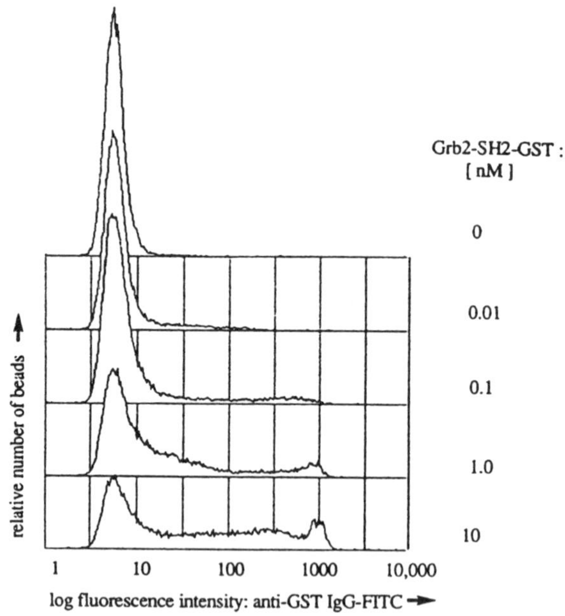

Figure 7.13 Binding of Grb2-SH2-GST to library beads. Stacked overlays of fluorescence histograms from flow cytometric analysis of library beads that were incubated at the indicated concentrations of Grb2-SH2-GST. (Reproduced with permission from Muler et al., J. Biol. Chem., *1996, 271, 16502. Copyright© 1996, The American Society for Biochemistry and Molecular Biology.)*

scheme promises to be advantageous in cases where the receptors show only modest discrimination between the radioactivity levels after the equilibration with the radiolabeled receptor. The radiolabeling of substrates allows for the general tracing of receptor-ligand interactions during "on-bead" affinity assays, and the successful radioassay of a receptor shows that the microradiographic conditions do not affect the integrity of the molecular tags, nor do they induce artifacts in the screening.

7.5.3 High-Pressure Liquid Chromatography (HPLC)

The rapid deconvolution of combinatorial libraries prepared via solution-phase synthesis was demonstrated using HPLC (Griffey et al., 1998). Solution-phase combinatorial synthesis was used to prepare large quantities (20–100 mg) of libraries containing 25–1000 members. These large libraries were fractionated into smaller subsets of compounds using HPLC techniques, which are well suited for biological evaluation. The fractions, which demonstrate biological activity, can be separated further using an orthogonal HPLC method. In many cases, single compounds can be isolated and identified without the need for iterative synthesis, deconvolution, or tagging.

7.6 ENCODING METHODS

If an active compound was found through screening of a combinatorial library assembled using "split-and-pool" method, the structural elucidation of the active compound is a major challenge. Assuming the synthesis was successful in every step, the knowledge of the reaction sequence is equivalent to knowing the member's chemical structure. Therefore, various chemical encoding schemes have been developed to record the synthetic history of each library member. These methods have the potential to code very large numbers of components, but inevitably require extra steps to introduce tags, and also limit to varying degrees the scope of the chemistry available for ligand synthesis due to the required orthogonality with tagging chemistry, protection, and cleavage strategies. Recently, the use of radio-frequency tags as a nonchemical coding strategy has been reported (Nicolaou et al., 1995; Moran et al., 1995). Several encoding methods (for review see Czarnik, 1997a, b) have been used or proposed to be used in combinatorial chemistry.

Provided that sufficient material is available, direct analysis of selected ligands obviates the need for tagging, and in the case of peptides and oligonucleotides, this can be achieved by sensitive sequencing techniques. For smaller, nonoligomeric compounds, mass spectrometry (MS) provides a generally applicable technique for analysis of the small quantities (typically in the picomole range) of material available and has been applied to affinity-selection methods as well as to the identification of material released from single beads. Increasing library complexity can quickly blur MS information. MS analysis of a single bead used as an encoding method (Geysen et al., 1996) can only be applicable to very small or specially designed libraries that do not have isobaric complexity. NMR structural determination is unambiguous. However, the amount of pure compound from a single bead is not enough for such determination at present, although single-bead NMR has been studied. In the following, a brief account of three compound identification methods will be given:

- chemical encoding
- electronic encoding
- single-bead analysis

7.6.1 Chemical Encoding

The "split-and-pool" synthesis (also known as portion mixing-and-divide or couple-and-recombine methods) was a revolutionary advance in combinatorial synthesis (Furka et al., 1988, 1991; Lam et al., 1991; Houghten et al., 1991). Its creative power in producing large and diverse compound libraries has been unsurpassed. As HTS is miniaturized and the required amount of compound is further lowered, the split synthesis will become a powerful way to make discrete compounds at the single-bead level.

This method involves dividing a portion of resin into a number of lots, allowing each lot to react with a different synthon, and then mixing all the lots. The mixed resin lots can undergo a common addition or removal of a protecting group, a common transformation, or treatment. They are then redivided, and each new lot that contains all the products of the first step is allowed to react with a different synthon to produce all possible products. The total number of compounds prepared is the product of the number of lots into which each step is divided. For example, if the procedure were performed five times with division into lots of 10, 12, 9, 12, and 11 each, a total of $10 \times 12 \times 9 \times 12 \times 11 =$ 142,560 compounds would be produced in only five parallel reaction steps in ~11 reaction vessels.

Figure 7.14 is a schematic representation of the "split-and-pool" synthesis protocol of a "one bead one compound" encoded synthetic library. An independent coupling immediately follows each synthetic step, carried out on a particular portion of the bead population, with a tag component coding for the step. In principle, the tag allows us to retrace the history of the operations any particular bead has gone through and, therefore, to deduce the sequence of its corresponding ligand.

A library of $\sim 10^6$ different peptide sequences on small, spherical (10 μm diameter) beads by the combinatorial chemical coupling of both L-and D-amino acid building blocks was synthesized (Needels et al., 1993). To each bead are covalently attached many copies of a single peptide sequence and, additionally, copies of a unique single-stranded oligonucleotide that codes for that peptide sequence. The oligonucleotide tags are synthesized through a parallel combinatorial procedure that effectively records the process by which the decided peptide sequence is assembled. The collection of beads was screened for binding to a fluorescence-labeled anti-peptide antibody using a FACS instrument. Those beads to which the antibody bound tightly were isolated by fluorescence-activated sorting, and the oligonucleotide identifier attached to individual beads was amplified by the PCR. Sequences of the amplified DNAs were determined to reveal the

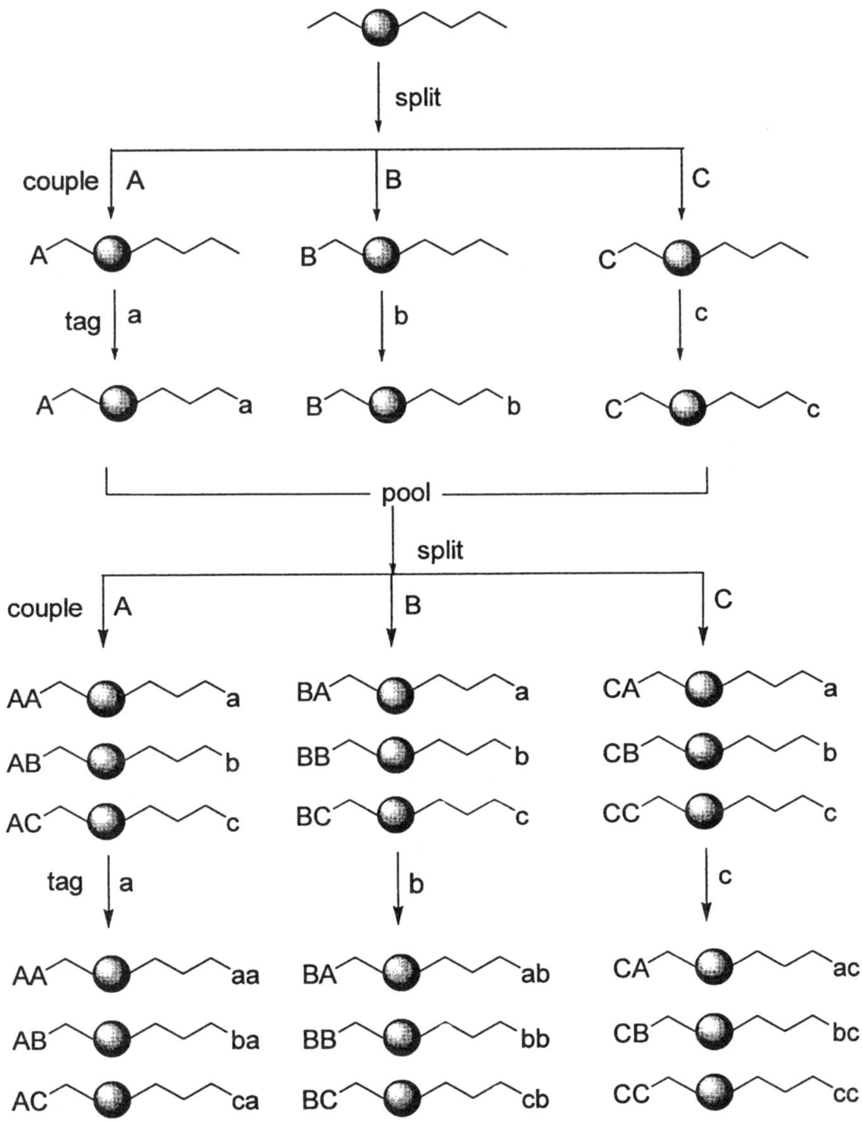

***Figure 7.14** Synthesis scheme for an encoded "split-and-pool" library.*

identity of peptide sequences that bind to the antibody with high affinity.

The limitations of this approach are that the use of a set of protecting groups on the amino acids and nucleotide building blocks is mutually orthogonal and that each of the polymer chains needs to be stable to reagents used in the synthesis and deprotection of the second chain.

A method for encoding non-natural components in a combinatorial library with standard amino acids is the parallel and alternating synthesis of two polymer chains (Kerr et al., 1993). The ability to decipher non-natural components in a library by Edman sequencing allows for applications in combinatorial synthesis.

The construction of a peptide combinatorial library, having 117,649 different members, synthesized on beads and indexed with inert chemical tags was reported (Ohlmeyer et al., 1993). These tags are used as a binary code to record the reaction history of each bead.

Split synthesis was used to construct a combinatorial library. During each step of synthesis, however, the beads were attached tagging molecules that encode both the step number and the chemical reagent used in that step. The array of tags used form a binary record of the synthetic steps for each bead. The tagging molecules are not sequentially connected; hence, no cosynthesis is required. With only 20 such tags, it is possible to uniquely encode 2^{20} = 1,048,576 different syntheses.

Capillary GC is a convenient analytical technique for separating and identifying tags. When tagging molecules with unusually electrophoric functionality (e.g., fluoro- or chlorocarbons) are used, the conjunction electron capture (EC) detection with GC separation enables them to analyze tags from a single 50-μm bead without significant interference from contaminants.

A set of 18 GC-separable tags (Diagram 7.3) with linkers that allowed them to be attached to, and then detached from, the synthesis beads were prepared. By using different lengths of the hydrocarbon chain (N = 1–10), and three different aromatic electrophores (Ar), more than 20 tagging molecules were prepared, well-separated by capillary GC, and selectively detected by EC at levels < 1 pmol.

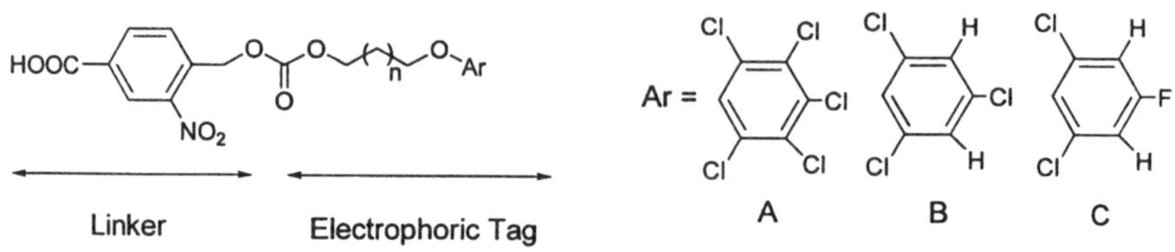

Diagram 7.3

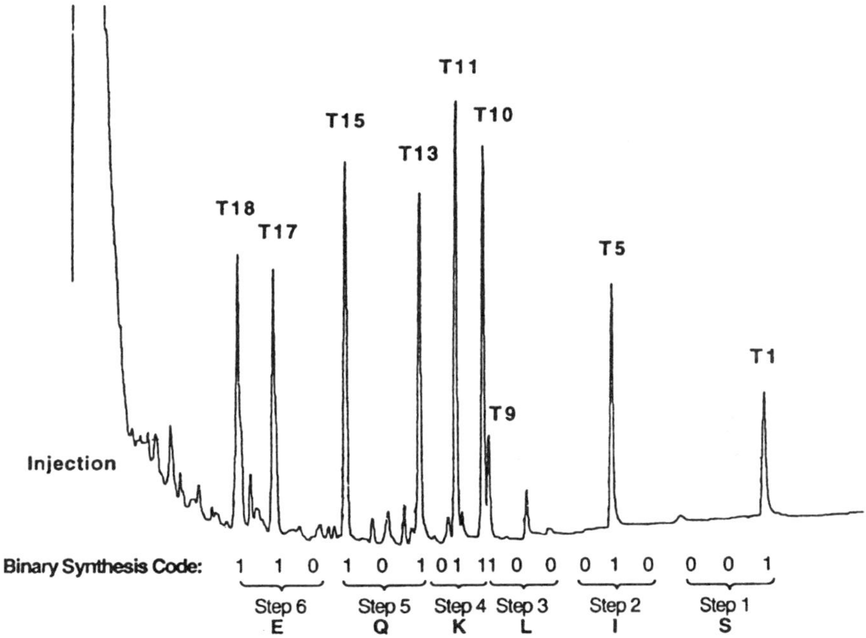

Figure 7.15 *Gas chromatogram of tags from EQKLISEEDLGGGG-Bead. The synthesis code from one stained synthesis bead was read by releasing and analyzing tags as described in the text. (Reproduced with permission from* Proc. Natl. Acad. Sci. USA, *1993, 90, 10926. Copyright© 1993, National Academy of Sciences, USA.)*

Given any bead, the attached tags could be detached by UV irradiation and unambiguously decoded by ECGC (e.g., see Figure 7.15). To verify that the codes corresponded to the actual peptide sequence present on the beads, two beads were picked at random, and the microsequencing results of the peptide sequences agree with the tags present on each bead read by ECGC.

A construct incorporates diazoketone functionality that can be converted to a reactive acylcarbene for direct attachment to a polymer matrix and a catechol diether moiety that can be cleaved oxidatively to effect tag release for analysis were introduced (Nestler et al., 1994). The acylcarbene tagging method is a major advance over previous tagging procedures. Not only are no specific functional groups required for tag attachment, but the tags and linkers are generally inert to the often-vigorous reaction conditions that are commonly used in organic synthesis.

An encoding methodology that uses chemically robust secondary amines as tags was described (Ni et al., 1996). These amines are incorporated into an N-[(dialkylcarbamoyl)methyl]glycine-coding oligomer through simple chemistry that is compatible with a wide range of polymer-supported transformations useful in combinatorial synthesis. In the decoding process, acidic hydrolysis of the tagging polymer regenerates the secondary amines, which, after dansylation, are resolved and detected at sub-picomole levels by reversed-phase HPLC with a fluorescence detector. RP-HPLC analysis of a small aliquot (2%) of the 18 dansylated secondary amines from a single bead (Figure 7.16) shows 18 well-resolved peaks of similar intensity. The versatility of this strategy is demonstrated by encoded syntheses of members of several representative heterocyclic compound classes, including β-lactams, 4-thiazolidinones, and pyrrolidines.

Capillary electrochromatography (CEC) has been coupled to a triple quadrupole mass spectrometer for the unambiguous decoding of active single beads from encoded combinatorial organic synthesis (Lane & Pipe, 1998). The high separation efficiency and plug flow profile of CEC coupled with the inherent specificity of a triple quadrupole mass spectrometer operating in the parent ion scanning mode allowed a highly specific and sensitive method to be developed for the unambiguous identification of a series of surrogate tags at the low femtomole level.

The high sensitivity of single-bead FTIR microspectroscopy (Yan et al., 1995) opens the possibility of using IR detectable groups as a molecular tag. Unlike most of the other encoding strategies, an advantage of this approach is the ability to access the encoding information while the tags are still attached to the resin, and the analysis of tags takes only minutes. Hochlowski et al. (1997) studied the potential applications of encoding combinatorial libraries with various IR and Raman sensitive tags using single-bead FTIR and single-bead Raman. The potential applications of FTIR and Raman microspectroscopy for analysis of organic reaction on solid support and for encoding combinatorial libraries has also been shown by Rahman et al. (1998). Organic groups that could be used as IR and Raman tags include $-C{\equiv}C{-}^2H$ (2585–2630cm^{-1}), $-C{=}C{-}^2H$ (2225–2335 cm^{-1}), $-C{-}^2H$ (2085–2255 cm^{-1}), $-C{=}C{-}C{\equiv}N$ (2215–2240 cm^{-1}), and $-N{=}N{=}N$ (2120–2160 cm^{-1}). Additionally, CH_2NO_2 (1375–1390 cm^{-1}), $ArNO_2$ (1310–1355 cm^{-1}), and C-S-S-C (505–515 cm^{-1}) could be used specifically as a Raman tag.

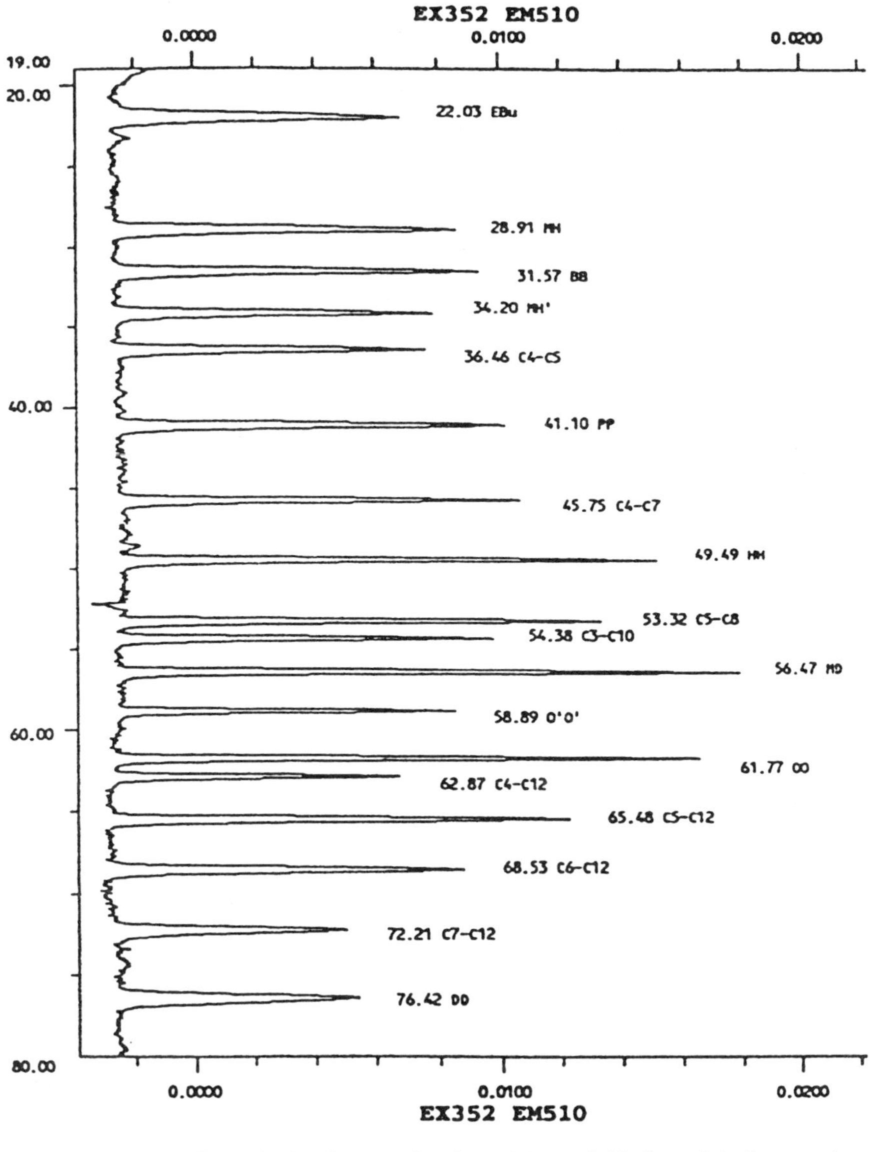

Figure 7.16 *HPLC analysis of an equimolar mixture of 18 dansylated secondary amine tags recovered from a single resin bead. (Reproduced with permission from* J. Med. Chem., *1996, 39, 1803. Copyright© 1996, American Chemical Society.)*

Another single-bead MS encoding technique is called partial termination method (Youngquist et al., 1994, 1995). Although it has been used for peptides, the idea can be potentially useful for organic compounds. During the synthesis of the library, a capping reagent is used to effect partial termination at each coupling step. In this way, each of the resin beads carries not only the full-length product, but also small amounts of sequence-specific termination products (totally 5%). The sequence of the full-length product is determined with MALDI-MS, which is used to measure mass differences between adjacent members of the termination series (Figures 7.17 and 7.18). In addition to rapid peptide sequencing, the technique also allows direct assessment of the quality of the synthetic library because deletion peptides, side-reaction products, and incomplete deprotection products are readily observed.

A method called "methionine scanning" developed by Bradley and coworkers encoded both regular peptides and "inverted peptides" (C-terminal exposed). The method is based on the incorporation of a small amount of Fmoc-methionine (totally 5–10%) with each amino acid during the synthesis of the peptide library. After synthesis, each bead carries a number of related peptides, the predominant species being the unmodified peptide as well as a series of peptides with methionine "scanned" into the sequence. In the mass spectrum, a "ladder" of peptides, each differing in mass from its neighbors by one amino acid residue, was released, allowing the peptide to be sequenced by mass differences (Davies & Bradley, 1997).

7.6.2 Electronic Encoding

Chemical tagging methods all have limitations. Attaching and detaching steps are introduced, and chemistry that can be used for assembling the library is limited to those that are not reactive to the tag. In addition, the time-consuming tag analysis is unavoidable. Thus, an alternative way to encode the library needs to be developed.

The split synthesis, although it allows for a fast synthesis of very large libraries, is usually slow in extracting structure/activity information even with the chemical encoding. A method called "mix and sort" synthesis can speed up the information extraction step. This method assigns a particular compound to each reactor unit and sorts them into the proper reactor after each "split-and-pool" synthesis step by reading encoded information on each reactor. It provides a method to make single compounds in milligram amounts using powerful "split-and-pool" strategy (Table 7.1). But, how to sort a large number of reactors?

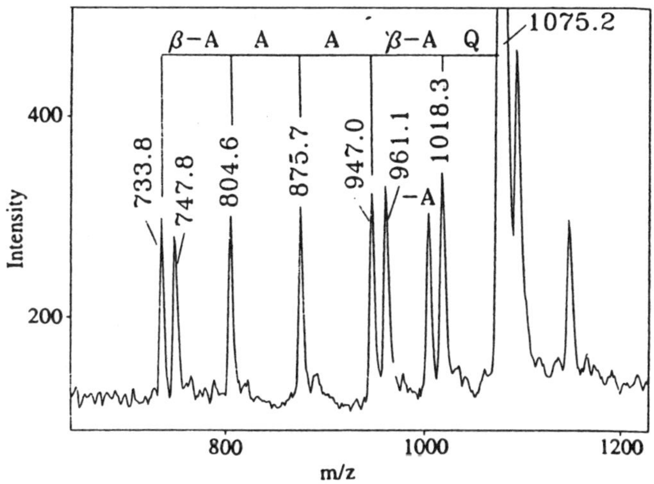

Figure 7.17 *MALDI mass spectrum of a peptide and its termination products using 5% of the material from a single bead. Each of the peaks had a smaller peak at higher mass (+18) that resulted from hydrolysis of the homoserine lactone. (Reproduced with permission from* J. Am. Chem. Soc., *1995, 117, 3902. Copyright© 1995, American Chemical Society.)*

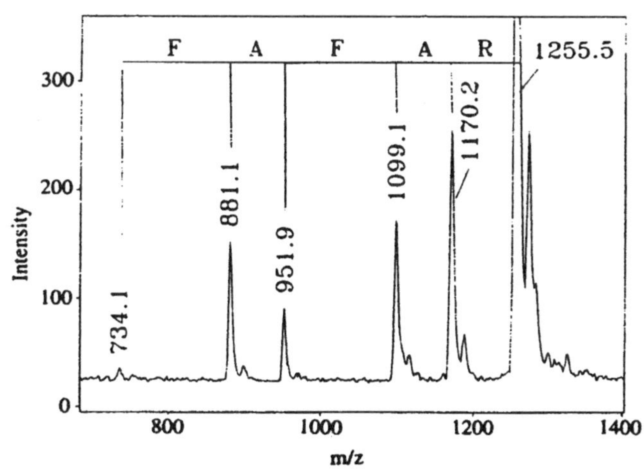

Figure 7.18 *MALDI mass spectrum of a coded peptide containing both alanine and β-alanine. β-Ala is coded with two terminating reagents to produce a doublet in the MALDI spectrum. The peak at 1005.5 was the result of an alanine deletion during synthesis. (Reproduced with permission from* J. Am. Chem. Soc., *1995, 117, 3902. Copyright© 1995, American Chemical Society.)*

Table 7.1 Comparison of combinatorial techniques.

Technique	Single/ Mixture	Speed of Synthesis	Information Retrieval	Application
Parallel synthesis	Single	Slow	Fast	Lead optimization
Split-and-pool	Mixture	Fast	Slow	Lead finding
Encoded split synthesis	Mixture	Fast	Moderate	Lead finding
Mix and sort	Single	Moderate	Fast	Lead finding and optimization

Two groups have reported the use of radiofrequency (rf) memory chip technology for the encoding of combinatorial libraries (Nicolaou et al., 1995; Moran et al., 1995). The encased microchip contains both memory, which is uniquely labeled, and antenna that receives and transmits rf signals from and to a computer-interfaced transceiver. By tagging each reactor with an rf memory chip, the code on the chip can be read simply by holding the chip over a transceiver unit. With an automated unit, it is possible to sort 10,000 reactors overnight. This technique has been widely used for "mix and sort" synthesis.

An interesting and much less expensive visual tagging method for mix and sort combinatorial chemistry was reported. The method employs only reactor caps with different colors and the addition of different colored beads into the reactor to track and control the synthesis of libraries with small to moderate sizes (Guiles et al., 1998).

7.6.3 Structural Determination on a Single Bead

Although the split synthesis is a success in terms of synthesizing a large number of diverse compounds, it poses challenges to both bioassay and the structural determination to identify the "active." The current answer to these two challenges are (1) while researchers are actively investigating the miniaturized assay format, the assays of a small pool of cleaved compound mixture or a large pool of bead-bound library are adopted; (2) various tagging methods have been developed to record the synthesis history. When the "active" is found, the tag is read, and the structure is revealed. However, if the structure of compound can be elucidated from the analysis of material on a single bead, the "active" bead can be decoded directly without resorting cleavage, tagging, or re-synthesis procedures. A variety of single-bead analysis methods have been recently reviewed (Swali & Bradley, 1997).

7.6.3.1 SINGLE-BEAD MS

The determination of the MW by analyzing a fraction of compound from a single bead has been frequently reported (Zambias et al., 1994;

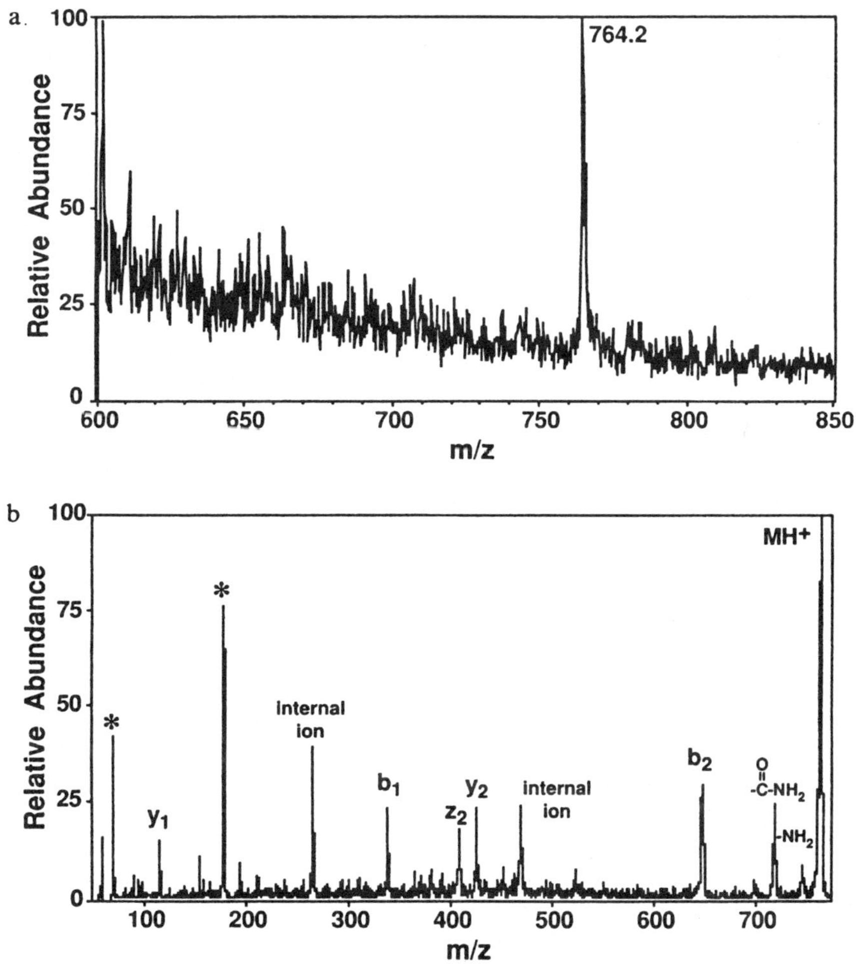

Figure 7.19 *(a) ESI mass spectrum of a selective active tripeptide following the irradiation of a single bead for 1 hr at 365 nm. (b) Tandem mass spectrum (MS/MS) from a single bead containing the tripeptide following photocleavage for 1 hr at 365 nm.*

Egner et al., 1995a, b; Haskins et al., 1995; Brummel et al., 1994, 1996; Yu et al., 1997; 1998). A tripeptide amide Fmoc-Asp-Arg(Tos)-Val-NH_2 was synthesized on resin with a photolabile linker (Brown et al., 1995). A fraction of this compound was released from resin and subjected to ESI-MS analysis. The molecular ion shows a strong peak at 764.2 (Figure 7.19). Tandem MS (MS/MS) was performed on the protonated parent ion, using the rest of the sample. From the fragmentation pattern, the sequence of the compound can be deduced without the need for encoding methodology.

A mass spectrometric strategy for the direct determination of structures of compounds on active beads selected from combinatorial mixtures by ultrahigh-throughput screening is presented (Blom et al., 1998). Sample preparation was expedited by performing sample cleanup on-line using a rapid, isocratic LC method. The complementary mass spectrometric data needed to achieve the required specificity in the determination (exact mass MS and MS/MS data were used) are obtained in a single analysis using automatic function switching. The use of API techniques with triple quadrupole and hybrid time-of-flight (TOF) instrumentation permitted full automation of the data acquisition procedure. Finally, the primary data analysis is with an automated computer procedure. Using concerted analyses of complementary mass spectrometric data enhances the specificity of the determination (i.e., the ability of the procedure to decrease the number of candidate structures to a minimum). Analyses of these data produced unequivocal identifications of the structure in 70% of the cases. The automatic computer analyses generated definitive solutions in 80% of the analyses. The remainder of the beads contained mixtures of components or components not in the enumerated library (either byproducts or degradants of expected components).

High-loading resin beads such as dendrimers were prepared. In such resins, a single bead gives sufficient material for NMR characterization, mass spectrometry, and repeated conventional HPLC analysis (Wells et al., 1998; Pursch et al., 1996).

7.6.3.2 SINGLE-BEAD NMR

Macro bead was produced to hold thousandfold scale-up of the amount of substance per bead (10–65 nmole per bead), with a diameter of 400–750 μm. The increased capacity of a single bead will facilitate the assay of compound released from the single bead and the structural elucidation on this amount of material. Such capability was demonstrated by a series of MAS ^1H NMR experiments to monitor a synthetic reaction step by step (Figure 7.20) using regular CP/NMR (solid-state) probe (Pursch et al., 1996). It takes only six to eight minutes to record a spectrum with 128–160 scans. The linewidths of ^1H NMR signals are in the order of 5–8 Hz. In Figure 7.20, ^1H NMR spectra of the polymer support and the addition of an acid-sensitive linker (AC linker) are shown in a and b. In the next step, phenylalanine-Fmoc is bound to the supported linker, as seen by the proton signals of the

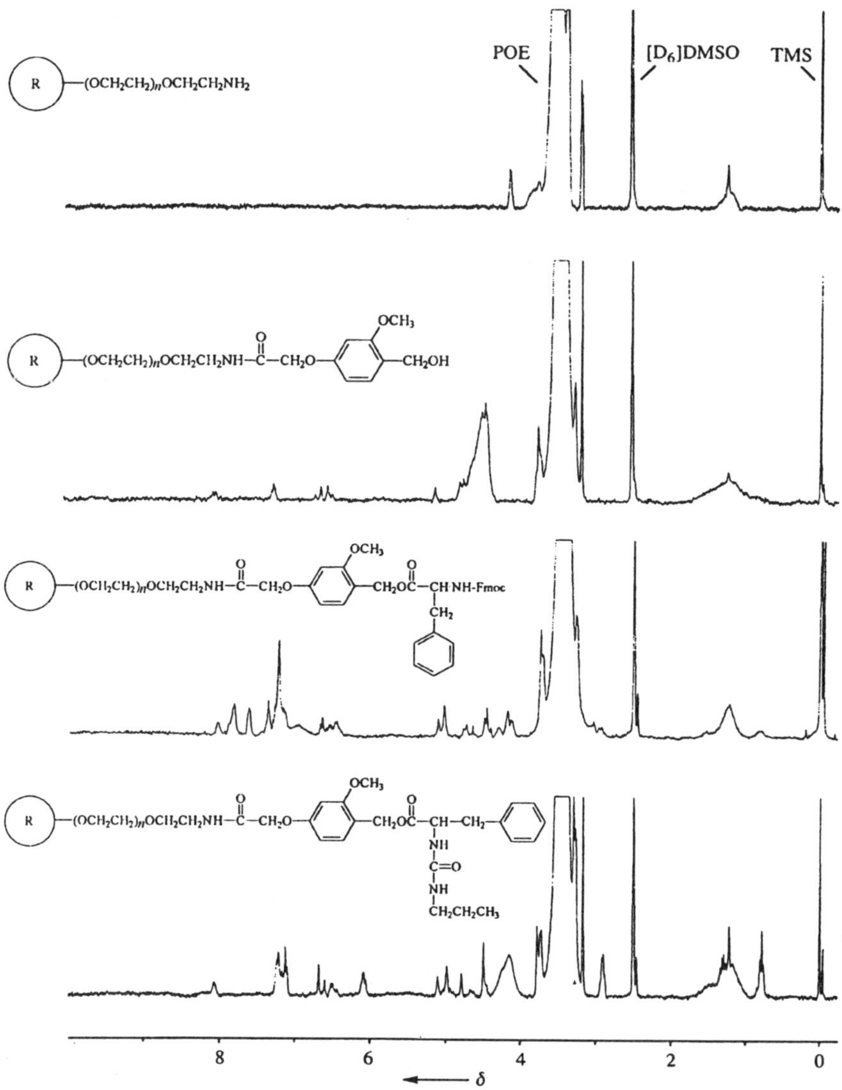

Figure 7.20 *^{1}H MAS NMR spectra of a suspension of individual resin particles in the hydantoin reaction sequence: spectra of the support-bound polyethyleneglycol spacer (PEG, see a), with acid-labile linker (b), after treatment with phenylalanine—Fmoc (c), and after scission of the Fmoc group and treatment with isocyanate (d). (Reproduced with permission from Pursch, et. al.,* Angew. Chem. Int. Engl., *1996, 35, 2868. Copyright© 1996, Wiley-VCH Verlag GmbH.)*

Fmoc unit in the aromatic region (c). Then Fmoc is cleaved, and the activated resin is treated with n-propylisocyanate. The Fmoc signals in the aromatic region decrease, and signals in the aliphatic region appear (d). The high resolution of the ^1H NMR signals is particular remarkable. In addition, a 2D-MAS NMR spectrum (COSY) was obtained in 40 minutes, facilitating structural investigation.

An alternative to scale-up the single-bead loading is the use of isotope labeling to increase the NMR sensitivity (Sarkar et al., 1996). A [3,5-dimethoxy-^{13}C] benzoic acid was coupled to 100-μm-diameter Wang resin beads (Diagram 7.4). Using the isotope-filtered NMR to selectively observe the proton resonance of the 3,5-dimethoxy^{13}C group, a 1D ^{13}C-filtered ^1H NMR spectrum of the compound and the corresponding unfiltered spectrum are shown in Figure 7.21. A 2D ^1H-^{13}C HMQC spectrum was also obtained to facilitate structural determination.

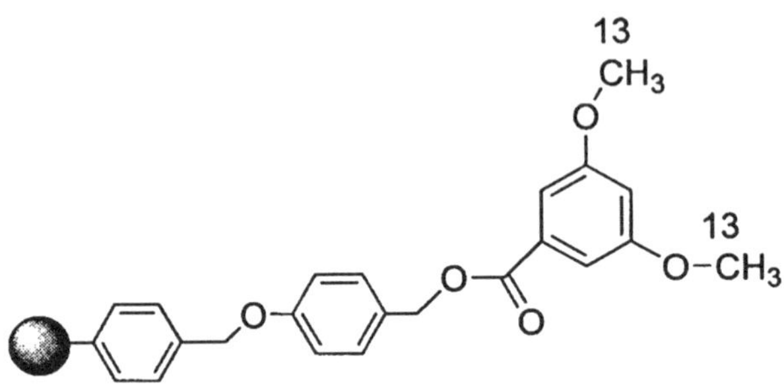

Diagram 7.4

7.7. SUMMARY

Majorities of screening assays are carried out in solution using modern high-throughput screening (HTS) technologies. So far, various affinity selection studies are limited to model proteins and small compound mixtures. They are not the main tools for lead selection at present. However, they may play a more important role when these methods go beyond the model protein stage.

Because the trend of HTS is to further increase throughput by miniaturization, compounds released from a single bead will be plenty for assays in future. Therefore, one-bead-one-compound libraries will provide highly diverse discrete compounds. If no mixture assay will be required, the affinity selection will become unnecessary someday. On

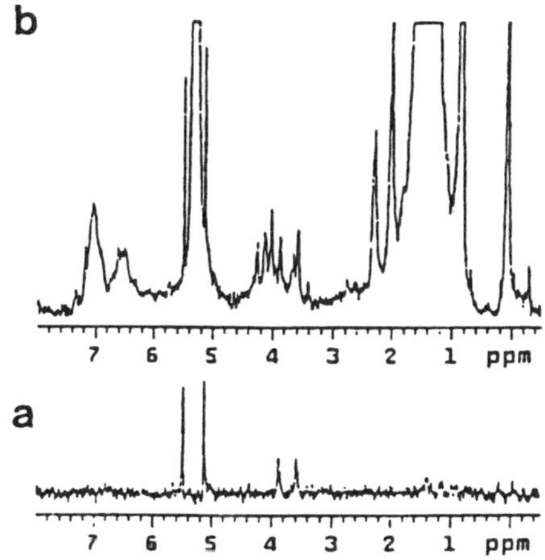

Figure 7.21 *(a) 500-MHz ^{13}C-filtered ^{1}H NMR spectrum of the compound in Diagram 4 coupled to Wang resin obtained on a single bead using the decoupler coil of a standard, commercially available heteronuclear Nano NMR probe. The single bead was suspended in 30 μL of dichloromethane-d2 in a standard (40 μL) nanoprobe cell and spun at the magic angle (54.7°) at ~2kHz. This proton spectrum was acquired using a one-dimensional version of the HMQC sequence run without ^{13}C decoupling but with pre-saturation (11-Hz field strength for 1s) of the residual protonated dichloromethane resonance at 5.32 ppm. Total experiment time was 170 min (5000 transients, 1.02-s acquisition time, 1.0-s pre-saturation delay). (b) ^{1}H NMR spectrum obtained using a one-pulse experiment. Total experiment time was 250 min (5000 transients, 2.05-s acquisition time, 1.0-s pre-saturation delay). Both spectra were processed with 4 Hz of exponential weighting (the dichloromethane ^{13}C satellites have 2-Hz line widths while the less mobile 3,5-dimethoxy-^{13}C resonance has 20-Hz line widths). (Reproduced with permission from* J. Am. Chem. Soc., *1996, 118, 2305. Copyright© 1996, American Chemical Society.)*

the other hand, various single-bead-based structure elucidation methods will become essential to decode the identified hits.

However, before highly miniaturized assays become available, compounds attached to beads cannot be handled individually. Chemical tagging on individual beads eventually allows decoding of the selected hits, but it also introduces additional synthesis steps and additional chemical moieties that may be sensitive to reaction conditions to be used. Radio-frequency tagging is a better alternative. If the size and the cost of these tags can be reduced, this tagging method may become more powerful in the future.

7.8 REFERENCES

Avila, L. Z., Chu, Y.-H., Blossey, E. C., Whitesides, G. M. Use of affinity capillary electrophoresis to determine kinetic and equilibrium constants for binding of arylsulfonamides to bovine carbonic anhydrase. *J. Med. Chem.* 1993, 36, 126.

Berlin, K., Jain, R. K., Tetzlaff, C., Steinbeck C., Richert, C. Spectrometrically monitored selection experiments: quantitative laser desorption mass spectrometry of small chemical libraries. *Chem. & Biol.* 1997, 4, 63.

Blackburn, C., Pingali, A., Kehoe, T., Herman, L. W., Wang, H., Kates, S. A. Libraries of angiotensin converting enzyme inhibitors: solid-phase synthesis and affinity selection. *Bioorg. & Med. Chem. Lett.* 1997, 7, 823.

Blom, K. F., Combs, A. P., Rockwell, A. L., Oldenburg, K. R., Zhang, L.-H., Chen, T. Direct mass spectrometric determination of bead bound compounds in a combinatorial lead discovery application. *Rapid Commun. Mass Spectr.* 1998,12, 1192.

Brown, B. B., Wagner, D. S., Geysen, H. M. A single-bead decode strategy using electrospray ionization mass spectrometry and a new photolabile linker: 3-Amino-3-(2-nitrophenyl) propionic acid. *Mol. Diversity.* 1995, 1, 4.

Bruce, J. E., Anderson, G. A., Chen, R., Cheng, X., Gale, D. C., Hofstadler, S. A., Schwartz, B. L., Smith, R. D. Bio-affinity characterization mass spectrometry. *Rapid Comm. Mass Spectr.* 1995, 9, 644.

Brummel, C. L., Lee, I. N. W., Zhou, Y., Benkovic, S. J., Winograd, N. A mass spectrometric solution to the address problem of combinatorial libraries. *Science.* 1994, 264, 399.

Brummel, C. L., Vickerman, J. C., Carr, S. A., Hemling, M. E., Roberts, G. D., Johnson, W., Weinstock, J., Gaitanopoulos, D., Benkovic, S. J., Winograd, N. Evaluation of mass spectrometric methods applicable to the direct analysis of non-peptide bead-bound combinatorial libraries. *Anal. Chem.* 1996, 68, 237.

Chen, A., Shapiro, M. J. NOE pumping: A novel NMR technique for identification of compounds with binding affinity to macromolecules. *J. Am. Chem. Soc.* 1998, 120, 10258.

Cheng, X., Chen, R., Bruce, J. E., Schwartz, B. L., Anderson, G. A., Hofstadler, S. A., Gale, D. C., Smith, R. D. Using electrospray ionization FTICR mass spectrometry to study competitive binding of inhibitors to carbonic anhydrase. *J. Am. Chem. Soc.* 1995, 117, 8859.

Chu, Y.-H., Avila, L. Z., Biebuyck, H. A., Whitesides, G. M. Use of affinity capillary electrophoresis to measure binding constants of ligands to proteins. *J. Med. Chem.* 1992, 35, 2915.

Chu Y.-H., Cheng, C. C. Affinity capillary electrophoresis in biomolecular recognition. *Cell. Mol. Life Sci.* 1998, 54, 663.

Chu, Y.-H., Dunayevskiy, Y. M., Kirby, D. P., Vouros, P., Karger, B. L. Affinity capillary electrophoresis—mass spectrometry for screening combinatorial libraries. *J. Am. Chem. Soc.* 1996, 118, 7827.

Czarnik, A. W. Encoding strategies in combinatorial chemistry. *Proc. Natl. Acad. Sci. USA.* 1997a, 94, 12738.

Czarnik, A. W. Encoding methods for combinatorial chemistry. *Current Opinion in Chem. Biol.* 1997b, 1, 60.

Davies, M., Bradley, M. Inverted peptide-single bead analysis by methionine scanning and mass spectrometry. *Tetrahedron Lett.* 1997, 49, 8565.

Dunayevskiy, Y. M., Lai, J.-J., Quinn, C., Talley, F., Vouros, P. Mass spectrometric identification of ligands selected from combinatorial libraries using gel filtration. *Rapid Comm. Mass Spectr.* 1997, 11, 1178.

Dunayevskiy, Y. M., Lyubarskaya, Y. V., Chu, Y., Vouros, P., Karger, B. L. Simultaneous measurement of nineteen binding constants of peptides to vancomycin using affinity capillary electrophoresis-mass spectrometry. *J. Med. Chem.* 1998, 41, 1201.

Egner, B. J., Cardno, M., Bradley, M. Linkers for combinatorial chemistry and reaction analysis using solid phase *in situ* mass spectrometry. *J. Chem. Soc. Chem. Commun.* 1995a, 2163.

Egner, B. J., Landley, G. J., Bradley, M. Solid phase chemistry: Direct monitoring by matrix-assisted laser desorption/ionization time of flight mass spectrometry. A tool for combinatorial chemistry. *J. Org. Chem.* 1995b, 60, 2652.

Furka, A., Sebestyen, F., Asgedom, M., Dibo, G. In *Highlights of Modern Biochemistry: Proceedings of the 14th International Congress of Biochemistry.* VSP, Utrecht, The Netherlands, 1988, Vol. 5, 47.

Furka, A., Sebestyen, F., Asgedom, M., Dibo, G. General method for rapid synthesis of multicomponent peptide mixtures. *Int. J. Pept. Protein Res.* 1991, 37, 487.

Gafni, A., Cohen, Y. Complexes of Macrocycles with γ-cyclodextrin as deduced from NMR diffusion measurements. *J. Org. Chem.* 1997, 62, 120.

Gao, J., Cheng, X., Chen, R., Sigal, G. B., Bruce, J. E., Schwartz, B. L., Hofstadler, S. A., Anderson, G. A., Smith, R. D., Whitesides, G. M. Screening derivatized peptide libraries for tight binding inhibitors to carbonic anhydrase II by electrospray ionization-mass spectrometry. *J. Med. Chem.* 1996, 39, 1949.

Geysen, H. M., Wagner, C. D., Bodnar, W. M., Markworth, C. J., Parke, G. J., Schoenen, F. J., Wagner, D. S., Kinder, D. S. Isotope or mass encoding of combinatorial libraries. *Chem. Biol.* 1996, 3, 679.

Griffey, R. H., An, H., Cummins, L. L., Gaus, H. J., Haly, B., Herrmann, R., Cook, P. D. Rapid deconvolution of combinatorial libraries using HPLC fractionation. *Tetrahedron.* 1998, 54, 4067.

Guiles, J. W., Lanter, C. L., Rivero, R. A. A visual tagging process for mix and sort combinatorial chemistry. *Angew. Chem. Int. Ed.* 1998, 37, 926.

Haskins, N. J., Hunter, D. J., Organ, A. J., Rahman, S. S., Thom, C. Combinatorial chemistry: direct analysis of bead surface associated materials. *Rapid Commun. Mass Spectr.* 1995, 9, 1437.

Hochlowski, J., Grillot, A.-L., Sowin, T. Encoding of combinatorial chemistry libraries with infrared and Raman active tags. Pacific Conference on Chemistry and Spectroscopy. 1997.

Houghten, R. A., Pinilla, C., Blondelle, S. E., Appel, J. R., Dooley, C. T., Cuervo, J. H. Generation and use of synthetic peptide combinatorial libraries for basic research and drug discovery. *Nature.* 1991, 354, 84.

Hughes, I. Design of self-coded combinatorial libraries to facilitate direct analysis of ligands by mass spectrometry. *J. Med. Chem.* 1998, 41, 3804.

Kaur, S., Huebner, V., Tang, D., McGuire, L., Drummond, R., Csetjey, J., Stratton-Thomas, J., Rosenberg, S., Figliozzi, G., Banville, S., Zukermann, R., Dollinger, G. In Proceedings 43rd Annual Conference on Mass Spectrometry and Allied Topics, Atlanta, GA, May 21–26, 1995, 30.

Kaur, S., McGuire, L., Tang, D., Dollinger, G., Huebner, V. Affinity selection and mass spectrometry-based strategies to identify lead compounds in combinatorial libraries. *J. Protein Chem.* 1997, 16, 505.

Kaur, S., Tang, D., Simon, R. J., Goff, D., Zukermann, R. N. In Proceedings 42nd Annual Conference on Mass Spectrometry and Allied Topics, Chicago, IL, May 29–June 3, 1994, 408.

Kerr, J. M., Banville, S. C., Zuckermann, R. N. Encoded combinatorial peptide libraries containing non-natural amino acids. *J. Am. Chem. Soc.* 1993, 115, 2529.

Lam, K. S., Salmon, S. E., Hersh, E. M., Hruby, V. J., Kazmierski, W. M., Knapp, R. J. A new type of synthetic peptide library for identifying ligand-binding activity. *Nature.* 1991, 354, 82.

Lane, S. J., Pipe, A. Unambiguous bead decoding by microelectrospray capillary electrochromatography tandem mass spectrometry of dansylated secondary amine tags from encoded combinatorial organic synthesis. *Rapid Commun. Mass Spectr.* 1998, 12, 667.

Lin, M., Shapiro, M. J., Wareing, J. Diffusion-edited NMR—Affinity NMR for direct observation of molecular interactions. *J. Am. Chem. Soc.* 1997a, 119, 5249.

Lin, M., Shapiro, M. J., Wareing, J. Screening mixtures by affinity NMR. *J. Org. Chem.* 1997b, 62, 8930.

Mckendrick, J. E., Frormann, S., Luo, C., Semchuck, P., Vederas, J. C., Malcolm, B. A. Rapid mass spectrometric determination of preferred irreversible proteinase inhibitor in combinatorial libraries. *Int. J. Mass Spectr.* 1998, 176, 113.

Moran, E. J., Sarshar, S., Cargill, J. F., Shahbaz, M. M., Lio, A., Mjalli, A. M. M., Armstrong, R. W. Radio frequency tag encoded combinatorial library method for discovery of tripeptide-substituted cinnamic acid inhibitors of the protein tyrosine phosphatase PTP1B. *J. Am. Chem. Soc.* 1995, 117, 10787.

Morris, K. F., Johnson Jr., C. S. Diffusion-ordered two-dimensional nuclear magnetic resonance spectroscopy. *J. Am. Chem. Soc.* 1992, 114, 3139.

Muller, K., Gombert, F. O., Manning, U., Grossmuller, F., Graff, P., Zaegel, H., Zuber, J. F., Freuler, F., Tschopp, C., Baumann, G. Rapid identification of phosphopeptide ligands for SH2 domains. *J. Biol. Chem.* 1996, 271, 16500.

Nedved, L. M., Habibi-Goudarzi, S., Ganem, B., Henion, J. D. Characterization of benzodiazepine "combinatorial" chemical libraries by on-line immunoaffinity extraction, coupled column HPLC-ion spray mass spectrometry-tandem mass spectrometry. *Anal. Chem.* 1996, 68, 4228.

Needels, M. C., Jones, D. G., Tate, E. H., Heinkel, G. L., Kochersperger, L. M., Dower, W. J., Barrett, R. W., Gallop, M. A. Generation and screening of an oligonucleotide-encoded synthetic peptide library. *Proc. Nat. Acad. Sci. USA.* 1993, 90, 10700.

Nestler, H. P., Bartlett, P. A., Still, W. C. A general method for molecular tagging of encoded combinatorial chemistry libraries. *J. Org. Chem.* 1994, 59, 4723.

Nestler, H. P., Wennemers, H., Sherlock, R., Dong, D. L.-Y. Microautoradiographic identification of receptor-ligand interactions in bead-supported combinatorial libraries. *Bioorg. & Med. Chem. Lett.* 1996, 6, 1327.

Ni, Z.-J., Maclean, D., Holmes, C. P., Murphy, M. M., Ruhland, B., Jacobs, J. B., Gordon, E. M., Gallop, M. A. Versatile approach to encoding combinatorial organic syntheses using chemically robust secondary amine tags. *J. Med. Chem.* 1996, 39, 1601.

Nicolaou, K. C., Xiao, X. Y., Parandoosh, Z., Senyei, A., Nova, M. Radio frequency encoded combinatorial chemistry. *Angew. Chem. Int. Ed. Engl.* 1995, 34, 2289.

Ohlmeyer, M. H. J., Swanson, R. N., Dillard, L. W., Reader, J. C., Asouline, G., Kobayashi, R., Wegler, M., Still, W. C. Complex synthetic chemical libraries indexed with molecular tags. *Proc. Natl. Acad. Sci. USA.* 1993, 90, 10922.

Pursch, M., Schlotterbeck, G., Tseng, L.-H., Albert, K., Rapp, W. Monitoring the reaction progress in combinatorial chemistry: ^{1}H MAS NMR investigations on single macro beads in the suspended state. *Angew. Chem. Int. Ed. Engl.* 1996, 35, 2867.

Rahman, S. S., Busby, D. J., Lee, D. C. Infrared and Raman spectra of a single resin for analysis of solid-phase reactions and use in encoding combinatorial libraries. *J. Org. Chem.* 1998, 63, 6196.

Rigler, R. Fluorescence correlation, single molecule detection and large number screening, Applications in biotechnology. *J. Biotech.* 1995, 41, 177.

Rymden, R., Carlfors, J., Stilbs, P. Substrate binding to cyclodextrin in aqueous solution: a multicomponent self-diffusion study. *J. Inclusion. Phenom.* 1983, 1, 159.

Sarkar, S. K., Garigipati, R. S., Adams, J. L., Keifer, P. A. An NMR method to identify nondestructively chemical compounds bound to a single solid-phase-synthesis bead for combinatorial chemistry applications. *J. Am. Chem. Soc.* 1996, 118, 2305.

Stankova, M., Issakova, O., Sepetov, N. F., Krchnák, V., Lam, K. S., Lebl, M. Application of one-bead one-structure approach to identification of nonpeptidic ligands. *Drug Dev. Res.* 1994, 33, 146.

Swali, V., Bradley, M. Analytical techniques for single bead analysis in combinatorial chemistry. *Anal. Commun.* 1997, 34, 15H.

Tu, J., Yu, Z., Chu, Y. Combinatorial search for diagnostic agents: Lyme antibody H9724 as example. *Clinical Chem.* 1998, 44:2, 232.

Wells, N. J., Davies, M., Bradley, M. Cleavage and analysis of material from single resin beads. *J. Org. Chem.* 1998, 63, 6430.

Wieboldt, R., Zweigenbaum, J., Henion, J. Immunoaffinity ultrafiltration with ion spray HPLC/MS for screening small-molecule libraries. *Anal. Chem.* 1997, 69, 1683.

Wigger, M., Nawrocki, J. P., Watson, C. H., Eyler, J. R., Benner, S. A. Assessing enzyme substrate specificity using combinatorial libraries and electrospray ionization-Fourier transform ion cyclotron resonance mass spectrometry. *Rapid Commun. in Mass Spectr.* 1997, 11, 1749.

Wu, Q., Gao, J., Joseph-Mcarthy, D., Sigal, G. B., Bruce, J. E., Whitesides, G. M., Smith, R. D. Carbonic anhydrase-inhibitor binding: From solution to the gas phase. *J. Am. Chem. Soc.* 1997, 119, 1157.

Yan, B., Kumaravel, G., Anjaria, H., Wu, A., Petter, R., Jewell Jr., C. F., Wareing, J. R. Infrared spectrum of a single resin bead for real-time monitoring of solid-phase reactions. *J. Org. Chem.* 1995, 60, 5736.

Youngquist, R. S., Fuentes, G. R., Lacey, M. P., Keough, T. Matrix-assisted laser desorption ionization for rapid determination of the sequences of biologically active peptides isolated from support-bound combinatorial peptide libraries. *Rapid Commun. Mass Spectr.* 1994, 8, 77.

Youngquist, R. S., Fuentes, G. R., Lacey, M. P., Keough, T. Generation and screening of combinatorial peptide libraries designed for rapid sequencing by mass spectrometry. *J. Am. Chem. Soc.* 1995, 117, 3900.

Yu, Z., Tu, J., Chu, Y.-H. Confirmation of cross-reactivity between lyme antibody H9724 and human heat shock protein 60 by a combinatorial approach. *Anal. Chem.* 1997, 69, 4515.

Yu, Z., Yu, X. C., Chu, Y.-H. MALDI-MS determination of cyclic peptidomimetic sequences on single beads directed toward the generation of libraries. *Tetrahedron Lett.* 1998, 39, 1.

Zaegel, H., Zuber, J. F., Freuler, F., Tschopp, C., Baumann, G. Rapid identification of phosphopeptide ligands for SH2 domains: screening of peptide libraries by fluorescence-activated bead sorting. *J. Biol. Chem.* 1996, 271, 16500.

Zambias, R. A., Boulton, D. A., Griffin, P. R. Microchemical structural determination of a peptoid covalently bound to a polymeric bead by matrix-assisted laser desorption ionization time-of-flight mass spectrometry. *Tetrahedron Lett.* 1994, 35, 4283.

Zhao, Y., Breemen, R. B. V., Nikolic, D., Huang, C., Woodbury, C. P., Schilling, A., Venton, D. L. Screening solution-phase combinatorial libraries using pulsed ultrafiltration/electrospray mass spectrometry. *J. Med. Chem.* 1997, 40, 4006.

8

Optimization of Lead Compounds

COMBINATORIAL CHEMISTRY, IN a broad sense, includes not only the synthesis of a large number of compounds, but also the selection of lead compounds and the optimization of these compounds to generate drug candidates or lead compounds for other applications. Combinatorial chemistry has already been applied in optimizing the *in vitro* activity of lead compounds. Both combinatorial and parallel analog synthesis has made great impact on this process. In this chapter, selected analytical advancements in the optimization of *in vitro* activity and physicochemical properties are discussed. Before these discussions, arguments are presented for what constitutes a lead compound.

8.1 LEAD COMPOUNDS AND LEAD OPTIMIZATION

In a combinatorial library, it is rare to find a compound that possesses all the suitable properties for a desired application, such as a drug. However, there are always some compounds that possess one of the desired properties, such as *in vitro* activity, physicochemical properties, or lower toxicity. These compounds should be selected for further development to improve properties that have not yet met with the desired standard. This lead optimization process is known as very time-consuming.

In most drug-discovery programs, lead compounds are narrowly defined as those with only acceptable *in vitro* activity. Usually, only these compounds are selected for optimizing their physicochemical properties, bioavailability, and pharmacokinetic profile in the development stage. Compounds with optimal drug-like properties other than *in vitro* activity are not even selected in the current programs. If these compounds can be selected for optimization of their *in vitro* activity, they have an equal chance to become drugs. Therefore, a lead compound should be defined as a compound that possesses one or more of the desired properties for a drug, and other properties of this compound can be optimized.

The current high-throughput screening based only on the *in vitro* activity misses a large portion of lead compounds, such as the compound with good drug-like solubility, permeability, and toxicity profile, but a low *in vitro* activity.

8.2 OPTIMIZING THE *IN VITRO* ACTIVITY: SAR BY NMR

When the target biological molecule is known, a specially designed, small library can be synthesized based on knowledge on the target molecule. This approach can be effectively assisted by protein structural studies. A successful example is SAR by NMR, meaning structure-activity relationships by NMR spectroscopy. This approach offers the advantage of combining rational drug design of certain target molecules and combinatorial chemistry with ^{15}N-enriched proteins of molecular weight up to about 30 kDa for which assignment of the signals for proteins and nitrogen atoms of the amide group is possible. The ^{15}N-^{1}H pairs of folded protein usually give well-resolved 2D ^{15}N heteronuclear single quantum correlation (HSQC) spectra. Addition of an inhibitor yields a shift of the HSQC NMR signals for all groups next to the binding site.

"SAR by NMR" is a linked-fragment approach wherein ligands are constructed from building blocks that have been optimized for binding to individual protein subsites (Figure 8.1). In the first step of this process, a library of organic compounds is screened to identify molecules that bind to the protein, which is ^{15}N labeled. Once a lead molecule is identified, analogs are screened to optimize binding to this site (Figure 8.1, step 2). Next, a ligand is sought that interacts with a nearby site. Binding to a second pocket is determined by observing changes in a different set of amide chemical shifts in a screen conducted in the presence of the first ligand (Figure 8.1, step 3). The binding of this ligand is optimized, and its approximate location is determined by protein structural studies. When two "lead" fragments have been selected, their location and orientations in the ternary complex are determined experimentally by NMR spectroscopy. On the basis of this structural information, compounds are synthesized in which the

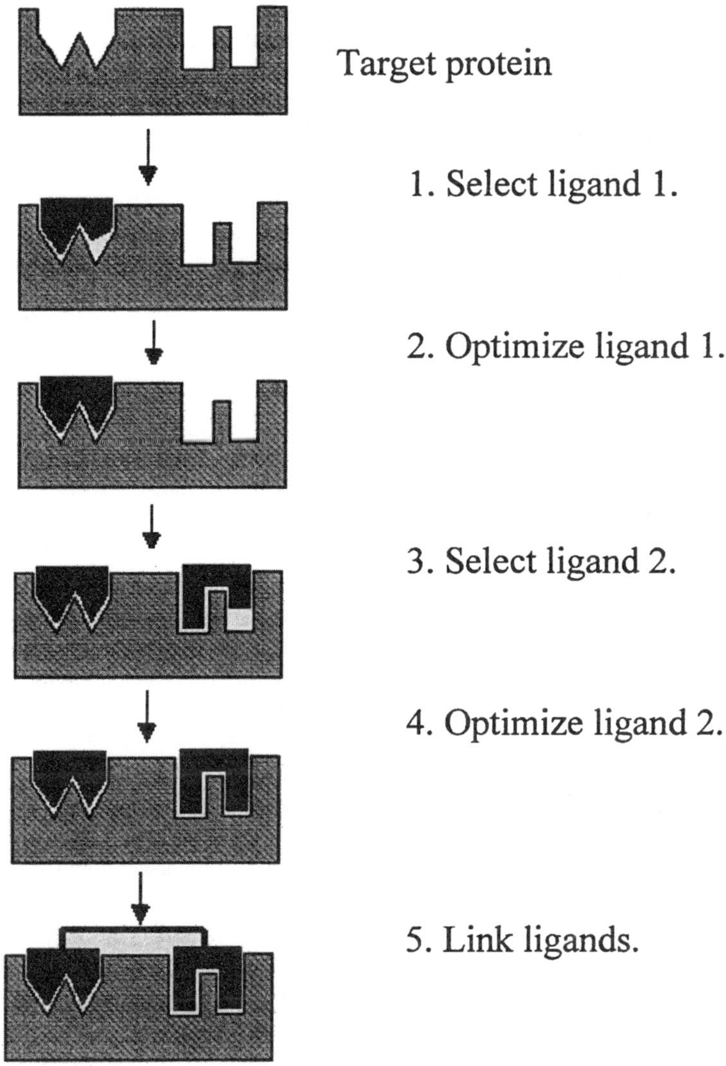

Figure 8.1 *The principle of SAR by NMR approach.*

two fragments are linked together (Step 5). SAR by NMR discovers high-affinity ligands with experimentally derived information. Although the small fragment molecules might only bind in the micromolar to millimolar range to the target protein, the affinity of a linked molecule is, in principle, the product of the binding constants of the individual fragments plus a term that accounts for the changes in binding affinity that are due to linking (Jencks, 1981). In principle, molecules with submicromolar affinity can be obtained. This principle was demonstrated by discovering a high-affinity ligand to FKBP (FK506 binding protein) (Shuker et al., 1996). In a period of two months, two identified fragment molecules with binding affinities of 2 and 100 μM were linked to produce a potent inhibitor with an affinity of 19 nM.

With the use of "SAR by NMR" approach, potent (IC_{50} < 25 nM) nonpeptide inhibitors of the matrix metalloproteinase stromelysin (MMP-3) were also discovered (Hajduk et al., 1997b). Two ligands that bind weakly to stromelysin (acetohydroxamic acid, K_D=17 mM; 3-cyanomethyl-4′-hydroxybiphenyl, K_D = 0.02 mM) were first identified. On the basis of NMR-derived structural information, the two fragments were connected to produce a 15-nM inhibitor of this enzyme (Figure 8.2). This inhibitor was discovered within six months and requires only a minimal amount of synthesis. The rapid discovery of the inhibitor was benefited from some known structural information, such as the ligand (acetohydroxamic acid), for the first site was chosen based on the fact that many of the known MMP inhibitors contain a hydroxamine moiety. The knowledge that the second site may be a hydrophobic interaction pocket guided a focused screening of only hydrophobic compound mixture. After the two fragment ligands were found, the three-dimensional structure of a ternary complex was determined by NMR spectroscopy to aid in the design of linked compounds. After the high-affinity-linked inhibitor was discovered, the 3-D structure of stromelysin complexed to the inhibitor was determined by NMR again to determine whether the linked compound binds to stromelysin as designed. The linked molecule binds to stromelysin in a similar manner as the unlinked compounds.

It is important to note that this approach was successful when a screen of 115,000 compounds using an enzyme inhibition assay failed to discover any lead compounds with potency greater than 10 μM. Biological screens depend on finding a potent molecule that already contains the necessary functional groups in their proper spatial orientation for tight binding to the protein. In contrast, using the SAR by NMR method, pieces that are important for binding are identified and optimized prior to the construction of the linked compounds. Information on the SAR is obtained prior to linking that will dictate which linked molecules should be prepared.

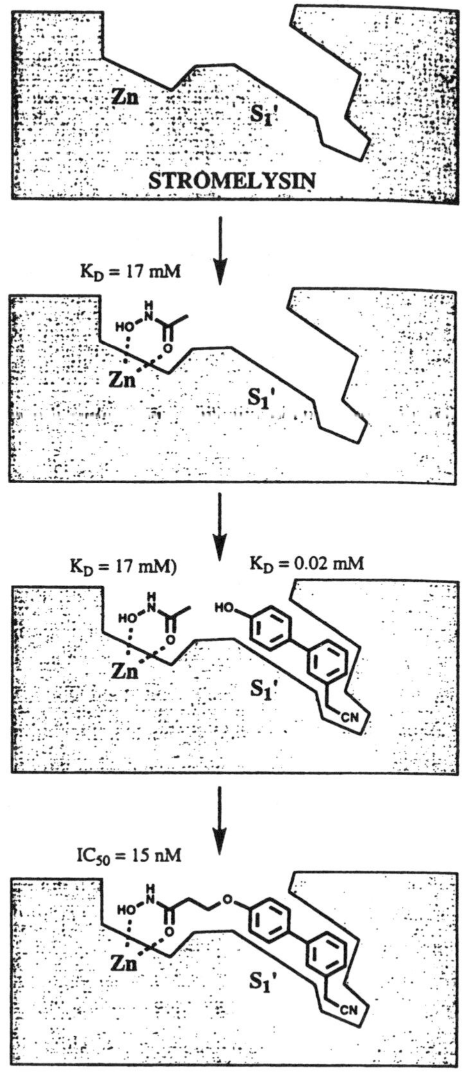

Figure 8.2 *A summary of the SAR by NMR method as applied to discovery of stromelysin inhibitors. (Reprinted with permission from* J. Am. Chem. Soc., *1997, 119, 5818–5827. Copyright© 1997 American Chemical Society.)*

To circumvent the limitation of the requirement for small proteins (<30kDa) in SAR by NMR approach, alternative methods were developed. Many NMR-sensitive parameters change upon complex formation, such as chemical shifts, relaxation time, and diffusion rates. Using FKBP and stromelysin as model systems, Fesik and coworkers investigated methods that exploit the changes in either the relaxation rates or diffusion rates of a small compound that occur upon binding to the biomolecule (also see Section 7.4) and thus eliminate the need for isotopically labeled targets (Hajduk et al., 1997a). Because this assay system is much simpler compared to other assay systems, the identification of false positives caused by unrelated mechanisms is minimized. Furthermore, because these methods rely on detecting the NMR signals of uncomplexed molecules rather than bound ligands or macromolecules, the techniques are applicable even to very large proteins.

A one-dimensional, relaxation-edited approach is applied to the FKBP. A Carr-Purcell-Meiboom-Gill (CPMG) spin-lock time is needed to reduce or eliminate the signals of the protein and bound ligands, without significantly affecting the signals of unbound ligands. If signal intensities of the protein and free ligand are plotted as a function of spin-lock time, the average proton T_2 relaxation times for FKBP and the ligand 2-phenylimidazole are 0.04 and 2.1 s. Therefore, for a spin-lock time of 400 ms, greater than 99% of the protein signal intensity was eliminated, whereas the ^1H NMR signals corresponding to the ligand were only reduced by 17%.

First, a relaxation-edited spectrum of the test compounds in the absence of FKBP was obtained (Figure 8.3a). The signals corresponding to all of the nine compounds (Diagram 8.1) in the mixture appear in this spectrum. Next, relaxation-edited spectra of FKBP alone and the test compounds in the presence of FKBP were obtained and subtracted to produce a spectrum (b) that only contains the resonance of those compounds whose transverse relaxation rates were not reduced in the presence of FKBP. By subtracting these two spectra, the spectrum of the bound compound appears (c). The bound ligand can be readily identified from the analysis of the chemical shifts that correspond to those of the free molecule (d). Figure 8.3e depicts a control experiment performed on a separate mixture containing only eight compounds that do not bind to FKBP.

The diffusion-edited detection of ligand binding is based on the fact that, at a low gradient strength of 1.2 G/cm, the NMR signals of the protein and the compounds are observed. However, at a high gradient strength of 48 G/cm, the signals from the free compounds are completely eliminated, while those from the protein are only reduced by 50% (see also Section 7.4).

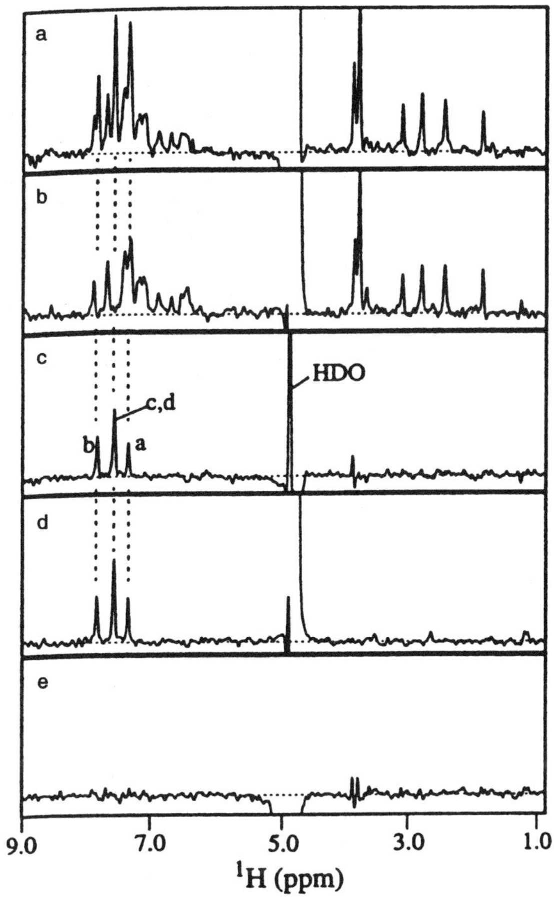

Figure 8.3 *Analysis of ligand binding to FKBP with use of a relaxation-edited approach. (a) Relaxation-edited ^{1}H NMR spectrum of a mixture of nine compounds* ***(1, 3–10)*** *in the absence of FKBP. (b) Relaxation-edited ^{1}H NMR spectrum of a mixture of nine compounds* ***(1, 3–10)*** *in the presence of FKBP after correcting for residual protein signals by subtracting an analogous spectrum of FKBP alone. (c) A difference spectrum obtained by subtracting the spectrum in b from a. The resonances of* ***1*** *occur at 7.89 (b), 7.63 (overlap of c and d), and 7.41 ppm (a) in the absence of protein and are indicated by the vertical dashed lines. (d) A reference spectrum of 1 alone. (e) A difference spectrum obtained in an analogous fashion to the spectrum shown in C, but on a mixture of eight compounds* ***(3–10),*** *which do not bind to FKBP. All relaxation-edited spectra utilized a CPMG spin-lock time of 400 ms. (Reprinted with permission from* J. Am. Chem. Soc., *1997, 119, 12257–12261. Copyright© 1997 American Chemical Society.)*

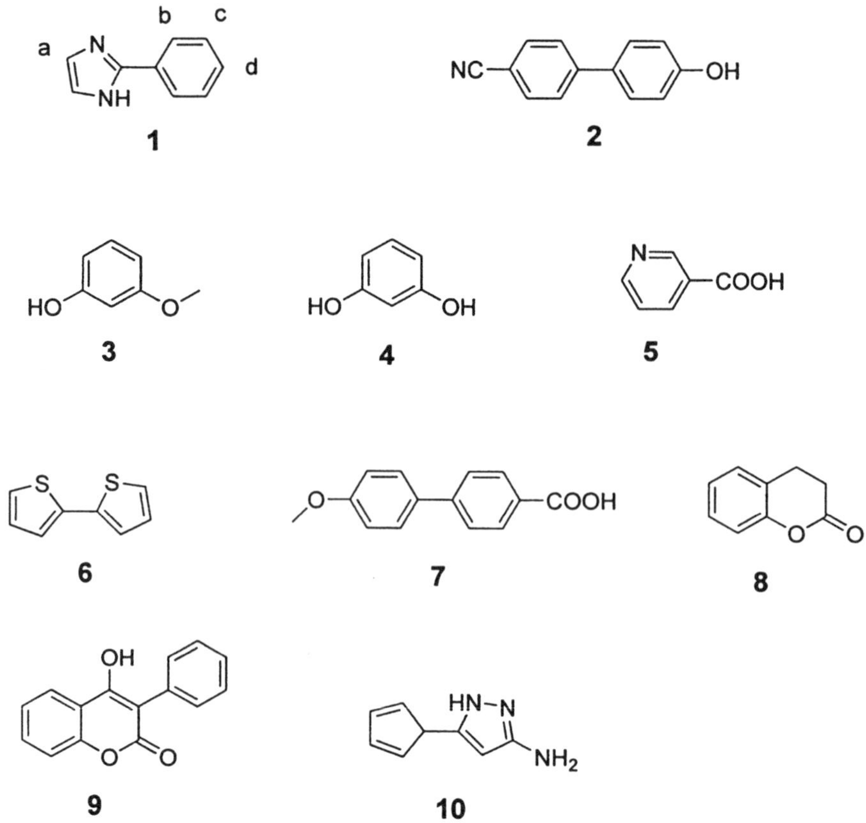

Diagram 8.1

To identify compounds that bind to stromelysin from a mixture of nonbinding compounds using a diffusion-edited approach, a series of pulse field gradient stimulated echo spectra (PFG-STE) were recorded and analyzed (Hajduk et al., 1997a). First, a PFG-STE spectrum of the nine compounds (Diagram 8.1), including one ligand in the absence of protein, was acquired at a low gradient strength (1.2 G/cm) (Figure 8.4a). At this gradient strength, the resonance from all of the compounds is observed. Next, PFG-STE spectra of the test compounds in the presence of stromelysin were obtained at low and high gradient strengths and were subtracted to produce a spectrum (Figure 8.4b) that only contains the signals of the compounds that do not bind to the protein. By subtracting the spectra shown in Figures 8.4a and b, the ^1H NMR signals of the compound that binds to stromelysin are

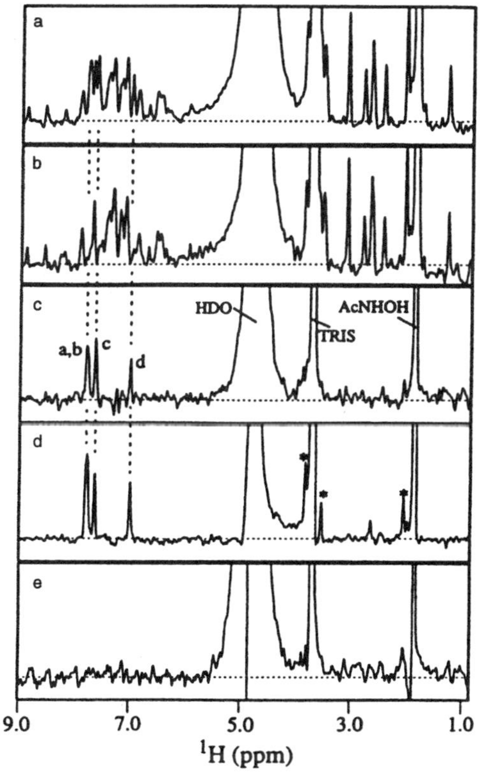

Figure 8.4 *Analysis of ligand binding to the catalytic domain of stromelysin by using a diffusion-edited approach. (a) PFG-STE spectrum of a mixture of nine compounds* ***(2–10)*** *in the absence of stromelysin with use of low gradient strengths. (b) PFG-STE spectrum of a mixture of nine compounds* ***(2–10)*** *in the presence of stromelysin with use of a low-gradient strength, after removal of protein signals by subtracting a PFG-STE spectrum of the same sample obtained at high gradient strengths. (c) A difference spectrum obtained by subtracting the spectrum in b from that in a. The resonances of* ***2*** *occur at 7.84 (overlap of a and b), 7.70 (c), and 7.06 ppm (d) in the absence of protein and are indicated by the vertical dashed lines. The signals from TRIS (3.74 ppm) and acetohydroxamic acid (AcNHOH, 1.94 ppm) were significantly attenuated in the difference spectrum, but not eliminated. (d) A reference spectrum of* ***2*** *alone. Signals arising from impurities in the buffer are indicated with an asterisk (*) and do not appear in the difference spectrum (Figure 8.2c) because they do not bind to the protein. (e) A difference spectrum obtained in an analogous fashion to the spectrum shown in c, but on a mixture of eight compounds* ***(3–10)*** *that do not bind to stromelysin. (Reprinted with permission from* J. Am. Chem. Soc., *1997, 119, 12257–12261. Copyright© 1997 American Chemical Society.)*

readily identified from an analysis of the chemical shifts that correspond to those of the free molecules (Figure 8.4d). No ligand signals were observed in a control experiment performed on a separate mixture containing only eight compounds that do not bind to stromelysin (Figure 8.4e).

Although these methods can identify compounds that bind to a specific protein target, they do not provide information on the three-dimensional structure of the ligands in the bound state.

The tumbling time, τ_c, is a physical parameter that distinguishes free and bound molecules. Small molecules exhibit positive, no, or very small negative NOEs. When a small molecule binds to a large protein, relaxation is governed by the slow tumbling time of the protein, resulting in strong negative NOEs, so-called transfer NOEs. The discrimination between transfer NOEs and NOEs of the ligand in solution is facilitated by the fact that transfer NOEs build up much faster. The selection of transfer NOEs over regular NOEs should be optimal at short mixing times.

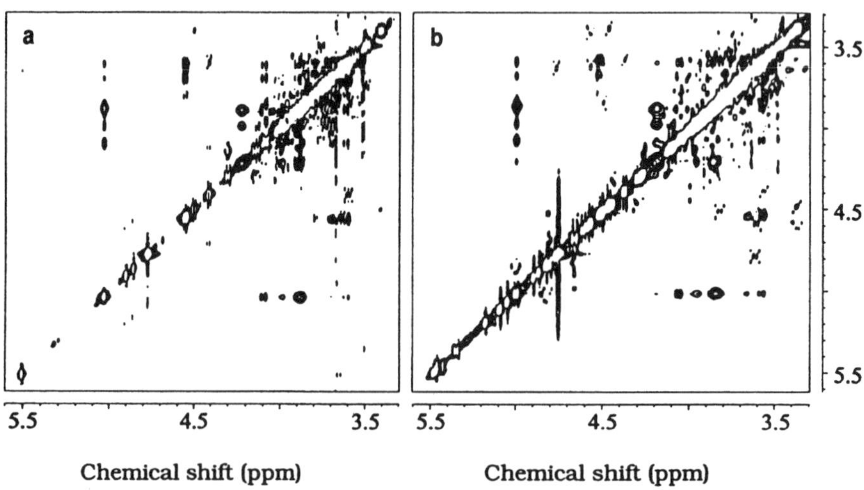

Figure 8.5 *Two-dimensional NOESY spectra for the oligosaccharide libraries A (a) and B (b) in D_2O solution in the presence of* A. aurantia agglutinin *displaying the transfer NOE pattern of α-L-Fuc-(1→6)-β-D-GlcNAc-OMe. Transfer NOE cross-peaks are of the same sign as the diagonal signals (negative NOEs). The spectra were obtained at 500 MHz and a temperature of 306 K. A spin lock filter was applied after the first π/2 pulse to remove disturbing protein signals. (Reprinted with permission from* Eur. J. Biochem., *1997, 246, 705–709. Copyright© 1997, Springer-Verlag.)*

Two libraries, A and B, consisting of six and 15 individual carbohydrate derivatives, respectively, were selected to test the binding affinity with *A. aurantia agglutinin* (Meyer et al., 1997). In each library, the disaccharide α-L-Fuc-(1→6)-β-D-GlcNAc-OMe is supposed to bind to *A. aurantia agglutinin.* It was unknown whether the other oligosaccharides have binding affinity. Two-dimensional NOESY spectra of the saccharide libraries in the presence of *A. aurantia agglutinin* clearly show the transfer NOE signal patterns that identify the bound library compounds (Figure 8.5). To illustrate this, Figure 8.6 shows a NOESY spectrum of pure α-L-Fuc-(1→6)-β-D-GlcNAc-OMe complexed with *A. aurantia agglutinin* for comparison. From a comparison with Figure 8.6, it is seen that the transfer NOE signal pattern is conserved in the NOESY spectra of library A and B in the presence of the protein. The signals of the library components that do not bind are suppressed. In

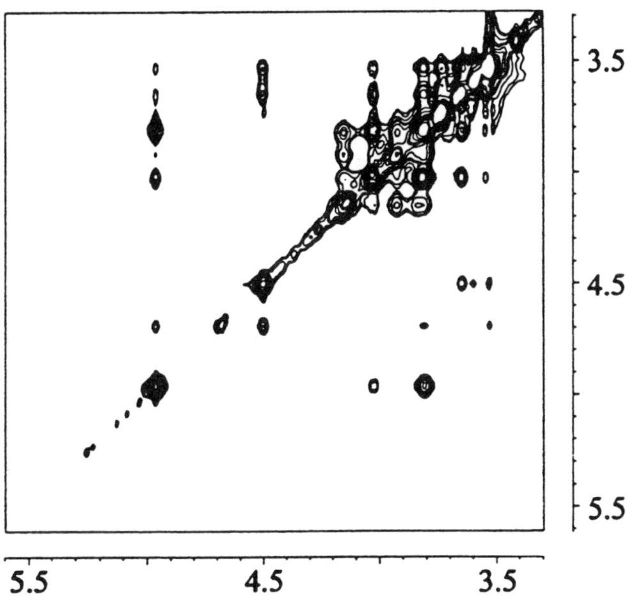

Figure 8.6 *Two-dimensional NOESY spectrum of α-L-Fuc-(1→6)-β-D-GlcNAc-Ome complexed with* A. aurantia agglutinin. *The spectrum was obtained at 600 MHz and a temperature of 303 K. The mixing time was set at 150 ms. The transfer NOE pattern is almost identical with the ones in Figure 8.5 proving that only α-L-Fuc-(1→6)-β-D-GlcNAc-Ome is bound by the lectin. (Reprinted with permission from* Eur. J. Biochem., *1997, 246, 705–709. Copyright© 1997, Springer-Verlag.)*

this case, the library members and their NMR spectra are known, and, therefore, it is straightforward to define the identity of the bound saccharide from the observed transfer NOE pattern. If library members and their ^{1}H-NMR spectra are unknown as the cases in combinatorial chemistry, identification of the bound component will normally require additional steps in the NMR experiment such as TOCSY, COSY, or HMQC.

In principle, transfer NOEs provide information on the three-dimensional structure of the ligands in the bound state. It is possible to perform a conformational analysis of the bound ligand in its bound state from a mixture of compounds. In practice, it is advantageous to perform the conformational analysis in a separate experiment where the pure compound with binding affinity is analyzed in the presence of the protein. Due to the NMR time scale, this method is limited to detect ligands that have dissociation constants from approximately 10^{-3} to 10^{-7}. It is not applicable to detect tight-binding ligands.

8.3 OPTIMIZING PHYSICOCHEMICAL PROPERTIES

Besides the *in vitro* activity discussed above, a drug must be soluble in the gastrointestinal (GI) fluid to be orally active. The rate and extent of dissolution depend critically upon water solubility. In addition, the drug must have a certain amount of hydrophobicity to traverse the biological membranes of the GI tract as well as target organelles. The hydrophobicity of a drug can be measured by its partition coefficient in octanol/water and permeability across the cell membranes.

The majority of drugs are intended for oral therapy, and improving bioavailability is time consuming and is sometimes more difficult than the optimization of *in vitro* activity. Oral activity prospects are improved through increased potency, but improvements in solubility or permeability can also achieve the same goal. Despite increasingly sophisticated formulation approaches, deficiencies in physicochemical properties may represent the difference between failure and success in the development of an oral drug product.

8.3.1 Drug Design Principles

Acceptable drug absorption depends on the triad of potency, solubility, and permeability. Many marketed drug examples shows that one parameter can compensate for another. For example, a very high aqueous solubility of 50 mg/ml more than counterbalances a very low absorption rate in the intestine loop of 0.001 min^{-1}. Poorer permeability in orally active peptidic-like drugs is usually compensated by very high solubility.

Principles for designing orally bioavailable drugs have been discussed recently (Navia & Chaturvedi, 1996; Lipinski et al., 1997). Lipinski et al. (1997) studied some key physicochemical properties of small molecule drugs and new chemical entities and found that four parameters are globally associated with solubility and permeability: namely, molecular weight; log P; the number of H-bond donors; and the number of H-bond acceptors. They called it the "rule of 5" because the cutoffs for each of the four parameters were all close to 5 or a multiple of 5. The "rule of 5" states that poor absorption or permeation are more likely when

- there are more than 5 H-bond donor (expressed as the sum of OHs and NHs)
- the molecular weight is more than 500
- the Log P is more than 5 (or MLog P is more than 4.15)
- there are more than 10 H-bond acceptors (expressed as the sum of Ns and Os)

Compound classes that are substrates for biological transporters are exceptions to the rule. The calculation procedure according the "rule of 5" has been incorporated into the compound registration system in a number of pharmaceutical companies. Although this knowledge is useful after compounds have been synthesized, it is important that such calculations be performed for every library member when designing a combinatorial library synthesis.

Current screening methodologies have three limitations:

1. Due to the very nature of screening, high molecular weight, high lipophilicity compounds usually cannot be detected and discarded. The physicochemical property profile of "hits" largely depends on the artificial criteria. The experience is that more lipophilic and, therefore, generally less soluble leads are generated.
2. Currently, only the *in vitro* activity screening is in practice in lead discovery stage. The lack of proper physicochemical screening definitely will miss a significant portion of lead compounds that have a favorable physicochemical property profile, but lower *in vitro* activity.
3. Because the lead compounds selected are not based on any physicochemical criteria, poor physicochemical properties would predominate in the many compounds that enter into and fail to survive preclinical stages and Phase I safety evaluation.

Obtaining oral activity can easily become a rate-limiting step, and, hence, methods that allow physicochemical predictions from molecular structure are badly needed. Computation methods in the early discovery setting need to deal with large numbers of compounds and serve as filters that direct chemistry SAR toward compounds with greater probability of oral activity. These computation methods become particularly important as experimental studies become more difficult, because compounds are available for physicochemical screening in only very small quantities and in non-traditional formats, such as those from combinatorial synthesis. Many studies on calculation methods for Log P and solubility have been reported. In most instances, these calculations help to remove bad compounds, not necessarily select good ones. The best bets still come from physicochemical screening assays.

8.3.2 High-Throughput Solubility Measurement

Thermodynamic solubility measurements are important in preclinical development and clinical development stages, not necessarily in the lead discovery stage. In most HTS assays, compounds were not solubilized in aqueous media under thermodynamic equilibrating conditions. Rather, compounds were dissolved in dimethyl sulfoxide (DMSO) as stock solution, and then serially diluted into 96-well plates for assays. Because of many interactions between compounds and assay components, proteins, membranes, and walls, even very insoluble compounds can be assayed. The apparent "solubility" in the HTS screen is always higher than the true thermodynamic solubility. In the lead discovery stage, the measurement of the true thermodynamic aqueous solubility is not very relevant to the screening manner in which leads are detected.

High-throughput solubility measurement has been developed by Lipinski et al. (1997) and has now been incorporated into drug discovery programs in a number of pharmaceutical companies. Compound is dissolved in DMSO at a concentration of 10 μg/μL of DMSO, which is close to the 30 mM DMSO stock concentration used in biological laboratories. The additions of DMSO are spaced a minute apart. A total of 14 additions are made. These correspond to solubility increments of <5 μg/mL to a top value of >65 μg/mL if the buffer volume is 2.5 mL. Precipitation can be quantified by an absorbance increase due to light scattering by precipitated particulate material in a dedicated diode array UV instrument. Increased UV absorbance from light scattering is measured in the 600–820 nm range because most compounds have

UV absorbance well below this range. The precipitation point is calculated from a bilinear curve fit to the absorbance-concentration plot. The upper solubility limit is based on the premise that, for most projects, permeability is not a major problem and that solubility assays will most often be requested for poorly soluble compounds. In the absence of poor permeability, solubilities above 65 μg/mL suggest that if bioavailability is poor, solubility is not the problem. The efficiency gain in an early discovery stage solubility assay lies in the SAR direction provided to chemistry and in the more efficient application of drug metabolism and pharmaceutical sciences resources once oral activity is detected. However, the solubility measurements are not a substitute for careful thermodynamic solubility measurements on well-characterized crystalline drugs and should not be used for decision making in a development setting.

8.3.3 High-Throughput Log P Measurement

The correlation of lipophilicity of an organic compound with the permeability of the compound has been a subject of study for years. The permeability of the blood-brain barrier (logPS) to a chemically diverse series of compounds was measured using a short-duration vascular perfusion method. Although LogP cannot always correlate with the membrane permeation property, LogPS data were correlated with calculated solute descriptors and octanol-water partition coefficients ($\log P_{oct}$) (Gratton et al., 1997). A realistic form of the disposition function is derived to describe the kinetics of the intracellular distribution of extraneous compounds on the basis of their hydrophobicity, acid-base properties, affinity to proteins, and the rate of elimination (Balaz et al., 1990). These studies suggest that LogP is an important property for a compound to be eligible as a drug candidate. Currently, no high-throughput method is available to screen compounds at the lead discovery stage. Calculation methods, such as the Pomona College Medical Chemistry program (CLogP) and Moriguchi method (MLogP), are heavily used to provide some insights. However, discrepancies are often found when comparing calculated and experimental LogPs. High-throughput experimental approaches are needed.

High-performance liquid chromatography in reversed-phase separation mode has long been recognized as a potential method for lipophilicity determination. A new chromatographic hydrophobicity index (CHI, φ_0) has been suggested as a measure of the lipophilic character of compounds in reversed-phase HPLC (Valko & Slegel, 1993). The parameter φ_0 is defined as the organic-phase concentration (methanol or acetonitrile) in the mobile phase, which is required for log $k' = 0$ (retention time is double the dead time), that is, the molar fraction of the compound is identical in the mobile and the stational phase. The φ_0 values, therefore, range from 0 to 100%, and the higher the value, the more hydrophobic the compound. It is shown that the value of φ_0 is characteristic for a compound and depends only on the type of organic modification, pH, and temperature. It is independent of the column type and column length, flow rate, and the mobile phase compositions where the actual retention measurements are carried out. The φ_0 values for more than 500 compounds were calculated and are presented on the basis of their published retention data. Significant correlations were found between the φ_0 values and the logarithm of 1-octanol-water partition coefficients (log P).

The φ_0 values can be used as part of a protocol for high-throughput (50–100 compounds/day) physicochemical property profiling for rational drug design (Valko et al., 1997). The index is derived from retention times (t_R) observed in a fast gradient, reversed-phase HPLC method. The isocratic retention factors (log k′) were measured for a series of 76 structurally unrelated compounds by using various concentrations of acetonitrile in the mobile phase. By plotting the log k′ as a function of the acetonitrile concentration, the slope (S) and the intercept (log k'_w) values were calculated. The previously validated index of hydrophobicity φ_0 was calculated as $-\log k'_w/S$. A good linear correlation was obtained between the gradient retention time values, t_R, and the isocratically determined φ_0 values for the 76 compounds. The constants of this linear correlation can be used to calculate CHI. CHI values can be measured using acidic, neutral, or slightly basic eluents. Values corresponding to the neutral form of molecules could be measured for 52 of the compounds and showed good correlation ($r = 0.851$) to the calculated octanol/water partition coefficient (Clog P) values.

8.3.4 High-Throughput Permeability Measurements

Identifying orally active drugs is typically a lengthy process of eliminating numerous potential compounds to identify a few putative leads that have appropriate oral bioavailability characteristics.

Passive permeation through a biological membrane depends to a large extent on three interdependent physicochemical properties, namely lipophilicity, polarity, and molecular size. *In vitro* and *in situ* experimental models that are descriptive of drug absorption *in vivo* are valuable tools in the discovery of new chemical entities that are bioavailable after oral administration. A method is based on studies of drug diffusion across monolayers of human colon adenocarcinoma cell line (CACO-2). For organic molecules and peptidomimetic analogs, good correlations were obtained between CACO-2 cell monolayers and rat single-pass intestinal perfusion measurements (Stewart et al., 1995).

CACO-2 cell measurement has been automated to increase the throughput. A Hamilton Microlab 2200 system was modified to conduct CACO-2 diffusion studies in 48 diffusion chambers simultaneously (Mandagere et al., 1996). Permeability of compound mixtures was measured and compared with individual measurement for five different therapeutic classes (GI drugs, analgesics, water soluble vitamins, steroids, and respiratory drugs). The rank order permeability of all compounds in each series, whether run individually or as a mixture, showed the same pattern (McCarthy et al., 1996). Mixtures from combinatorial libraries were also evaluated in the CACO-2 model of intestinal absorption (Taylor et al., 1996).

An HT parallel artificial membrane permeation assay (PAMPA) was reported (Kansy et al., 1998). This assay permits the fast determination of artificial membrane permeation properties of compounds, related to transcellular *in vivo* absorption processes. The method shows trends in the ability of a compound to permeate membranes by passive diffusion. In addition, PAMPA can deliver information on the lipophilicity, the ionization state, and the solubility of a compound. The 96-plate format and the microplate photometer for UV measurement allow the screening of large combinatorial libraries.

8.3.5 High-Throughput Pharmacokinetics and Drug Metabolism Measurements

Combinatorial chemistry potentially can provide dramatically increasing numbers of lead compounds for optimization. Traditional methods used to determine the pharmacokinetics (oral bioavailability, clearance, volume of distribution, and half-life) of molecules have presented a bottleneck in current drug-discovery programs.

Selection of a new drug candidate from a series of compounds requires a means of performing rapid analytical method development and sensitive quantitation of each drug in serum, plasma, or other biological matrices. New approaches have been adopted to solve these problems. One approach is to screen a mixture of compounds in animals using LC-API MS/MS analysis. The greatest potential utility of the mixture approach is to screen larger numbers of compounds and to access SAR information quickly. A simultaneous evaluation of pharmacokinetic parameters of five compounds in the dog was reported (Berman et al., 1997). In the same year, a simultaneous determination of plasma concentration of up to 12 drug candidates over a concentration range of 1–1000 ng/mL in a single analysis was also reported (Olah et al., 1997). In a study of a series of farnesyl protein transferase (FPT) inhibitors, LC-API MS/MS method was found two to four times faster compared with GC method with nitrogen-phosphorus detection (Bryant et al., 1997).

8.4 SUMMARY

The concept of lead compound and the selection criteria need to be revised in order to increase the number of lead compounds and the chance of success. Lead compound is a compound with one or more desired properties, and the rest of the properties can be further optimized. The lead selection based only on *in vitro* activity is incomplete. Many compounds with good physicochemical properties lose the opportunity to enter the optimization pipeline.

Current physicochemical screening methods are not high throughput, and only compounds with certain *in vitro* activity are tested. Ideally, all designed library members should pass the calculated physicochemical profile, and all synthesized compounds should be subjected to high-throughput physicochemical property screening.

The optimization of *in vitro* activity requires significant rational approaches, not just empirical methods. SAR by NMR is a good example of the combination of rational and combinatorial methods. The lead optimization requires applications of many analytical techniques and developments of novel analytical approaches. The trend is still to increase the throughput by adopting high-speed serial analysis and, ideally, parallel analysis. There are significant difficulties and also great opportunities for breakthrough and innovation.

8.5 REFERENCES

Balaz, S., Wiese, M., Chi, H.-L., Seydel, J. K. Subcellular pharmacokinetics and quantitative structure/time/activity relationships. *Anal. Chim. Acta.* 1990, 235, 195.

Berman, J., Halm, K., Adkison, K., Shaffer, J. Simultaneous pharmacokinetic screening of a mixture of compounds in the dog using API LC/MS/MS analysis for increased throughput. *J. Med. Chem.* 1997, 40, 827.

Bryant, M. S., Korfmacher, W. A., Wang, S., Nardo, C., Nomeir, A. A., Lin, C.-C. Pharmacokinetic screening for the selection of new drug discovery candidates is greatly enhanced through the use of liquid chromatography-atmospheric pressure ionization tandem mass spectrometry. *J. Chromatogr. A.* 1997, 777, 61.

Gratton, J. A., Abraham, M. H., Bradbury, M. W., Chadha, H. S. Molecular factors influencing drug transfer across the blood-brain barrier. *J. Pharm. Pharmacol.* 1997, 49, 1211.

Hajduk, P. J., Olejniczak, E. T., Fesik, S. W. One-dimentional relaxation- and diffusion-edited NMR methods for screening compounds that bind to macromolecules. *J. Am. Chem. Soc.* 1997a, 119, 12257.

Hajduk, P. J., Sheppard, G., Nettesheim, D. G., Olejniczak, E. T., Shuker, S. B., Meadows, R. P., Steinman, D. H., Carrera Jr., G. M., Marcotte, P. A., Severin, J., Walter, K., Smith, H., Gubbins, E., Simmer, R., Holzman, T. F., Morgan, D. W., Davidsen, S. K., Summers, J. B., Fesik, S. W. Discovery of potent non-peptide inhibitors of stromelysin using SAR by NMR. *J. Am. Chem. Soc.* 1997b, 119, 5818.

Jencks, W. P. On the attribution and additivity of binding energies. *Proc. Natl. Acad. Sci. USA.* 1981, 78, 4046–4050.

Kansy, M., Senner, F., Gubernator, K. Physicochemical high throughput screening: parallel artificial membrane permeation assay in the description of passive absorption processes. *J. Med. Chem.* 1998, 41, 1007.

Lipinski, C. A., Lombardo, F., Dominy, B. W., Feeney, P. J. Experimental and computational approaches to estimate solubility and permeability in drug discovery and development settings. *Adv. Drug Delivery Rev.* 1997, 23, 3.

Mandagere, A. K., Correll, M. A., Mooney, J. P., Poole, J. C., Hwang, K.-K., Cheng, L. K. Application of automation to CACO-2 drug diffusion studies. *Pharma. Res.* 1996, 13, S237.

McCarthy, S., Osborne, S., Kuhfeld, M. Automated permeability analysis of mixtures across CACO-2 cell monolayers. *Pharma. Res.* 1996, 13, S242.

Meyer, B., Weimar, T., Oeters, T. Screening mixtures for biological activity by NMR. *Eur. J. Biochem.* 1997, 246, 705.

Navia, M. A., Chaturvedi, P. R. Design principles for orally bioavailable drugs. *Drug Discovery Today.* 1996, 1, 179.

Olah, T. V., McLoughlin, D. A., Gilbert, J. D. The simultaneous determination of mixtures of drug candidates by liquid chromatography/atmospheric pressure chemical ionization mass spectrometry as an *in vivo* drug screening procedure. *Rapid Commun. Mass Spectr.* 1997, 11, 17.

Shuker, S. B., Hajduk, P. J., Meadows, R. P., Fesik, S. W. Discovering high-affinity ligands for proteins: SAR by NMR. *Science.* 1996, 274, 1531.

Stewart, B. H., Chan, O. H., Lu, R. H., Reyner, E. L., Schmid, H. L., Hamilton, H. W., Steinbaugh, B. A., Taylor, M. D. Comparison of intestinal permeabilities determined in multiple *in vitro* and *in situ* models: relationship to absorption in humans. *Pharma. Res.* 1995, 12, 693.

Taylor, E. W., Gibbons, J. A., Braechman, R. A. Screening of chemical mixtures from combinatorial libraries in the CACO-2 model of intestinal absorption. *Pharma. Res.* 1996, 13, S242.

Valko, K., Bevan, C., Reynolds, D. Chromatographic hydrophobicity index by fast-gradient RP-HPLC: A high-throughput alternative to log P/log D. *Anal. Chem.* 1997, 69, 2022.

Valko, K., Slegel, P. New chromatographic hydrophobicity index (φ_0) based on the slope and the intercept of the log k′ versus organic phase concentration plot. *J. Chromatography.* 1993, 631, 49.

9

Final Thoughts and Future Perspectives

PREVIOUS CHAPTERS HAVE provided an overview of the current analytical methods in combinatorial chemistry. Examples and highlights of many recently published results have shown that tremendous progress has already been made in these areas. These citations are by no means complete, but they do represent where we are. Because summaries have been given at the end of each chapter, this chapter will suggest selected issues that need to be solved and some trends for future developments.

Last summer, I went to Zurich for the ETH Workshop on Analytical Challenges in Combinatorial Chemistry organized by Drs. Beck-Sickinger, Bovermann, Pretsch, and Zenobi. At the speakers' dinner, Dr. R. Zenobi asked a very interesting question: "Up to now, is there any breakthrough in analytical technologies for combinatorial chemistry?" Unfortunately, no one could point out a single major breakthrough. Various analytical tools with some innovative improvements have been applied to solve combinatorial chemistry problems. However, the essence of combinatorial chemistry is "parallel" instead of "serial." The analytical methods so far have not achieved "parallel analysis." Then, what is the expected major breakthrough? I firmly believe that the mismatch between the parallel synthesis and the serial analysis is a major issue. Routine parallel analysis methods will be the major breakthrough for years to come. Optimization of hundreds of organic reactions on solid-phase is another challenge for years to come. Additionally, we also see increasingly miniaturized synthesis on a chip. Are analytical chemists ready for that? Technical advances for the application of HT NMR have been made. Is an automated NMR interpretation package ready to deal with the high volume of data? Some recent research in these areas will be highlighted in the following.

9.1 REACTION OPTIMIZATION EFFORTS

If solid-phase organic reactions are not optimized, the libraries made using those synthetic routes will contain a high percentage of impurities. Although the quality for lead discovery libraries can be set at a lower standard, a synthesis well containing less than 80% of the expected compound will be problematic for interpreting the assay result. A recent trend is to purify every compound from the final library. This will cause further reduction of the library size. The operation is prohibitorily costly and time consuming. Although the purification approach is a workable solution for very small libraries, it is unthinkable for medium to large libraries. A way to significantly improve the quality of combinatorial libraries is to intensify the effort in reaction optimization.

Despite many progresses, the obstacles in the reaction optimization are still formidable. There are several areas requiring intensive research. I will point out a few of them:

1. The effects of solid support on hundreds of organic reactions.
2. Effective on-support yield determination based on many organic functional groups to facilitate multi-step synthesis.
3. The influence of solvents on hundreds of solid-phase organic reactions.
4. Effective parallel instead of serial analytical methods for the total quality control of trial libraries.

At present, we are going through a difficult period as pointed out by Joyce, Still, and Chapman (1997): "Perhaps the most disturbing trend in combinatorial chemistry is the willingness of many in the field to give up on highly diverse compound libraries. The size of the typical library has been decreasing at a rate of tenfold per six months for the past several years. There is a serious danger that combinatorial chemistry will degenerate into the robotic synthesis of individual analogs." However, we accept the temporary setback, never the defeat. The key to our success is to pay attention to these fundamental problems. More and more involvement from academic researchers is crucial."

There are two aspects of combinatorial chemistry: the commercial aspect and the scientific aspect. It is the scientific aspect that plays the decisive role in this combinatorial chemistry era. Any ignorance of the basic scientific research in combinatorial chemistry will not only jeopardize this revolutionary science, but also the commercial benefit for many.

9.2 MINIATURIZED ANALYSIS

For lead discovery, only minute amounts of compounds are needed in the increasingly miniaturized high-throughput screening assays. The split synthesis may become one of the major methods to provide discrete compounds when compounds released from a single bead are enough for assaying and structure identification. Now, high-throughput organic synthesis of discrete compounds can also be performed on a chip. Up to 50 nmol of a single compound can be generated in each 700 nL reaction well using microfluidics (DeWitt et al., 1998). Are analytical methods ready for the miniaturized synthesis?

The sensitivity of NMR has been improved greatly recently. On-line stopped-flow and static ^{1}H NMR spectra were acquired with capillary columns of 75–315 μm i.d. With detection over a length of 1.2 cm, cell volumes cover a range of 50–900 nL.

NMR spectra with resolution about 3 Hz was obtained using a 50 nL detection cell to measure 1.1 nmol of dansylated γ-aminobutyric acid under static conditions in a 75 μm i.d. capillary. Such small-volume cells allow on-line NMR detection for capillary electrophoresis separations (Behnke et al., 1996; Olson et al., 1998). In the near future, deuterium lock channels, multinuclear probes, and higher field strengths will add to the capabilities of microcoil NMR. Recent efforts have included coupling standard HPLC/NMR to MS. This combination will likely be demonstrated soon on the nanoliter scale.

Photolithographic and chemical (wet) etching techniques to prepare chemical analysis systems that use capillary electrophoresis on glass substrates may provide multispectroscopic analytical systems on a single chip (Wu et al., 1994).

The sensitivity of MS is further improving. A "picospray" ESI source demonstrates both picoliter sample volume and picoliter/min flow rates for high-sensitivity electrospray ionization (Valaskovic et al., 1995). Good quality MS/MS data can be obtained at the level of

100 fmol of injected sample using automated procedures (Davis et al., 1995). This technique will aid greatly the structural elucidation on a single bead.

Miniaturized synthesis and screening pose more challenge to the sensitivity of analytical technologies. Techniques like MS, FTIR, GC, and CE are highly sensitive. The improvement of sensitivity of other analytical techniques is still an area of research.

9.3 PARALLEL ANALYSIS

Serial analysis can never match the demand from massive parallel synthesis in combinatorial chemistry. Parallel analytical techniques are in high demand for the characterization and purification of combinatorial libraries.

Recently, a new method called capillary array electrophoresis (CAE) was reported where multiple CE separations are performed in parallel using bundles of microcapillaries. Capillary array electrophoresis is advantageous because the capillaries can be separated and independently manipulated at the inlet for ease of rapid, parallel loading. It is currently feasible to run more than 100 capillaries in an array (Mathies & Huang, 1992). The detection is based on a laser-excited, confocal-fluorescence scanner. This scanner uses an epi-illumination format where the laser is focused on the sample by a microscope objective and the fluorescence is gathered by the same objective followed by confocal detection (Huang et al., 1992).

Micromachining, a well-known photolithographic technique for structures in the micrometer range, was introduced to allow capillary electrophoresis to be performed on a chip-like structure (Manz et al., 1992). Up to ~80 independent separation and loading channels on a single chip have been fabricated (Woolley et al., 1994). Micromachined columns have been fabricated for gas chromatography, liquid chromatography, and capillary electrophoresis and have been evaluated in the cases of gas chromatography and capillary electrophoresis (Jacobson et al., 1994). A microfabricated multichannel chip was also successfully interfaced to ESI-MS. A series of sequential experiments were conducted by infusing various samples from different channels, demonstrating that rapid analysis with high-throughput can be implemented (Xue et al., 1997). Such miniaturized systems (Ramsey, 1998; Ramsey et al., 1998) have great potential in purifying and analyzing combinatorial chemistry samples.

Imaging techniques based on MS (Brummel et al., 1994), IR (Holzwarth et al., 1998; Reetz et al., 1998; Haap et al., 1998), and fluorescence may also be applicable in parallel characterization of libraries.

9.4 HT NMR DATA INTERPRETATION

HT NMR allows the acquisition of a spectrum in two to three minutes. This method has been coupled with computer programs for spectral data reduction and classification using principal component analysis. Each spectrum (δ 10.0–δ 0.24) was segmented into 256 equal chemical shift regions using AMIX software (version 2.1.3, Bruker), and the total integrated intensity in each region was determined to provide a series of descriptors of the spectra normalized to the total integral of each spectrum to remove concentration effects. These data were autoscaled to give a mean of zero and a variance of ±1 for each descriptor and were subjected to PC analysis using the software package PIR-OUETTE (version 2.03, Infometrix Inc, Seattle) running on an IBM-compatible personal computer (Spraul et al., 1997).

9.5 LC-NMR-MS

LC-NMR-MS is a powerful tool for obtaining structural information on metabolites from complex biological matrices in a single experiment (Burton et al., 1997). This technique may also be applied to combinatorial chemistry. The mass spectral data usually provide good molecular mass information but limited structural information, as little fragmentation occurs. The ability to obtain UV, NMR, and MS data for individual components in a complex mixture within a single LC run is advantageous, particularly when sample quantity is limited. A number of problems associated with the operation of a quadrupole mass spectrometer close to a high-field magnet have been investigated (Holt et al., 1997). One of the major problems, a 100-fold loss of sensitivity, may be improved by shielding the mass spectrometer from the magnet. This might be achieved if the ion optics and analyzer housing were manufactured from μ-metal.

9.6 REFERENCES

Behnke, B., Schlotterbeck, G., Tallarek, U., Strohschein, S., Tseng, L.-H., Keller, T., Albert, K., Bayer, E. Capillary HPLC-NMR coupling: high-resolution ^{1}H NMR spectroscopy in the nanoliter scale. *Anal. Chem.* 1996, 68, 1110.

Brummel, C. L., Lee, I. N. W., Zhou, Y., Benkovic, S. J., Winograd, N. A mass spectrometric solution to the address problem of combinatorial libraries. *Science.* 1994, 264, 399.

Burton, K. I., Everett, J. R., Newman, M. J., Pullen, F. S., Richards, D. S., Swanson, A. G. On-line liquid chromatography coupled with high field NMR and mass spectrometry (LC-NMR-MS): a new technique for drug metabolite structure elucidation. *J. Pharm. & Biomed. Anal.* 1997, 15, 1903.

Davis, M. R., Stahl, D. C., Hefta, S. A., Lee, T. D. A microscale electrospray interface for on-line, capillary liquid chromatography/tandem mass spectrometry of complex peptide mixtures. *Anal. Chem.* 1995, 67, 4549.

DeWitt, S. H., Chiang, W., Chen, Y., Hamilton, T., Kelley, S., Leber, H., McBride, S., Moroney, R., O'Mara, K. Synthesis on a chip. 1998. Washington, DC, American Chemical Society. 216th ACS National Meeting, Boston.

Haap, W. J., Walk, T. B., Jung, G. FT-IR mapping—a new tool for spatially resolved characterization of polymer-bound combinatorial compound libraries with IR microscopy. *Angew. Chem. Int. Ed.* 1998, 37, 3311.

Holt, R. M., Newman, M. J., Pullen, F. S., Richards, D. S., Swanson, A. G. High-performance liquid chromatography/NMR spectrometry: further advances in hyphenated technology. *J. Mass Spectr.* 1997, 32, 64.

Holzwarth, A., Schmidt, H.-W., Maier, W. F. Detection of catalytic activity in combinatorial libraries of heterogeneous catalysis by IR thermography. *Angew. Chem. Int. Ed.* 1998, 37, 2644.

Huang, X. C., Quesada, M. A., Mathies, R. A. Capillary array electrophoresis using laser-excited confocal fluorescence detection. *Anal. Chem.* 1992, 64, 967.

Jacobson, S. C., Hergenröder, R., Koutny, L. B., Warmack, R. J., Ramsey, J. M. Effects of injection schemes and column geometry on the performance of microchip electrophoresis devices. *Anal. Chem.* 1994, 66, 1107.

Joyce, G. F., Still, W. C., Chapman, K. T. Combinatorial chemistry. Searching for a winning combination. *Current Opinion in Chem. Biol.* 1997, 1, 3–4.

Manz, A., Harrison, D. J., Verpoorte, E. M. J., Fettinger, J. C., Paulus, A., Lüdi, H., Widmer, H. M. Planar chips technology for miniaturization and integration of separation techniques into monitoring systems. *J. Chromatog.* 1992, 593, 253.

Mathies, R. A., Huang, X. C. Capillary array electrophoresis: an approach to high-speed, high-throughput DNA sequencing. *Nature* 1992, 359, 167.

Olson, D. L., Lacey, M. E., Sweedler, J. V. The nanoliter niche. *Anal. Chem.* 1998, 70, 257A.

Ramsey, J. M. *Microdomain chemical analysis.* Book of Abstracts, 216th ACS National Meeting, Boston, August 23–27, 1998. Publisher: American Chemical Society, Washington, DC.

Ramsey, J. M., Jacobson, S. C., Kutter, J. P. *Microchip separations: Small and fast.* Book of Abstracts, 216th ACS National Meeting, Boston, August 23–27, 1998. Publisher: American Chemical Society, Washington, DC.

Reetz, M. T., Becker, M. H., Kuhling, K. M., Holzwarth, A. Time-resolved IR-thermographic detection and screening of enantioselectivity in catalytic reactions. *Angew. Chem. Int. Ed.* 1998, 37, 2647.

Spraul, M., Hofmann, M., Ackermann, M., Nicholls, A. W., Damment, S. J. P., Haselden, J. N., Schockcor, J. P., Nicholson, J. K., Lindon, J. C. Flow injection proton nuclear magnetic resonance spectroscopy combined with pattern recognition methods: implications for rapid structural studies and high throughput biochemical screening. *Anal. Commun.* 1997, 34, 339.

Valaskovic, G. A., Kelleher, N. L., Little, D. P., Aaserud, D. J., McLafferty, F. W. Attomole-sensitivity electrospray source for large-molecule mass spectrometry. *Anal. Chem.* 1995, 67, 3802.

Woolley, A. T., Mathies, R. A. Ultra-high-speed DNA fragment separations using microfabricated capillary array electrophoresis chips. *Proc. Natl. Acad. Sci.* 1994, 91, 11348.

Wu, N., Peck, T. L., Webb, A. G., Magin, R. L., Sweedler, J. V. ^{1}H-NMR spectroscopy on the nanoliter scale for static and on-line measurements. *Anal. Chem.* 1994, 66, 3849.

Xue, Q., Foret, F., Dunayevskiy, Y. M., Zavracky, P. M., McGruer, N. E., Karger, B. L. Multichannel microchip electrospray mass spectrometry. *Anal. Chem.* 1997, 69, 426.

Index